Gender, Environment, and Human Rights:

An Intersectional Exploration

Swati Chakraborty
Concordia University, India

Subrata S. Satapathy
Christ Academy Institute of Law, India

IGI Global
Publishing Tomorrow's Research Today

Vice President of Editorial — Melissa Wagner
Managing Editor of Acquisitions — Mikaela Felty
Managing Editor of Book Development — Jocelynn Hessler
Production Manager — Mike Brehm
Cover Design — Phillip Shickler

Published in the United States of America by
IGI Global Scientific Publishing
701 East Chocolate Avenue
Hershey, PA, 17033, USA
Tel: 717-533-8845
Fax: 717-533-8661
Website: https://www.igi-global.com E-mail: cust@igi-global.com

Library of Congress Cataloging-in-Publication Data

Names: Chakraborty, Swati, 1985- editor.
Title: Gender, environment, and human rights : an intersectional exploration / Swati Chakraborty, Subrata Satapathy.
Description: Hershey, PA : IGI Global Scientific Publishing, 2025. | Includes bibliographical references and index. | Summary: "This book explores various themes that reflect this interconnectedness. From the role of women in environmental conservation to the human rights implications of climate change, and from corporate accountability to the integration of indigenous knowledge, each section provides a nuanced view of how gender, environment, and human rights are inextricably linked"-- Provided by publisher.
Identifiers: LCCN 2024061195 (print) | LCCN 2024061196 (ebook) | ISBN 9798369360699 (hardcover) | ISBN 9798369360705 (paperback) | ISBN 9798369360712 (ebook)
Subjects: LCSH: Women's rights. | Women's rights--India--21st century. | Sustainable development. | Sustainable development--India--21st century. | Human rights. | Human rights--India--21st century.
Classification: LCC HQ1236.5.I4 G385 2025b (print) | LCC HQ1236.5.I4 (ebook) | DDC 304.2082--dc23/eng/20250225
LC record available at https://lccn.loc.gov/2024061195
LC ebook record available at https://lccn.loc.gov/2024061196

British Cataloguing in Publication Data
A Cataloguing in Publication record for this book is available from the British Library.

Table of Contents

Detailed Table of Contents

In a world full of technology and modernization, we still live in a society where menstruation and periods are a dirty and negative thing that we hesitate to talk about. Where females are still burdened under the myths and taboos. This research is aimed towards breaking the walls of period related myths and taboos and Introducing the IT involvement in the ground level of this field. We can use technology and create huge machines and algorithms that can detect period related diseases within a second. But what is the use of all these if females will yet not be comfortable in coming out and ask for the technology to test them. It's necessary to start working from the ground level so that females can get everything they need to know about periods without feeling uncomfortable or ashamed. This research is aimed towards understanding and looking at the social and technical aspects of menstruation. This project is a combination of society and technology.

The right to the best possible level of physical and mental health is protected by a number of international legal instruments, such as the International Covenant on Economic, Social, and Cultural Rights. Both liberties and entitlements are included in this right and this includes the protection against torture and against non-consensual medical treatment or experimentation, which is especially important for people with disabilities. The entitlement is the freedom to obtain first-rate medical treatment without encountering prejudice. A human rights-based perspective on health requires nations to set up health systems that respect human rights and are efficient, gender-transforming, integrated, and accountable. It also urges the adoption of public health initiatives targeted at enhancing the fundamental factors that influence health, such access to hygienic conditions and clean water. This chapter deeply dives in the healthcare and human rights perspectives with artificial intelligence applications.

The plight of female prisoners is not unknown globally; the UN Rules for the Treatment of Women Prisoners and Non-custodial Measures for Women Offenders, popularly known as "Bangkok Rules", is the primary international instrument dedicated to incarcerated women. The Indian National Prison Manual, 2016 is based upon the principles of these rules along with another UN-led "Nelson Mandela Rules". However, the Ministry of Women & Children Development led an exclusive study in 2018 titled "Women in Prisons" that recorded the major issues & human rights violations of such women behind bars due to various reasons like inadequate female staff, accommodation, medical access along with basic right of healthy food. This paper hypothesized that according to the historical & current scenario, prisons in the country specifically still do not cater to the necessities for the subsistence of the incarcerated women. The study intends to analyze major concerns like obligations of the Indian Prison Administration vis-à-vis the progress made till now, specifically concerning Female Prisoners.

This research paper explores the profound connection between Indigenous women's spirituality and their roles in environmental protection in Canada. By examining traditional ecological knowledge, spiritual practices, and contemporary environmental activism, the study aims to highlight the integral role of Indigenous women in safeguarding their ancestral lands. The paper will delve into case studies of specific Indigenous communities, analyze the impact of colonization and industrialization, and offer insights into how Indigenous women's spiritual perspectives can inform broader environmental conservation efforts.

This comparative study of Mariama Bâ's "Une Si Longue Lettre" and Ashapurna Devi's "Pratham Pratishruti (The First Promise)" explores the feminist consciousness of these two pioneering authors. Despite being from different continents, eras, and cultures, their novels highlight similar struggles of women within patriarchal societies. During times when global focus was on wars and colonization, these authors brought attention to the overlooked domestic struggles of women. The study demonstrates how Bâ and Devi's protagonists break free from the confines of patriarchy, finding their paths to freedom. The research aims to uncover the overlooked narratives of women in rural India and Africa, emphasizing the domestic realms depicted in these contemporary literary works.

Women can be empowered by laws that safeguard and promote their sexual and reproductive rights. The concept of reproductive rights specifies the rights of women to make decisions about their own bodies and health, highlighting the right of choice in reproductive matters. The social and cultural context in India presents a complex and multifaceted view of the concept of reproductive rights of women. Childbearing is viewed as a societal responsibility of women in Indian society. This restricts women's right to make decisions about their own lives by putting pressure on them to have children even if they are not ready. In India, abortion is also frequently seen as a women's issue, and women who choose to have an abortion are stigmatized and discriminated against. Authors propose actionable policy recommendations for an equitable and just society where women's rights are both legally protected and culturally esteemed.

"We are still living with the results of millennia of patriarchy that excludes women and prevents their voices from being heard," - António Guterres. In underprivileged countries including Ghana, Egypt, Costa Rica, and Malawi, women face such challenges and risks regularly as per the IMF Report. Unique challenges have often been faced by women and girls when it comes to climate disasters because of deep-rooted inequalities, especially towards girls and women which make them much more vulnerable to its effects. Issues in health care, food security, earning for living, and being forced to migrate often show up because of disasters. Hence, it would be correct to say tackling climate change needs to be equal from the ground up, or everyone suffers. Including women and girls in the climate policies will make solutions more practical and relevant. It will consider the needs of everyone, not just a select few. Empower girls and women to be not just victims, but as an active participant in shaping a sustainable future and will make the UN Climate Change Conference (COP 26 & 33) successful.

This chapter explores the essential role of women in environmental conservation, focusing on their challenges and successes. It examines how gender, socioeconomic factors, and cultural contexts impact women's leadership in conservation. Using a comprehensive literature review, case studies, and original research, the chapter highlights the barriers women face—such as discrimination, resource limitations, and cultural norms—and showcases their innovative contributions to conservation. It underscores the need to empower women in leadership roles to achieve sustainable environmental management and advance human rights. The study offers valuable insights for policymakers, conservation organizations, and activists seeking to promote gender equality in environmental stewardship.

Rivers have historically served as lifelines for civilizations, providing essential resources and opportunities for livelihood. However, they also present significant challenges, particularly when turned into disaster propelled by human interventions. The very qualities that make rivers indispensable to the dependent communities also render them vulnerable in many ways. Among the myriad challenges facing riverine communities, river bank erosion stands out as a particularly pressing issue, with far-reaching social, economic, and environmental implications. In recent time, well-being of riparian people in West Bengal, India, is seriously threatened by river bank erosion in the Ganges river basin, especially in parts of Malda and Murshidabad districts have [un]intentionally gotten worse since the Farakka Barrage was built in 1972.The article emphasises how these elements are interrelated and have consequences for the welfare of communities by looking at them through the lenses of geopolitics, feminism, and human rights.

Women, constituting nearly half of India's population, are crucial for the nation's economic growth. Gender equality as a core global agenda, is encapsulated in the United Nations' Sustainable Development Goals (SDGs), particularly SDG 5, which focuses on gender equality and the empowerment of all women and girls. Despite educational advancements, women in India continue to face significant barriers in other areas. Therefore, it is imperative to implement relevant legal instruments and policies to support environment-friendly agricultural practices, benefiting both ecological sustainability and women's economic participation. Judicial pronouncements related to women also contribute to sustainability and women's participation in labour.

Child sexual abuse is a social evil that is frequently around the globe. They are becoming more numerous every day, which makes it difficult to evaluate the situation. The chance of negative psychosocial and mental health effects associated with child sexual abuse is unquestionably increased, but even stability processes identified a number of protective factors, including as family support. The perpetrators of this crime take advantage of their position of vulnerability, trust, and power in order to commit child sexual abuse or attempt to do so for financial gain or other purposes. Given how similar the terms "child sexual exploitation" and "child sexual abuse" are, it can be challenging to distinguish between them. Children who experience child sexual abuse suffer grave consequences. It serves as a huge roadblock to growth and advancement. Many health behaviours brought on by child sexual abuse result in negative mental, physical, and social outcomes throughout life. There are numerous circumstances that can contribute to child sexual abuse.

Climate change disproportionately affects vulnerable people, exacerbates inequalities and threatens human rights. In this chapter the author has explores the complex interplay between climate change and human rights, focusing on rights to life, health, food, water, shelter and culture highlighting the challenges faced by Indigenous peoples, women, children and poor meet on climate change adaptation emphasis. The chapter discusses the role of international instruments such as the UNFCCC and human rights treaties in addressing these issues and explores the need to create stronger legal safeguards, greater public participation and focus on stronger groups to build a just and equitable future

The particular focus of this study is on bringing to light the differential nature of climate change effects across the population, as seen through the lens of Amartya

Sen's crucial thesis on 'missing women.' We argue that there is a need for economists to map how climate change makes invisible women, particularly in the labour force. The precarious accessibility women have, and the freedom women enjoy by virtue of gaining access to education, housing, credit and so on, is doubly at risk from the effects of climate change. It is essential to bring to light the susceptibility of the labor market at large, and the susceptibility of women workers in particular, to the vagaries of nature. Be it in the form of heat stress, or global warming disrupting large scale supply chains, or even wiping out small industries, the impact of climate change on the labor market is substantially large and deserving of closer study, especially through a fruitful application of Sen's essential ideas on human development and gender parity.

The feminization of agriculture in India, driven by male migration to urban areas, intersects with the growing challenges of climate change, creating a complex landscape of both opportunities and challenges. Women, increasingly responsible for agricultural work, face heightened vulnerabilities due to their reliance on natural resources and limited access to adaptive technologies, credit, and decision-making power. Climate change exacerbates these challenges, intensifying workloads and threatening food security as crop yields decline and traditional farming practices become less viable. However, this convergence also presents opportunities for empowering women through targeted interventions such as land ownership, access to climate-resilient technologies, and involvement in decision-making processes. By recognizing and supporting women's critical roles in agriculture India can enhance agricultural productivity, promote gender equality, and build climate resilience in rural communities.

This chapter provides a concise overview of the integration of children's rights into corporate operations, highlighting the growing importance of this issue in the globalized business environment. As companies become more interconnected with global supply chains, they must address the direct and indirect impacts of their activities on children, particularly in areas like child labor prevention, fair wages, education, and healthcare. By embedding children's rights into their strategies, businesses not only

Preface

As global citizens confronting the myriad challenges of the 21st century, we are increasingly aware that the issues of gender equality, environmental sustainability, and human rights are not isolated problems but are deeply intertwined facets of a complex global tapestry. This realization has inspired the creation of '*Gender, Environment, and Human Rights*: An Intersectional Exploration' a pioneering collection that delves into the intersections of these critical dimensions.

In curating this volume, we—Swati Chakraborty and Subrata S. Satapathy—sought to weave together a diverse array of perspectives and insights that illuminate the intricate relationships among gender, environmental issues, and human rights. Our aim is not merely to present a compilation of articles but to foster a profound understanding and inspire action in the face of these intersecting challenges.

Our world is at a crossroads where the need for integrative and holistic solutions has never been more pressing. Environmental degradation, climate change, and human rights violations are not abstract concerns but have tangible, far-reaching impacts on communities, particularly those most marginalized. Women and other gender minorities, in particular, bear disproportionate burdens as environmental stewards and human rights defenders. Recognizing and addressing these inequities is vital for creating a just and sustainable future.

The chapters within this book explore various themes that reflect this interconnectedness. From the role of women in environmental conservation to the human rights implications of climate change, and from corporate accountability to the integration of indigenous knowledge, each section provides a nuanced view of how gender, environment, and human rights are inextricably linked. We delve into the stories of grassroots activists, the challenges faced by women leaders, and the critical need for inclusive policies that consider these intersections.

Our hope is that this volume serves as a catalyst for dialogue and transformation. By highlighting the experiences and contributions of those at the forefront of these issues, we aim to empower individuals, communities, and policymakers alike. This book is designed to inspire action and foster collaboration, offering both theoretical insights and practical strategies to address systemic inequalities and environmental degradation.

In presenting this compilation, we invite readers from all walks of life—scholars, activists, policymakers, and the general public—to engage deeply with the material. It is our aspiration that the knowledge contained within these pages will not only enhance understanding but also drive meaningful change towards a more equitable and sustainable world.

We are profoundly grateful to all contributors for their dedication and expertise. Together, we hope to bridge gaps in understanding and inspire collective efforts towards achieving justice and sustainability for all.

CHAPTER 1: INTERWOVEN REALITIES - GENDER, ENVIRONMENT, AND HUMAN RIGHTS

In this foundational chapter, we address the urgent need to bridge the gap between climate change and human rights. It argues that climate change is not only an ecological crisis but also a profound human rights issue, particularly affecting the most vulnerable populations. The chapter advocates for a comprehensive perspective that incorporates human rights into climate policy decisions, emphasizing the importance of equitable and enduring solutions. It serves as a call to action for researchers, activists, and policymakers to expand their focus beyond economic and environmental concerns to include fundamental human rights, which are increasingly at risk due to climate-related challenges.

CHAPTER 2: WOMEN'S LEADERSHIP IN ENVIRONMENTAL DIALOGUE - A HUMAN RIGHTS APPROACH

This chapter investigates the critical intersection of women's leadership, environmental conservation, and human rights. It explores how integrating gender perspectives into conservation efforts can enhance both environmental and human rights outcomes. By reviewing existing literature and applying a human rights framework, the chapter highlights the significant yet often overlooked contributions of women in environmental stewardship. It also identifies barriers women face in these roles and proposes strategies to advance gender equality in environmental policy and practice, underscoring the dual benefits of promoting sustainable development and advancing women's rights.

CHAPTER 3: CELLS TO SOCIETY: LET'S TALK ABOUT MENSTRUATION

Addressing the societal taboos surrounding menstruation, this chapter aims to demystify and destigmatize menstrual health through a combination of social and technological perspectives. It highlights the persistent myths and discomfort associated with menstruation and argues for the use of technology to improve menstrual health diagnostics. The chapter advocates for breaking down barriers to open discussion and education, emphasizing the need for technological advancements to be paired with efforts to normalize menstruation and improve women's access to health resources without stigma.

CHAPTER 4: FEATURING HEALTHCARE, ENVIRONMENT AND HUMAN RIGHTS: APPLICATIONS OF ARTIFICIAL INTELLIGENCE IN HEALTH DOMAIN

This chapter delves into the intersection of healthcare, human rights, and environmental issues through the lens of artificial intelligence (AI). It discusses how AI can enhance healthcare delivery while respecting human rights and addressing gender disparities. The chapter examines international legal frameworks that protect health rights and proposes how AI can contribute to more effective, equitable, and human rights-compliant health systems. It highlights the potential of AI to address public health challenges and improve access to healthcare in underserved populations.

CHAPTER 5: DECODING WOMEN'S REPRODUCTIVE AUTONOMY - A CRITICAL EXAMINATION OF ABORTION LAWS IN INDIA

Focusing on reproductive autonomy, this chapter critically examines the history and impact of abortion laws in India. It provides a comprehensive analysis of the Medical Termination of Pregnancy (MTP) Act of 1971 and subsequent policies, evaluating their effectiveness in safeguarding women's reproductive rights. The chapter contrasts India's abortion laws with global standards and explores the legal and societal factors affecting women's access to safe and legal abortions, offering insights into both the progress made and the ongoing challenges.

CHAPTER 6: TRADITIONAL KNOWLEDGE SYSTEM OF THE COLD DESERT OF LEH-LADAKH (INDIA) FOR ENVIRONMENTAL MANAGEMENT

Exploring the Cold Desert of Leh-Ladakh, this chapter highlights the traditional ecological knowledge (TEK) of indigenous communities in managing their harsh environment. It documents the rich cultural heritage and environmental practices developed by the Ladakhi, Tibetan, and Changpa peoples. The chapter emphasizes the importance of preserving these traditional practices amidst modern challenges and globalization, advocating for the documentation and integration of TEK into contemporary environmental management strategies.

CHAPTER 7: HUMAN RIGHTS OF INCARCERATED WOMEN IN INDIA: EMPTY RHETORIC OR EVOLUTIONARY PROGRESS?

This chapter examines the human rights conditions of female prisoners in India, assessing whether recent reforms have led to meaningful improvements. It critiques the implementation of international standards such as the Bangkok Rules and the Nelson Mandela Rules in the context of Indian prisons. The chapter highlights ongoing issues such as inadequate facilities and healthcare, offering a critical analysis of the progress made and identifying areas requiring further attention to ensure the rights and well-being of incarcerated women.

CHAPTER 8: SACRED LANDSCAPES - THE INTERSECTION OF INDIGENOUS WOMEN'S SPIRITUALITY AND ENVIRONMENTAL PROTECTION IN CANADA

This research paper explores the vital role of Indigenous women's spirituality in environmental protection within Canada. It investigates how traditional ecological knowledge and spiritual practices contribute to environmental conservation efforts. By examining case studies of Indigenous communities, the chapter highlights the impact of colonization and industrialization on spiritual and environmental practices and proposes ways to incorporate Indigenous perspectives into broader environmental conservation strategies.

CHAPTER 9: THE OVERLOOKED NARRATIVES OF WOMEN IN RURAL INDIA AND AFRICA, AS DEPICTED IN CONTEMPORARY LITERATURE

This comparative study analyzes the feminist themes in Mariama Bâ's *Une Si Longue Lettre* and Ashapurna Devi's *Pratham Pratishruti*. It reveals how these authors, despite differing contexts, address the domestic struggles of women in patriarchal societies. The chapter emphasizes the narratives of women's resistance and empowerment as depicted in their literature, shedding light on the overlooked experiences of women in rural settings and the ways in which these literary works challenge and redefine gender norms.

CHAPTER 10: EMPOWERMENT THROUGH LEGISLATION: A LEGAL ANALYSIS OF WOMEN'S SEXUAL AND REPRODUCTIVE RIGHTS IN INDIA

This chapter explores the role of legislation in promoting and protecting women's sexual and reproductive rights in India. It critiques the social and cultural barriers that impact women's autonomy over their reproductive choices and proposes legal reforms to enhance these rights. By analyzing existing laws and societal attitudes, the chapter aims to provide actionable policy recommendations for creating a more equitable legal framework that supports women's reproductive autonomy and challenges societal stigmas.

CHAPTER 11: WOMEN AT THE CROSSROADS OF CLIMATE CHANGE - ADVANCING GENDER EQUALITY THROUGH CLIMATE ACTION

Addressing the unique challenges faced by women in the context of climate change, this chapter argues for the inclusion of gender perspectives in climate policies. It highlights how climate change exacerbates existing inequalities, particularly for women and girls in developing countries. The chapter advocates for empowering women as active participants in climate action, emphasizing that gender-inclusive policies are crucial for effective and equitable climate solutions.

CHAPTER 12: WOMEN'S LEADERSHIP IN ENVIRONMENTAL CONSERVATION - CHALLENGES AND SUCCESS STORIES

This chapter examines the vital contributions of women to environmental conservation, focusing on both the obstacles they face and their successes. Through a detailed literature review and case studies, the chapter discusses how cultural, socio-economic, and gender-related factors influence women's leadership in conservation efforts. It highlights the need to support and empower women leaders to advance sustainable environmental management and human rights.

CHAPTER 13: WOMEN, GENDER EQUALITY AND ENVIRONMENTAL JUSTICE

Exploring the intersection of environmental justice and gender equality, this chapter emphasizes the need for an inclusive approach to addressing environmental and social inequities. It discusses how environmental degradation disproportionately affects marginalized communities, particularly women, and advocates for integrating gender perspectives into environmental policies. The chapter calls for interdisciplinary research and action to create resilient and equitable societies capable of confronting environmental challenges.

CHAPTER 14: WOMEN AS ENVIRONMENTAL CHANGE-MAKERS IN CIVIL SOCIETY

This chapter explores the transformative roles women play as environmental leaders within civil society. It highlights how women's participation in environmental decision-making fosters both gender equality and effective environmental action. The chapter examines the benefits of women's leadership in shaping environmental policies and encourages greater involvement of women in both local and global environmental initiatives.

CHAPTER 15: UNDERSTANDING DISASTERS, VULNERABILITIES, AND RESILIENCE - IMPACT OF RIVER BANK EROSION ON RIPARIAN COMMUNITIES

Focusing on the issue of river bank erosion in West Bengal, India, this chapter investigates the socio-economic and environmental impacts on riparian communities. It explores how human interventions and natural factors contribute to the erosion crisis and its consequences for local populations. The chapter uses a multidisciplinary approach to analyze the complex interplay of geopolitics, feminism, and human rights in understanding and addressing the challenges faced by these communities.

CHAPTER 16: WOMEN LABOUR FORCE PARTICIPATION IN INDIA AND SUSTAINABLE DEVELOPMENT GOALS - A LEGAL FRAMEWORK

This chapter examines the relationship between women's labor force participation in India and the Sustainable Development Goals (SDGs). It discusses how legal frameworks can support women's economic empowerment and align with the SDGs. The chapter analyzes existing policies and proposes legal reforms to enhance women's participation in the workforce, contributing to broader goals of gender equality and sustainable development.

CHAPTER 17: CHILDHOOD AT RISK - EXAMINING THE EFFECTIVENESS OF THE FRAMEWORK IN PROTECTING CHILDREN FROM SEXUAL OFFENSES

Addressing the issue of child sexual abuse, this chapter evaluates the effectiveness of existing legal frameworks in protecting children. It discusses the psychological, social, and physical impacts of abuse and examines the strengths and limitations of current protective measures. The chapter provides recommendations for improving legal and support systems to better safeguard children and address the pervasive issue of child sexual exploitation.

CHAPTER 18: THE UNEQUAL BURDEN - CLIMATE CHANGE AND VULNERABLE POPULATIONS

This chapter explores how climate change disproportionately affects vulnerable populations, exacerbating existing inequalities and threatening human rights. It focuses on the rights to life, health, food, water, and shelter, highlighting the specific challenges faced by Indigenous peoples, women, children, and the poor. The chapter discusses the role of international agreements and proposes strategies for strengthening legal protections and public participation to build a just and equitable future amidst climate challenges.

CHAPTER 19: CLIMATE CHANGE, WOMEN'S RIGHTS AND THE WAY FORWARD: NOTES ON THE INDIAN PERSPECTIVE

In this chapter it can be seen that the particular focus of this study is on bringing to light the differential nature of climate change effects across the population, as seen through the lens of Amartya Sen's crucial thesis on 'missing women.' We argue that there is a need for economists to map how climate change makes invisible women, particularly in the labour force. The precarious accessibility women have, and the freedom women enjoy by virtue of gaining access to education, housing, credit and so on, is doubly at risk from the effects of climate change. It is essential to bring to light the susceptibility of the labor market at large, and the susceptibility of women workers in particular, to the vagaries of nature. Be it in the form of heat stress, or global warming disrupting large scale supply chains, or even wiping out small industries, the impact of climate change on the labor market is substantially large and deserving of closer study, especially through a fruitful application of Sen's essential ideas on human development and gender parity.

CHAPTER 20: CLIMATE CHANGE AND FEMINIZATION OF AGRICULTURE IN INDIA: MAPPING THE OPPORTUNITIES AND CHALLENGES

This chapter delves into the key aspects of climate change and feminization of agriculture. The feminization of agriculture in India, driven by male migration to urban areas, intersects with the growing challenges of climate change, creating a complex landscape of both opportunities and challenges. Women, increasingly responsible for agricultural work, face heightened vulnerabilities due to their reliance on natural resources and limited access to adaptive technologies, credit, and decision-making power. Climate change exacerbates these challenges, intensifying workloads and threatening food security as crop yields decline and traditional farming practices become less viable. However, this convergence also presents opportunities for empowering women through targeted interventions such as land ownership, access to climate-resilient technologies, and involvement in decision-making processes. By recognizing and supporting women's critical roles in agriculture India can enhance agricultural productivity, promote gender equality, and build climate resilience in rural communities.

CHAPTER 21: THE CONSIDERATION OF CHILDREN'S RIGHTS IN CORPORATE ACTION-CURRENT DEVELOPMENTS IN GERMANY

This chapter provides a concise overview of the integration of children's rights into corporate operations, highlighting the growing importance of this issue in the globalized business environment. As companies become more interconnected with global supply chains, they must address the direct and indirect impacts of their activities on children, particularly in areas like child labor prevention, fair wages, education, and healthcare. By embedding children's rights into their strategies, businesses not only fulfill legal and ethical obligations but also enhance their reputations and contribute positively to society, facing both challenges and opportunities in this endeavor.

As we conclude this seminal volume, *Gender, Environment, and Human Rights: An Intersectional Exploration*, it is evident that the interconnected challenges of gender equality, environmental sustainability, and human rights require a comprehensive and integrated approach. In assembling this diverse collection of insights and analyses, our goal has been to illuminate the complex and often overlooked intersections between these critical areas.

Each chapter in this book contributes to a broader understanding of how gender disparities, environmental crises, and human rights violations are not isolated issues but are deeply interwoven aspects of our global reality. From the critical examination of climate change's impact on human rights to the vital role of women's leadership in environmental conservation, the research presented here underscores the urgent need for intersectional solutions. The contributions span a range of topics, including the role of traditional knowledge in environmental management, the impact of legislation on reproductive rights, and the challenges faced by incarcerated women. These diverse perspectives not only highlight the breadth of the issues but also the depth of the systemic connections that need to be addressed.

In exploring the unique challenges faced by marginalized groups, particularly women, this volume advocates for a more inclusive and equitable approach to policy-making and advocacy. The chapters collectively argue that effective solutions must consider the diverse experiences and needs of those most affected by these intersecting issues. Whether addressing the stigmas surrounding menstrual health, analyzing the effectiveness of legal frameworks for child protection, or examining the role of women in civil society, the book provides a rich tapestry of knowledge aimed at fostering deeper understanding and driving meaningful change.

Our hope is that this book will serve not only as a resource for scholars, activists, and policymakers but also as a catalyst for action. By bridging gaps in understanding and offering practical strategies for addressing systemic inequalities and environmental degradation, we aspire to inspire collaborative efforts that move us toward a more just and sustainable world.

We extend our heartfelt gratitude to the contributors for their invaluable expertise and commitment to these critical issues. Together, through continued dialogue and action, we can address the complex challenges of our time and work towards a future where gender equality, environmental sustainability, and human rights are upheld for all.

Swati Chakraborty
Concordia University, Canada

Subrata S. Satapathy
Christ Academy Institute of Law, India

Chapter 1
Interwoven Realities: Gender, Environment, and Human Rights

Mumtaz Alam
http://orcid.org/0000-0002-0945-2447
Fiji National University, Fiji

Kunal Avishek Gounder
https://orcid.org/0000-0003-2790-6307
Fiji National University, Fiji

ABSTRACT

This chapter aims to narrow the existing information gap between climate change and human rights. This chapter serves as proof that climate change is a significant anthropogenic catastrophe and ecological calamity. To prioritise the interests of the most vulnerable individuals, it argues that decisions on climate change should consider human rights. This chapter emphasises the need for researchers, activists, and political leaders to have a broader perspective on climate change that extends beyond the apparent economic and environmental risks. Moreover, it is imperative to identify a feasible, equitable, and enduring resolution, as this issue poses a threat to fundamental human rights.

"Emphasizing that climate change-related impacts have a range of implications, both direct and indirect, for the effective enjoyment of human rights, including, inter alia, the right to life, the right to adequate food, the right to the highest attainable standard of health, the right to adequate housing, the right to self-determination and the right to safe drinking water and sanitation…"

DOI: 10.4018/979-8-3693-6069-9.ch001

HUMAN RIGHTS COUNCIL RESOLUTION 18/22

Climate change poses a significant threat to human rights, particularly impacting the basic rights of marginalized and vulnerable populations such as the elderly, women, children, and individuals living in poverty. Climate change has adversely affected essential needs such as sustenance, hydration, housing, and secure habitats, as well as causing increasing sea levels, erosion, severe weather events, and damage to biodiversity. Climate change poses a significant threat to several fundamental human rights, such as the right to life, the right to property, and the right to a sufficient means of subsistence. The imminent peril of submersion presents a grave menace to susceptible island nations, who risk losing their ancestral territory and cultural heritage if not forcibly relocated, potentially leading to the erosion of their distinctive way of life. Climate change poses a humanitarian dilemma of equal magnitude to its environmental impact. In order to safeguard justice and human rights, it is imperative to implement a climate action strategy that is centered around the principles of human rights. The integration of human rights concepts with climate action can lead to the attainment of transparency, justice, empowerment, and equitable treatment. This chapter aims to narrow the existing information gap between climate change and human rights. Climate change is a significant anthropogenic disaster and ecological calamity, as evidenced by this chapter. To prioritise the interests of the most vulnerable individuals, it argues that decisions on climate change should consider human rights. This chapter emphasises the need for researchers, activists, and political leaders to have a broader perspective on climate change that extends beyond the apparent economic and environmental risks. Moreover, it is imperative to identify a feasible, equitable, and enduring resolution as this issue poses a threat to fundamental human rights.

INTRODUCTION

Climate change poses a significant threat to human rights, particularly by infringing upon the fundamental rights of the most marginalised and vulnerable populations, such as the elderly, women, children, and those living in poverty. Climate change has had adverse effects on fundamental requirements like as sustenance, water, housing, and secure habitats, as well as on phenomena including escalating sea levels, erosion, severe weather events, and damage to biodiversity. Climate change has a significant impact on fundamental human rights such as property rights, the right to life, and the right to earn a living (Atapattu, 2015). Small island countries are in risk of being submerged, which might result in the loss of their native land and identity as well as the forced relocation that could cause a loss of culture. In addition to being an environmental problem, climate change is a humanitarian emergency. Therefore, in order to ensure that justice and the preservation of human rights are served, a rights-based approach must be applied in climate action. Human rights frameworks in climate action will guarantee accountability, non-discrimination, empowerment, and equality.

The concept of human rights-based approach to climate change was accepted by the international organizations after United Nations Human Rights Council (UNHRC) published "Report on the relationship between climate change and human rights" in January 2009 (Mahadew, 2021). The report was made based on the submissions made by the state and non-state actors. After the panel discussion on the report, it was evident that the governments were unable to agree on whether international human rights law require states to address the causes and effects of climate change or not. Hence, failure to agree to this in 2009 could be the reason behind the failure of international pledges, commitments and policies taking place today.

A wide range of human rights as affirmed by the Intergovernmental Panel on Climate Change (IPCC) and Human Rights Council resolution 41/21, are threatened by the negative effects of climate change phenomena, both directly and indirectly (Office of the United Nations High Commissioner for Human Rights, n.d.). These rights include the right to life, food, housing, health, safe drinking water and sanitation, culture, work and development. Vulnerable people and communities that are already in challenges are disproportionately affected by the negative effects of climate change, making them more fragile. These factors include geography, poverty, gender, age, disability and cultural or ethnic background. These groups have also contributed the least towards greenhouse gas emissions, yet being affected the most.

The aim of this chapter is to bridge the knowledge gap that exists between human rights and climate change. The chapter will emphasise that climate change is a human catastrophe as well as an environmental one, and that human rights must be taken into consideration when making decisions in order to prioritise the needs of the most vulnerable and marginalised groups of people. The chapter emphasises the significance of looking at climate change from a viewpoint that goes beyond the economic and environmental threats to researchers, activists, and political representatives. In addition, as a human problem posing a danger to fundamental human rights, making sure that a workable, just, and long-lasting solution is found.

MATERIALS AND METHODS

For this chapter, we focused on examples from around the world to support the implications of climate change on human rights and how a human rights-based approach could be utilized to ensure the protection of fundamental human rights. This research was conducted using secondary data from publications as well as websites and international organizations. The research project's objective was to identify the implications of climate change on human rights and what role a human rights-based approach could play. The research also identified how human-rights factors struggle in the international discussion that is linked to 17 SDGs.

CONCEPTUAL FRAMEWORK

In response to climate change and human rights, UNHRC has been stressing the application of a human rights-based approach to guide policy and strategy designs to deal with climate change. The human rights-based approach focuses on three main components:

- The main goal while drafting a new policy or strategy should be the achievement of human rights.
- To develop strategies to increase the capabilities of duty-bearers to fulfil their responsibilities and of rights-holders to make their claims, it is necessary to identify the rights-holders and their entitlements as well as the corresponding duty-bearers and their obligations.
- All policies and programs throughout the entire process should be guided by the values and norms found in international human rights legislation, particularly in the Universal Declaration of Human Rights and the fundamental human rights treaties.

HUMAN RIGHTS IMPLICATIONS OF CLIMATE CHANGE

The intensifying effects of climate change will have impacts on every individual around the globe, either directly or indirectly. Hence, it will have implications for their human rights. With climate change being on the international agenda for more than two decades, least actions have been taken to deal with the consequences efficiently and sustainably. Now, the recent topic among researchers is the human rights implications of climate change; the appealing question is what the policymakers will do to protect the basic human rights that are internationally recognised under International Human Rights Law. The International Court of Justice will hear a student-led project to decide the legality of climate change for the first time, starting as an extracurricular activity in a 2019 University of the South Pacific (U.S.P.) law class. An advisory opinion on the responsibilities of States under international law with regard to climate change, human rights, and intergenerational equality will be issued by the International Court of Justice, the highest court in the world (Rheeney, 2023).

RIGHT TO LIFE

According to the IPCC report (2022), "more frequent and intense extreme events have caused widespread adverse impacts and related losses and damages to nature and people, beyond natural climate variability." The hydrological cycle is being impacted by climate change, which is also making storms more frequent and intense. Over 90% of natural disasters include weather components, such as wildfires, flooding, drought and aridification (Climate Change and Water-related Disasters, n.d). The growing fatality rate emphasises the ongoing vulnerability of the population to natural disasters, particularly in low and middle-income nations, since there were 510,837 deaths and 3.9 billion people impacted by 6681 natural disasters between 2000 and 2019 (Kharb et al.,2022). Over the previous 50 years, extreme weather has resulted in 2 million fatalities and $4.3 trillion in economic damage, according to a United Nations assessment. The World Meteorological Organisation (WMO) of the UN has released fresh statistics showing that from 1970 to 2021, 11,778 weather-related catastrophes have occurred, with a significant increase throughout that time. According to the research, developing countries accounted for almost 90% of the global mortality attributed to these disasters. The majority of weather, climate, and water-related disasters, regrettably, affect the most vulnerable communities. According to a 2021 report by WMO, which covered disaster-related fatalities and losses from 1970 to 2019, there were over 50,000 such deaths worldwide at the start of the study period. The number of fatalities from disasters has fallen to less than 20,000 per year by the 2010s. Additionally, the WMO reported that 22,608 disaster deaths were reported worldwide in 2020 and 2021 combined in its update to that report (Al Jazeera, 2023).

In other words, the increase in the intensity and frequency of natural disasters due to climate change has increased the number of deaths and people affected unless there is an effective warning system to prepare people beforehand. Hence, it affects the right to life. It is expected that people with disabilities are vulnerable to natural events as they have difficulties in moving to safer places. The Convention on the Rights of Persons with Disabilities, specifically Article 11, mandates that signatory states must take all appropriate steps to guarantee the safety and protection of individuals with disabilities in risky situations, such as natural disasters (De Costa, 2014). Stein et al. (2023) state that over a billion people with disabilities are disproportionately affected by the climate issue. Disability-related deaths are two to four times higher worldwide, and even while their contributions to slow-onset climate change have been minimal, they are disproportionately vulnerable to its effects (Stein et al., 2023).

RIGHT TO HEALTH

Everyone has an obligation to the best possible level of both mental and physical wellness, according to Article 12 of the International Covenant on Economic, Social, and Cultural Rights (Atapattu, 2015). It goes into detail on the actions that governments need to do in order to fully realise this fundamental right; these actions include lowering the death rate of infants, enhancing industry and cleanliness in the environment, averting epidemics, and giving people access to essential medical care. Due to climate change, extreme weather conditions like El Nino and natural disasters have become frequent and intensified. Therefore, it results in human health problems such as malnutrition, injuries, death, stress hypertension, dengue, malaria, leptospirosis and an unsanitary environment. According to Nurse et al. (2014), morbidity and mortality are some of the current challenges for small states due to climate sensitivity issues, which extend further to food and waterborne diseases. Kapoor et al. (2021), also found that obesity, diabetes, and cardiovascular disease are examples of non-communicable diseases that climate change may indirectly affect. For example, hotter weather may make people less likely to exercise and encourage sedentary behaviour, which would increase the obesity rate. Not only that, Leal Filho et al., (2023) shed light on the considerable but overlooked psychological impacts of intensifying climate disasters like cyclones, floods and drought in remote Pacific communities. In these culturally-close-knit settlements where mental health issues remain highly stigmatized and professional psychological services are scarce, the trauma of losing lands, livelihoods and loved ones amplifies the lack of coping mechanisms for significant climate-linked anxiety, depression, and post-traumatic stress. According to the World Health Organisation (WHO), 3.6 billion people currently reside in regions that are extremely vulnerable to climate change. Climate change is predicted to result in an extra 250,000 fatalities year between 2030 and 2050, only from heat stress, malaria, diarrhoea, and undernourishment. (WHO, 2023).

Climate change not only results in illness but also leads to many other health-related situations. Currently, developing states face crippled healthcare services, especially due to a shortage of medical staff, poor facilities, and limited capacity (Alam & Ali, 2022). This is mainly due to financial constraints that the governments of developing states are unable to invest in. Hence, an increase in diseases due to climate change will increase the toll on patients, making access to health care even worse, and some might not even get access to the health care they might need urgently. Also, this might push people further into poverty as medical expenses will increase. Approximately 12% of the world's population spent at least 10% of their income on medical expenses (WHO, 2023).

Figure 1. An overview of climate-sensitive health risks, their exposure pathways and vulnerability factors. Climate change impacts health both directly and indirectly and is strongly mediated by environmental, social, and public health determinants.

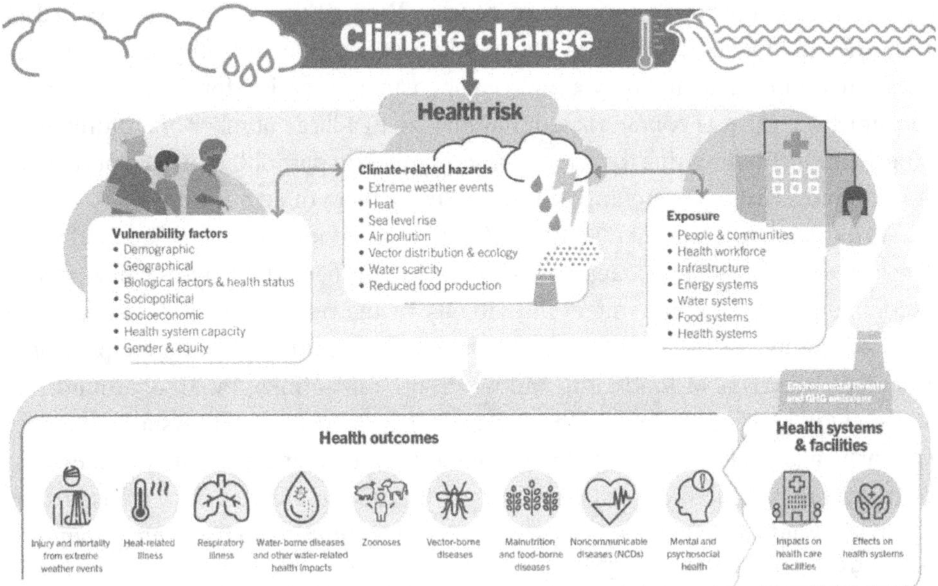

Source: *World Health Organization, 2023*

RIGHT TO FOOD AND WATER

Food and water are two basic human needs for a person to survive. The right to sufficient food was upheld by the Universal Declaration of Human Rights in 1948, and on July 28, 2010, the UN General Assembly recognised the right to clean water for drinking and sanitary conditions (United Nations Department of Economic and Social Affairs, n.d.). Climate change affects food and water security in many ways that endanger the right to food and security. Due to the rise in sea level, salinisation takes place when the quantity of salt in the soil increases, resulting in catastrophic impacts on agricultural food production as plants are intolerant to high salinity. Also, extreme weather events like drought and flooding damage agriculture production to an extreme level, risking food security and the right to food for vulnerable populations like the poor. Climate change also harms marine biodiversity, such as ocean acidification, damage to habitats and coral bleaching (Novianti et al., 2016). In simple economics concept, a decline in the supply of food will increase the price of food. Hence, the poor population will not be able to afford an adequate food supply to meet their needs. Hence it affects the right to food. Similarly, due to rising sea levels, the salinisation of fresh water is taking place, hence limiting the availability of fresh drinking water. In addition to that, long periods of drought are another factor that is contributing to water insecurity. Women and girls are compelled to go farther and longer to get water for their houses when the quantity and quality of water are diminished, and they are also responsible for caring for the infected who consume water that is contaminated. Families with a large number of children may experience difficulties due to water shortages; girls, in particular, may be pulled out of school to assist with fetching water, depriving them of an education.

WHO estimates indicate that over 600 million individuals experience foodborne diseases annually, whereas approximately 2 billion people do not have access to uncontaminated drinking water. Individuals in the age group of less than five years constitute 30% of fatalities caused by foodborne illnesses. Climate pressures exacerbate the risks of foodborne and waterborne infections. In 2020, around 770 million individuals encountered hunger, primarily concentrated in Asia and Africa. The influence of climate change on the quantity, quality, and diversity of accessible food exacerbates food and nutrition issues. In 2020, there was an increase of 98 million individuals experiencing food insecurity compared to the average number from 1981-2010. (WHO, 2023).

DISPROPORTIONATE IMPACT ON WOMEN, CHILDREN, AND INDIGENOUS COMMUNITIES

Gender inequality, together with the climate change crisis, has fuelled more challenges for women around the world, especially in low and middle-income nations. The impact of climate change is not balanced between genders; in fact, women bear the most impact. In low and middle-income countries, women are mostly involved in the agriculture sector to earn income for their families (UN Women, 2022). However, accelerating periods of drought and flood have been causing devastating effects on the agriculture sector, hence affecting the income of women. Low income increases the risk of other domestic issues, such as violence (Cooper, 2021). According to UN Women, climate change leads to many other issues, such as gender-based violence, sexual violence, human trafficking, child marriage and other types of violence. The routine tasks of women become more complex due to the scarcity of natural resources like water. Currently, women and girls have to walk for approximately 6 kilometres in developing countries and carry 20 litres of water every day (Otero, 2023). As climate change intensifies, water scarcity will worsen, forcing women and girls to walk farther to access water. Climate change also threatens women's health due to poor sanitation, increases in workload and poor access to health services. Studies have revealed that increased temperature and diseases such as dengue fever, malaria and Zika virus are directly linked to climate change, resulting in stillbirth and negative maternal outcomes (UN Women, 2022).

According to resolution 32/33 of the Human Rights Council, children are particularly susceptible to the effects of climate change, which might significantly affect their ability to maintain the best possible level of physical and mental health, as well as their access to school, sufficient food, decent housing, clean water and sanitary conditions (Agudo, n.d). Social problems such as poverty and lack of access to food and water in developing and least developing countries often result in children dropping out of school and assisting their families with tasks such as farming, carrying water, and hunting for food. Children have been mostly ignored during climate change decision-making despite being the most fragile group and future leaders. Only 2.4% of climate finance was allocated for projects incorporating child-responsive activities (UNICEF, 2023). Whereby UNICEF's 2021 landmark Children Climate Risk Index (CCRI) stated that 1 billion children are at high risk of climate change (UNICEF, 2023). A safe environment and livelihood are the basic needs of all children to grow healthy and happy. Thus, climate change has turned children's lives into struggles and fights for survival. As of 2022, 739 million children were subjected to high or extremely high-water shortages, and 436 million children reside in places that have high or extremely high-water vulnerability (UNICEF, 2023).

Indigenous communities are the first to feel the effects of climate change because they are situated in the most vulnerable places. The majority of indigenous groups are found in remote places, making them vulnerable to natural disasters like droughts. Alternatively, they may reside close to the coast, where they face increased risks from sea level rise, salinisation, and erosion. Under the Paris Agreement, countries are required to uphold, advance and take into account the rights of indigenous peoples while addressing climate change (Agudo, n.d). Indigenous communities are often victims of forced displacement. For example, the indigenous communities in Africa's Kalahari Desert are forced to live near the bore drilled by the government to support them with the supply of water as the severe temperature rise and desertification have affected their water source (United Nations, n.d). The survival lifestyle of indigenous communities is also affected by climate change, as the communities mostly rely on their agricultural produce and fishing to meet their food security. Due to climate change, this is at risk as long periods of drought and salinisation affect the agricultural produce. Similarly, ocean acidification and warming of ocean temperature have resulted in a decline in fisheries.

STRATEGIES FOR A RIGHTS-BASED APPROACH TO CLIMATE ACTION, ENSURING JUSTICE FOR ALL.

In order to plan for and adapt to climate decisions, it is essential to implement a human-based approach to climate change. First of all, the rights-based approach recognises that climate change poses a risk to human aspects in addition to the environment. Basic human rights, including the right to food, water, shelter, health, and many more, are in jeopardy because of the negative consequences of climate change. The fact that human rights are not taken into consideration while making decisions about climate action is even more alarming. Furthermore, the rights-based approach recognises that some categories of people are disproportionately affected by climate change, including women, children, the elderly, indigenous people, persons with disabilities, and the poor (Jodoin et al., 2021).

Several tactics for a rights-based approach to climate action include the following:

INTEGRATION OF HUMAN RIGHTS PRINCIPLES

Human rights considerations must be included in climate policy and initiatives in order to have a human dimension. This involves enabling communities and vulnerable groups to participate in the decision-making process for climate action. Their efforts and obstacles in dealing with the effects of climate change on a daily basis will serve as a guide for the creation of effective and efficient policies and initiatives. Additionally, Limon (2009) argues that because international organisations have not succeeded in humanising climate change, there has been a failure to comprehend global human suffering. To guarantee that there is no prejudice of any kind during the participation process, it is necessary to adhere to certain factors, such as equality and non-discrimination. Public opinion will benefit from civic organisations' participation in national and international decision-making.

There is still a number of important concerns that remain unsolved in the application of rights-based approach framework from the perspective of international human rights law. Mahadew (2021) shed light that a human right-based strategy may be used to address climate change has not altered much, not even with the OHCHR report. Therefore, it appears inappropriate for people to rely on international human rights legislation when suing for compensation in climate change cases. But there is an example from Europe that could turn the table around. A group of women claim the Swiss government did not take adequate precautions to protect their health during increasingly severe heatwaves linked to climate change, and the European Court of Human Rights (ECHR) has found in their favour. After their domestic judicial system rejected their lawsuit, the "Swiss grandmothers" first sailed across the Rhine in 2020 to bring it to the court in Strasbourg. According to the ECHR, the system failed to take into account the overwhelming body of scientific information about climate change and gave the matter due consideration. Numerous European nations may be significantly impacted by the ruling, which may put those nations that have not met their climate commitments on notice. Switzerland has not met with its obligations under the European Convention on Human Rights concerning climate change, according to the ECHR, which cited "critical gaps" in domestic regulations and the country's inability to meet its targets for reducing greenhouse gas emissions. The decision was made as global temperatures broke previous records last year and are about to cross a key point beyond which increasingly dire circumstances are predicted (Letzing & Sung, 2024).

ACCESS TO INFORMATION AND INSTITUTION

Access to information and institutions is a necessity for the sturdy achievement of a rights-based approach to climate action. Information regarding climate change should be available to everyone as it poses a risk to everyone, either directly or indirectly. Community members need to be aware of the problem and the dangers posed by climate change, especially the most vulnerable groups. Information access will help the community make sense of the situation and take appropriate action. Public access to national plans, policies, and strategies addressing the climate crisis should be available at the state level. It should be the responsibility of the government to notify and include its citizens in the decisions made. The public will be able to learn about their nation's position in the battle against climate change and offer suggestions for more effective and efficient measures. Similarly, the public should have access to how climate funds received by their country are utilised and the implications of climate-related projects.

In addition, institutions promote inclusive, diverse and cooperative decisions that facilitate the vital action adopted by many countries. But often, the decisions are in favour of cooperation and politics. Therefore, considering human rights-related institutions like UNHRC and civil society will bring a platform of critical debate on adapting to more effective policies and frameworks that focus on people rather than business ideologies and political agendas. Usually, it is visible that large groups of people protest outside the venue of institution meetings, and their voices are often not heard in the decision making and also leads to violence to some extent. For instance, UN Special Rapporteurs voiced concerns over harassment and intimidation of environmental defenders, civil society, human rights advocates, and climate campaigners at the close of COP27. They verified many claims that these actors—including indigenous had been stopped and questioned. They also expressed worries about possible post-COP27 retaliation against human rights advocates who are taking part (Vahí, 2022).

CAPACITY BUILDING INITIATIVES

Initiatives aimed at increasing capacity are crucial for marginalised groups because they provide them the chance to participate in human rights and climate decision-making processes, as well as to share their traditional knowledge that may be used to implement mitigation strategies. By teaching marginalised groups about climate catastrophe, we want to increase their knowledge of the issue and encourage them to actively participate in creating plans and policies that will directly impact their way of life. By facilitating the sharing of traditional mitigating techniques, including farming expertise and food preservation techniques, capacity development initiatives would increase the security of food and water. Sessions aimed at developing capacity may teach women about skills for mitigating risks, help communities create emergency plans to get ready for catastrophic disasters, and teach them how to help the disabled. Therefore, capacity building sessions are a grassroots effort to help communities comprehend the problems they encounter on a daily basis and equip them to deal with the climate catastrophe. This information will be sent along to the government for inclusion in their agenda.

RIGHTS BASED-APPROACH LEADS TO SUSTAINABLE DEVELOPMENT GOALS (SDG)

Both the SDGs and rights-based approach complement each other as both focuses on the sustainable outcome for the people. SDG integrates factors like people, planet, prosperity, peace and partnership to achieve the sustainable development. Several SDG goals are converted from objectives or ambitions into instantaneous rights. It ensures poverty reduction (SDG 1), prioritizing the needs of poor and vulnerable communities affected by adverse climate change impacts. Also, food security initiative (SDG 2) is being addressed which achieves "Right to Food". Prioritizing access to enough food for vulnerable groups due to destruction in food supply chain. "Good health and well-being" (SDG 3) and "Right to Health" aligns together as climate change increases the risk of health issues such as diseases, malnutrition and psychological effects. "Clean water and sanitation" (SDG 6) and "Right to Water" links together as well as it ensures availability and sustainability supply of water for people since it is a basic need for human survival. Last but not the least, "Climate Action" (SDG 13), requires to take urgent action to combat climate change and impacts associated. Similarly, rights-based approach focuses on urgent actions that helps to protect the human-rights that every human deserves. Especially the vulnerable group whose contributions is least towards climate crisis but faces adverse impacts.

To guarantee a human rights-based approach in national policies and programmes for the implementation, monitoring, and reporting of the Sustainable Development Goals (SDGs), human rights bodies at the local, regional, and international levels can be utilised. These organisations can act as a liaison between governments and marginalised communities and offer useful data for reporting and decision-making procedures. They may also provide knowledge on how to create connections between various rights and how to include organised civil society in SDG procedures. SDGs and human rights-based approach are interlinked, therefore the governments must consider the human rights-approach in decision making to achieve the SDGs. Increased cooperation across the strategy for sustainable development and the current human rights frameworks would guarantee consistency and prevent initiatives at the national level from being duplicated. Additionally, it will improve efficiency and guarantee that governments are using all of the resources at their disposal to uphold the fundamental liberties of the citizens they represent.

CONCLUSION

The implementation of a human rights-based approach in the decision-making at the national and international level is a way to fill gaps in dealing with this humanitarian crisis. Climate change has been discussed globally for more than two decades, with least actions taken to deal. Political and corporate agendas have hindered the implementation of effective strategies. Adopting a human rights-based approach in international law will be the ice breaker to allow effective actions by the government. So far, countries have only pledged and made commitments and policies, but no obligation has been committed where an action can be taken if they fail to fulfil their pledges, commitments, and policies. A human rights-based approach in international law will allow states to be obligated towards the victims of climate change and be compensated accordingly. This will pressurise the states to take effective actions in dealing with climate change so that fewer human rights are violated and fewer individuals are compensated. A human rights-based approach should be the weapon for bringing governments and international organisations effective. UNHRC and civil society are recommended to collaborate and work together in the international community to adapt to a human rights-based approach.

REFERENCES

Agudo, P. A. (n.d). *Climate change and the human rights to water and sanitation.* Office of the United Nations High Commissioner for Human Rights. https://www.ohchr.org/sites/default/files/2022-01/climate-change-2-friendlyversion.pdf

Al Jazeera. (2023, May 22). *Climate change causes 2m deaths in 50 years; poor suffer most: UN.* Al Jazeera. https://www.aljazeera.com/news/2023/5/22/climate-change-causes-2m-deaths-in-50-years-poor-suffer-most-un

Alam, , MAli, , M. F. (2022). Health and wellbeing and quality of life in the changing urban environment: Challenges and government response in Fiji island. *Journal of Positive School Psychology, 6*(9), 2853-2863.

Atapattu, S. (2015, October 16). *Human Rights Approaches to Climate Change.* Routledge. http://books.google.ie/books?id=A-e9CgAAQBAJ&pg=PT48&dq=https://doi.org/10.4324/9781315849683&hl=&cd=2&source=gbs_api

Climate change and water-related disasters. (n.d.). UNEP - UN Environment Programme. Retrieved April 2, 2024, from https://www.unep.org/topics/fresh-water/disasters-and-climate-change/climate-change-and-water-related-disasters

Cooper, B. (2021, October 18). *"the lower the income, the higher the prevalence for abuse": Domestic violence linked to poverty.* Indianapolis Recorder. https://indianapolisrecorder.com/the-lower-the-income-the-higher-the-prevalence-for-abuse-domestic-violence-linked-to-poverty/

Da Costa, K. (2014). Can the observance of human rights of individuals enhance their resilience to cope with natural disasters? *Procedia Economics and Finance,* 18, 62–70. DOI: 10.1016/S2212-5671(14)00914-9

Human Rights Council. (2011) Resolution 18/22 Human rights and climate change, A/HRC/ RES/18/22, (2011). In www.ohchr.org. UN Human Rights Council. Retrieved April 2, 2024, from https://www.ohchr.org/sites/default/files/Documents/Issues/ClimateChange/A.HRC.RES.18.22.pdf

IPCC. 2022: Summary for Policymakers [H.-O. Pörtner, D.C. Roberts, E.S. Poloczanska, K. Mintenbeck, M. Tignor, A. Alegría, M. Craig, S. Langsdorf, S. Löschke, V. Möller, A. Okem (eds.)]. In: *Climate Change 2022: Impacts, Adaptation, and Vulnerability.* Contribution of Working Group II to the Sixth Assessment Report of the Intergovernmental Panel on Climate Change [H.-O. Pörtner, D.C. Roberts, M. Tignor, E.S. Poloczanska, K. Mintenbeck, A. Alegría, M. Craig, S. Langsdorf, S. Löschke, V. Möller, A. Okem, B. Rama (eds.)]. Cambridge University Press, Cambridge, UK and New York, NY, USA, pp. 3-33, DOI: 10.1017/9781009325844.001

Jodoin, S., Savaresi, A., & Wewerinke-Singh, M. (2021). Rights-based approaches to climate decision-making. *Current Opinion in Environmental Sustainability*, 52, 45–53. DOI: 10.1016/j.cosust.2021.06.004

Kapoor, A., Alcayna, T., Boer, T. d., Gleason, K., Bhandari, B., & Heinrich, D. (2021). *Climate Change Impacts on Health and Livelihoods: Fiji Assessment.*https://www.climatecentre.org/wp-content/uploads/RCRC_IFRC-Country-assessments-FIJI.pdf

Kharb, A., Bhandari, S., Moitinho de Almeida, M., Castro Delgado, R., Arcos González, P., & Tubeuf, S. (2022, September 13). Valuing Human Impact of Natural Disasters: A Review of Methods. *International Journal of Environmental Research and Public Health*, 19(18), 11486. DOI: 10.3390/ijerph191811486 PMID: 36141766

Leal Filho, W., Krishnapillai, M., Minhas, A., Ali, S., Nagle Alverio, G., Hendy Ahmed, M. S., & Sharifi, A. (2023). Climate change, extreme events and mental health in the Pacific region. *International Journal of Climate Change Strategies and Management*, 15(1), 20–40. DOI: 10.1108/IJCCSM-03-2022-0032

Letzing, J., & Sung, M. (2024, April). *Is climate inaction a human rights violation?* World Economic Forum. https://www.weforum.org/agenda/2024/04/climate-progress-human-rights-violation

Limon, M. (2009). Human rights and climate change: Constructing a case for political action. *Harv. Envtl. L. Rev.*, 33, 439.

Mahadew, R. (2021). A Human-Rights-Based Approach to Climate Change. Revue juridique de l'Océan Indien, (31), 155-168.

Novianti, K., Warsilah, H., & Wahyono, A. (2016). Climate Change and Food Security on Coastal Community. *Jurnal Penelitian Kesejahteraan Sosial*, 15(3), 203–218.

Nurse, L. A., McLean, R. F., Agard, J., Briguglio, L. P., Duvat-Magnan, V., Pelesikoti, N., & Webb, A. (2014). *Small islands.* Cambridge University Press.

Office of the United Nations High Commissioner for Human Rights. (n.d.). *The impacts of climate change on the effective enjoyment of human rights.* Office of the United Nations High Commissioner for Human Rights. https://www.ohchr.org/en/climate-change/impacts-climate-change-effective-enjoyment-human-rights

Otero, M. (2023, May 17). *Six kilometers a day: Water and sanitation.* U.S. Agency for International Development. https://www.usaid.gov/six-kilometers-day

Rheeney, A. (2023, March 28). *U.S.P. Student initiative to go before I.C.J.* Samoa Observer. https://www.samoaobserver.ws/category/samoa/102759

Stein, P. J., Stein, M. A., Groce, N., & Kett, M. (2023). The role of the scientific community in strengthening disability-inclusive climate resilience. *Nature Climate Change*, 13(2), 108–109. DOI: 10.1038/s41558-022-01564-6

UNICEF. (2023, November 1). The climate-changed child. United Nations International Children's Emergency Fund. https://www.unicef.org/reports/climate-changed-child#download-the-report

United Nations. (n.d.). Climate change for indigenous peoples. United Nations. https://www.un.org/development/desa/indigenouspeoples/climate-change.html

United Nations Department of Economic and Social Affairs. (n.d.). *Water and food security*. United Nations. https://www.un.org/waterforlifedecade/food_security.shtml

Vahí, A. O. (2022). *COP27 and the centrality of human rights in Climate Action*. Universal Rights Group. https://www.universal-rights.org/cop27-human-rights/

Women, U. N. (2022, February 28). *Explainer: How gender inequality and climate change are interconnected*. UN Women. https://www.unwomen.org/en/news-stories/explainer/2022/02/explainer-how-gender-inequality-and-climate-change-are-interconnected

Chapter 2
Women's Leadership in Environmental Dialogue: A Human Rights Approach

Swati Chakraborty
https://orcid.org/0000-0003-0799-1954
GLA University, India

ABSTRACT

This research proposal explores the critical nexus between women's leadership, environmental conservation, and human rights. Recognizing the multifaceted challenges facing our planet, this study delves into the under-explored yet essential dimension of gender equality in conservation efforts. Through a comprehensive review of existing literature and the application of a human rights framework, the proposal aims to uncover the contributions of women to environmental stewardship, analyze the barriers they confront, and propose strategies for advancing gender equality in conservation. This paper underscores the importance of integrating a gender perspective into environmental policies and practices, highlighting how women's involvement in environmental advocacy not only promotes sustainable development but also advances gender equality and human rights.

INTRODUCTION

This research proposal explores the critical nexus between women's leadership, environmental conservation, and human rights. Recognizing the multifaceted challenges facing our planet, this study delves into the underexplored yet essential dimension of gender equality in conservation efforts. Through a comprehensive review of existing literature and the application of a human rights framework, the

DOI: 10.4018/979-8-3693-6069-9.ch002

proposal aims to uncover the contributions of women to environmental stewardship, analyze the barriers they confront, and propose strategies for advancing gender equality in conservation. By employing qualitative research methods, including interviews and case studies, the study seeks to amplify the voices of women leaders in the field, illuminate the intersectional dynamics shaping their experiences, and identify opportunities for transformative change. The findings of this research hold significant implications for academic scholarship, policy development, and practical interventions aimed at fostering inclusive, equitable, and rights-based approaches to environmental sustainability. This paper explores the critical role of women's leadership in environmental dialogue through the lens of human rights. It examines how women's involvement in environmental advocacy not only promotes sustainable development but also advances gender equality and human rights. By highlighting case studies and key figures, this paper underscores the importance of integrating a gender perspective into environmental policies and practices.

The intersection of gender, environmental issues, and human rights is increasingly recognized as essential for sustainable development. Women, who are often disproportionately affected by environmental degradation, have emerged as pivotal leaders in environmental advocacy. This paper aims to explore the significance of women's leadership in environmental dialogue and how it aligns with a human rights approach.

THE INTERSECTION OF GENDER, ENVIRONMENT, AND HUMAN RIGHTS

Gendered Impacts of Environmental Degradation

Environmental degradation often exacerbates existing gender inequalities. Women, especially in rural and indigenous communities, are primarily responsible for securing water, food, and energy for their families. Environmental issues such as deforestation, water pollution, and climate change disproportionately impact these responsibilities, thereby affecting women's health, economic opportunities, and social status.

All individuals are born equal and free, as affirmed by the Universal Declaration of Human Rights. Every person, irrespective of gender, is endowed with inherent human rights and dignity. However, climate change, biodiversity loss, habitat destruction, and pollution pose severe threats to lives, economies, and entire cultures and societies. These interconnected environmental crises endanger human rights, including the rights to life, health, food, water, sanitation, culture, self-determination, and more, with varying impacts based on gender.

Systemic gender-based discrimination and violence against women, along with stereotypes, resource constraints, differing nutritional needs throughout life (especially during infancy, childhood, pregnancy, and childbirth), and varying levels of exposure to pollution and hazardous substances, exacerbate the negative human rights effects of environmental degradation, including climate change. Despite these disproportionate impacts, many women lead environmental actions, managing resources, advocating sustainable lifestyles, and defending land, water, nature, and communities.

States, businesses, international organizations, and other actors have procedural and substantive obligations under international human rights and environmental laws, as well as international policy commitments, to address environmental crises. In doing so, they must also prevent the gendered impacts on human rights and ensure that actions to address environmental degradation are gender-responsive, equitable, non-retrogressive, non-discriminatory, and sustainable. This document outlines the key human rights obligations and responsibilities of states and other actors concerning gender and the environment.

Impact: Best or Worst

The worst impacts of environmental harm are often borne by those who have contributed the least to these crises, are in vulnerable situations, and have limited resources to cope with the consequences. The Paris Agreement, the Convention on Biological Diversity, the Convention to Combat Desertification, and various other multilateral environmental agreements include commitments to achieving gender equality, empowering women, and recognizing the importance of women's engagement in climate change responses, biodiversity conservation, and land preservation.

It is crucial that states take immediate and ambitious measures to respond to these interconnected environmental crises in a manner that is mindful of and seeks to address their gendered impacts. Effectively tackling land degradation, climate emergencies, mass extinction, zoonotic diseases, and the pollution crisis requires a gender-responsive, human rights-based approach. This approach must make room for the important but differentiated contributions of all genders, especially women and individuals with diverse gender identities.

Addressing the root causes of human rights harms, such as poverty, inequality, discrimination, and marginalization, will be key to effectively mitigating these environmental challenges.

Human Rights and Environmental Justice

The human rights framework provides a crucial lens for understanding environmental justice. Access to clean water, air, and land is not only an environmental issue but also a fundamental human right. Recognizing the gendered dimensions of these rights is essential for addressing the root causes of environmental injustice.

Gender-based discrimination is prohibited by every major international human rights treaty. The Convention on the Elimination of All Forms of Discrimination Against Women (CEDAW) requires states to take measures to amend or repeal laws, regulations, customs, and practices that constitute gender-based discrimination, including gender-based violence against women. Despite these commitments, women and individuals with diverse gender identities around the world continue to face historic, systemic, direct, and indirect discrimination. This discrimination manifests as an uneven distribution of power in society and within families, limited access to productive and natural resources, restricted access to information and decision-making authority, and discriminatory laws, social norms, and practices, often resulting in violence.

To address gender-based discrimination and violence in the environmental context, states should revise, amend, or repeal laws, policies, and practices that create or perpetuate gendered environmental impacts. They should work to reduce inequalities related to land ownership, tenure, and access to resources, and strengthen women's roles in conservation, stewardship, and management. ("HUMAN RIGHTS, THE ENVIRONMENT AND GENDER EQUALITY - UN Women")

In its General Recommendation No. 37, the Committee on the Elimination of Discrimination against Women acknowledges that environmental crises exacerbate existing gender inequalities and compound intersecting forms of discrimination. These intersecting forms of discrimination can be based on economic or social class; caste, ethnicity, race, religion, or indigeneity; disability; age; migration status; and sexual orientation, gender identity, or sex characteristics. Those experiencing such multiple and intersecting forms of discrimination are often disproportionately affected by environmental harms.

When crafting environmental laws, policies, and practices, states should aim to address these intersecting systemic inequities and take concrete steps to overcome the participation and leadership barriers faced by women who experience intersectional marginalization and discrimination. Such steps may include ensuring equal access to information, participation, and remedies; investing in gender-responsive social protection and care systems; and promoting empowerment and capacity-building initiatives.

Article 5 of CEDAW requires state action to eliminate ideas of inferiority or superiority of any sex and all gender-based stereotypes. In the context of environmental degradation, such stereotypes can create significant human rights risks. They may magnify the burden of unpaid care and domestic work that disproportionately falls on women and increase the threat of gender-based violence for women and individuals with diverse gender identities when using natural resources and ecosystem services, as well as in the context of both slow and sudden-onset environmental disasters. These stereotypes can also create barriers for women, girls, men, boys, and persons of diverse gender identities, hindering their participation in various forms of climate and environmental action that are viewed as gendered.

Consistent with their commitments under Article 5 of CEDAW, states should take concrete action to eliminate harmful gender stereotypes. This is essential not only to protect human rights but also to ensure a safe, healthy, clean, and sustainable environment for all.

International and regional human rights instruments, including the ICCPR, the Rio Declaration on Environment and Development, the UN Declaration on the Right to Development, the Aarhus Convention, and the Escazú Agreement, enshrine the right to equal, free, active, meaningful, and informed participation in decision-making processes. Women's involvement in environmental decision-making at all levels is also a strategic objective of the Beijing Platform for Action.

Several multilateral environmental agreements also incorporate procedural rights obligations. Women, including many indigenous women, possess essential knowledge, skills, and capabilities for effective environmental stewardship and action. They are already leading environmental and climate actions at all levels, but their voice, agency, and participation are under-supported, under-resourced, under-valued, and under-recognized, even though they enhance the effectiveness of environmental action.

States should undertake gender-responsive capacity-building, awareness, and monitoring efforts and institute affirmative measures to overcome traditional barriers to participation faced by women and individuals with diverse gender identities. Ensuring they are active participants and leaders in environmental action at all levels is crucial. Such measures should include ensuring gender balance and adequate gender expertise in environmental negotiation teams, implementing effective participatory processes that include women and their representative organizations in all their diversity when crafting environmental laws and policies, and promoting women to positions of power in all sectors of society.

The Declaration on Human Rights Defenders outlines the duties of not just states, but everyone, in the protection of human rights defenders. These duties include respecting, promoting, and implementing all human rights and ensuring that all persons under the jurisdiction of states can enjoy all rights and freedoms. Around the world, many women have taken action to protect lands, water, nature, communities, and human rights from environmental harms and climate impacts. Often, they do so at great personal risk, facing criminalization, silencing, threats, stigmatization, violence, and even death. Gender-based discrimination and stereotypes can lead to the marginalization of women environmental human rights defenders even within environmental movements, placing them at additional risk of gender-based violence.

States must respect, protect, and fulfill their human rights obligations to women environmental human rights defenders. This includes guaranteeing their rights to expression, assembly, and association in both virtual and physical spaces, and ensuring their participation in decision-making at local and international levels. Additionally, states must ensure the rights to redress and accountability for threats and harms against these defenders, in order to protect them, the communities and ecosystems they strive to defend, and the planet we all share.

Under the ICCPR, states are required to guarantee non-discrimination in the access of all persons, regardless of gender, to justice and to legal remedies for human rights violations. CEDAW Article 15 affirms the equal rights of women and girls before the law and in legal proceedings. Regional agreements, including the Aarhus Convention and the Escazú Agreement, specifically address access to justice in environmental matters. Additionally, a number of multilateral environmental agreements include provisions related to access to justice.

Across the world, women have been leaders in using legal mechanisms to seek environmental justice. However, barriers such as difficulty in claiming compensation and unequal recognition of women's legal capacity impede this pursuit. Effective, accessible, and gender-responsive accountability mechanisms are essential to addressing the human rights consequences of environmental harms.

States should remove barriers to access to justice in environmental matters by protecting women from reprisals and ensuring affordable or, where appropriate, free access to legal services, aid, and documents. They should also facilitate cooperation between states and non-state actors, including non-governmental and grassroots organizations, that play key roles in assisting women in accessing justice.

WOMEN'S LEADERSHIP IN ENVIRONMENTAL DIALOGUE

Climate change is an inevitable consequence of the historical global dependence on fossil fuels, which have been integral to economic growth since the Industrial Revolution. This reliance has been characterized by coal-fired power plants, oil wells, factories, and heavy-duty transport powered by internal combustion engines. These sources have already emitted enough carbon dioxide to potentially increase global temperatures by an additional 2 degrees Celsius this century. Currently, the world is experiencing unprecedented storms, wildfires, and severe droughts across various regions (Ashok Osai, 2022). By the 2030s, as temperatures continue to rise, climate hazards are expected to intensify, leading to more crippling heatwaves, worsening coastal flooding, and crop failures. At present warming levels, food production is starting to be strained. Although global food output is still increasing due to advancements in farming and crop technology, climate change has slowed this growth as the world's population exceeds eight billion.

Climate change poses more severe risks to countries in the Global South, particularly to poor and marginalized women whose vulnerabilities are exacerbated by existing societal inequalities. Globally, women are more likely to depend on climate-sensitive sectors for their livelihoods, such as agriculture, and they often have the least capacity to respond to climate extremes like droughts, floods, and storms. Consequently, women and the children they care for are 14 times more likely to die in disasters than men. By 2030, climate change could push 158 million more women and girls into poverty and cause hunger for 236 million more women worldwide. Climate-related displacement also increases women and girls' vulnerability to gender-based violence, human trafficking, and underemployment. (Evelyn Smail, 2023)

One way to equip women with the tools to tackle the effects of climate change is by building their entrepreneurial capacity. This essay explores the various elements of climate action that would greatly benefit from the leadership of women, particularly from marginalized communities in the developing world: climate finance, climate-smart agriculture, and climate education. With the right training and ecosystem support, innovation can be facilitated, and cutting-edge business models can be developed to disrupt traditional energy systems and drive the adoption of clean technology towards a net-zero future.

HISTORICAL CONTEXT AND CURRENT TRENDS

'Climate finance' refers to financial flows mobilized by developed nations' governments and private funds aimed at mitigating and adapting to climate change in developing countries (UNFCCC, n.d.). Achieving 'net zero' emissions by 2050 requires increasing low-carbon investments from US$900 billion in 2020 to US$5 trillion annually by 2030, with developing countries needing US$2 trillion annually—a fivefold increase from 2020 (UNFCCC, n.d.). Allocating at least 50 percent of this climate finance towards gender inclusion could facilitate the crucial transition to zero-carbon and climate-resilient development while promoting equitable social policies.

Despite these goals, many women in emerging economies still lack easy access to adequate funds to cover climate-related losses from events such as floods, droughts, heatwaves, and soil erosion (Osai, 2022). They also face challenges in accessing expensive climate adaptation technologies like flood safeguards, weather forecasting mechanisms, insurance tools, sensors, efficient irrigation systems, and water purification (Osai, 2022). Limited access to resources, particularly climate funding, hinders women's adoption of climate-resilient practices.

Cultural and social barriers in education, limited political participation, and lack of voice in decision-making processes contribute to these challenges. Gender disparities in ownership and access to resources such as land, credit, and technology further impoverish women, reduce their ability to adapt, and increase their vulnerability to climate risks (Osai, 2022). Given that women's livelihoods are often closely tied to climate-sensitive sectors, climate change disproportionately affects them compared to men. However, women's deep knowledge of community dynamics and their skills in natural resource management can significantly enhance the efficiency and resilience of climate change responses (Shiva, 1988).

Climate finance can address gender inequalities by prioritizing adaptation and mitigation projects that benefit women as a primary target group. For instance, investing in clean energy technologies can reduce women's dependence on fossil fuels like kerosene, which contribute to global carbon emissions. Globally, inefficient, fossil fuel-powered cookstoves affect 3 billion people, with women and girls using 90 percent of their incomes for the benefit of their families and communities (Maathai, 2004). Supporting startups that provide women access to and financing for clean cookstoves can alleviate health hazards caused by household air pollution and free up time for women to engage in education and income-generating activities (Maathai, 2004).

Major stakeholders like the Green Climate Fund and the Clean Investment Funds have integrated gender considerations into their frameworks. The Green Climate Fund, with a portfolio size of US$13.5 billion, implements transformative climate actions in over 120 countries and has a dedicated Gender Policy and Action Plan (UNFCCC, n.d.). Similarly, the Clean Investment Funds approved a Gender Action Plan in 2014 to support climate-smart development in developing countries (UNFCCC, n.d.).

Furthermore, the Climate Gender Equity Fund (CGEF), launched in November 2022, aims to promote climate solutions led by and benefiting women. Backed by USAID, Amazon, Reckitt, Visa Foundation, and The UPS Foundation, the CGEF has received initial commitments totaling US$20 million, with an additional US$5 million pledged by USAID during the COP28 conference in December 2023 (UN

Women, 2022). These funds will support grants to businesses, investment vehicles, accelerators, incubators, and grassroots organizations, emphasizing climate solutions that empower women. While these multilateral funds represent a fraction of public and private climate finance, their focus on gender equity sets a precedent for addressing systemic inequities and maximizing the impact of climate finance.

Historically, women have been at the forefront of environmental movements, from the Chipko movement in India to the Green Belt Movement in Kenya. Today, women continue to lead environmental initiatives globally, advocating for policies that address both environmental and gender injustices.

Furthermore, during COP28, several new gender initiatives were launched, including the Gender-Responsive Just Transitions & Climate Action Partnership. This partnership, supported by 68 countries, commits to enhancing gender-sensitive energy transitions and empowering women and girls in climate actions (COP28, 2023). The initiative focuses on three core pillars: enhancing data quality for informed transition planning, ensuring effective financial flows to regions most impacted by climate change, and promoting education, skills, and capacity-building for active participation in transition efforts.

Additionally, the US Government announced a significant investment of US$1.4 billion in the Women in the Sustainable Economy (WISE) Initiative. This initiative aims to bolster women's access to employment, training, leadership opportunities, and financial resources within green (agriculture) and blue (ocean) industries.

Key Figures and Case Studies

- **Wangari Maathai**: The founder of the Green Belt Movement in Kenya, Maathai's work in tree planting and environmental conservation also championed women's rights and community development.
- **Greta Thunberg**: While younger, Thunberg's leadership in the climate movement has inspired millions, highlighting the voices of young women in environmental advocacy.
- **The Chipko Movement**: Led by rural women in India, this movement against deforestation emphasized the critical role of women in sustainable resource management.

WOMEN IN CLIMATE-SMART AGRICULTURE

Significant portions of populations in the Global South depend on the agrarian economy, highlighting the importance of enhancing women's access to innovative climate-smart agricultural tools to narrow the climate mitigation gap. Globally, women constitute 48 percent of the agricultural workforce, with figures soaring to 69 percent in South Asia, and dominate the global fish processing sector at 85 percent, yet their contributions often entail insecure and poorly compensated work . As climate change escalates, the productive potential of these sectors is expected to diminish, placing increased economic strain on women and leading to reduced incomes. This decline in livelihood opportunities threatens both local communities and global food security, as essential natural resources dwindle.

Closing gender disparities in the agricultural sector to grant women access to productive resources has the potential to boost farming yields by 20–30 percent. Various initiatives worldwide serve as exemplars of how innovative technologies can be harnessed to integrate climate action into efforts aimed at gender inclusivity. For instance, in India, S4S Technologies addresses food wastage concerns by providing rural women farmers with solar-powered dehydration systems financed through loans. This includes a solar conduction dryer capable of processing over 45 types of food. The dehydration system extends the shelf life of produce up to a year without requiring cold storage. S4S Technologies purchases the processed food from women farmers, refining it into ingredients sold to the food and beverage industry.

CLIMATE ACTION AND EDUCATION

Inextricably linked to both climate finance and climate-smart agriculture is the imperative of incorporating a gender perspective into climate education and skill-building. This involves enhancing communities' adaptive capacity to climate change, promoting behavioral transformation, and expanding coping strategies. Historically, women have developed vital knowledge and skills related to water harvesting and storage, food preservation and rationing, and natural resource management. In Africa, for example, older women serve as repositories of wisdom with their inherited knowledge and expertise related to early-warning actions and mitigating the impacts of climate disasters.

EARLY EDUCATION FOR GREEN SKILLS

The foundation of this approach begins with early education, emphasizing the development of green skills for girls in the Global South. Such education allows them to increase climate resilience through consequential thinking and communication skills to understand and explain risks to others, systems thinking skills to comprehend the impacts of human activities like deforestation on landslides and flooding, and leadership skills for organizing collective action and driving systems change toward sustainability. There is a strong need for comprehensive climate change knowledge, including the science, ideas for mitigation and adaptation, and projections for future impacts. Essential skills include critical thinking, problem-solving, team collaboration, communication, and dialogue. Moreover, cultivating sustainable habits and behaviors, reducing carbon emissions, ensuring a just transition, and instilling respect for nature are crucial.

HEALTHCARE AND CLIMATE EDUCATION

Ingraining these practices in girls from an early age also equips them to address healthcare challenges exacerbated by climate change, such as worsening air quality, changes in the spread of infectious diseases, and threats to food and water quality. Educating girls in these areas not only helps build a resilient generation capable of tackling climate change but also ensures that they can lead their communities toward a sustainable future.

CHALLENGES AND OPPORTUNITIES

Barriers to Women's Participation

Despite their significant contributions, women face numerous barriers in environmental leadership, including limited access to education, financial resources, and decision-making platforms. ("PS wants more involvement of women in climate change decision-making") Cultural and societal norms often restrict women's participation in public discourse and policy-making.

Opportunities for Empowerment

- **Education and Capacity Building**: Enhancing women's access to education and training in environmental sciences and leadership can empower them to take active roles in environmental dialogue.
- **Inclusive Policy Frameworks**: Policies that explicitly incorporate gender perspectives and prioritize women's voices in environmental governance can lead to more effective and equitable outcomes.
- **Supportive Networks and Alliances**: Building networks of support among women leaders and fostering alliances with other social movements can amplify their impact.

INTEGRATING A HUMAN RIGHTS APPROACH

Legal and Policy Instruments

International frameworks such as the Convention on the Elimination of All Forms of Discrimination Against Women (CEDAW) and the United Nations Framework Convention on Climate Change (UNFCCC) provide mechanisms for integrating gender and human rights into environmental policies. National and local governments must align their policies with these frameworks to ensure comprehensive protection and empowerment of women.

Grassroots Movements and Advocacy

Grassroots movements play a critical role in advancing a human rights-based approach to environmental issues. Women-led organizations and community groups often drive these movements, advocating for both environmental sustainability and gender equality.

CONCLUSION

Women's leadership in environmental dialogue is not only vital for achieving sustainable development but also for advancing human rights. By recognizing and addressing the unique challenges faced by women in environmental advocacy, and by integrating a human rights approach, we can foster a more just and sustainable future for all. The importance of gender inclusion in climate action has never been more pronounced than it is today, as the world sits at a crucial juncture in the global transition to net-zero. As we strive to develop groundbreaking, climate-positive technologies in laboratories, green organizational supply chains, and educate the next generation in climate-friendly practices, it is essential to equip more women not only in policymaking roles but also in private enterprise. This includes roles in corporations, as scientists, technicians, and investors, and particularly as start-up entrepreneurs who can drive the transition away from fossil fuels. Start-ups are uniquely positioned to leverage a mix of talent and technology to develop agile market solutions that address significant challenges such as decarbonization along the gender axis.

Many initiatives across the areas of climate finance, climate-smart agriculture, and climate education are taking center stage in global forums and are showing promise. However, there is much more that can be implemented at scale, especially in countries of the Global South, to increase women's participation in the green transition.

Governments can play a crucial role by offering affordable, high-quality childcare to parents, increasing support for women-owned enterprises, expanding financial and networking opportunities for businesses managed and owned by women, and pursuing measures to eliminate the gender gap across multiple industries. By doing so, we can ensure that women are integral to the efforts in combating climate change and achieving a sustainable future.

REFERENCES

Agarwal, B. (2010). *Gender and Green Governance: The Political Economy of Women's Presence Within and Beyond Community Forestry*. Oxford University Press. DOI: 10.1093/acprof:oso/9780199569687.001.0001

Maathai, W. (2004). *The Green Belt Movement: Sharing the Approach and the Experience*. Lantern Books.

Shiva, V. (1988). *Staying Alive: Women, Ecology, and Development*. Zed Books.

Climate Impacts Tracker Asia. (2023). Gender Equality at COP28: Advancing Climate Justice for Women and Girls. Retrieved from https://www.climateimpactstrackerasia.org/gender-equality-cop28

COP28 Launches Partnership to Support Women's Economic Empowerment and Ensure a Gender-Responsive Just Transition at COP28 Gender Equality Day," COP28 News, December 2023

Osai, A. (2022). Women are Hit Hardest in Disasters, So Why are Responses Too Often Gender-Blind? United Nations Development Programme. Retrieved from https://www.undp.org/documents/women-hit-hardest-disasters

Smail, E. (2023). Gender Equality at COP28: Advancing Climate Justice for Women and Girls. Climate Impacts Tracker Asia. Retrieved from https://www.climateimpactstrackerasia.org/gender-equality-cop28

The Star. (2024, April 18). PS wants more involvement of women in climate change decision-making. Retrieved from https://www.the-star.co.ke/news/2024-04-18-ps-wants-more-involvement-of-women-in-climate-change-decision-making/

United Nations Framework Convention on Climate Change (UNFCCC). (n.d.). Gender and Climate Change. Retrieved from https://unfccc.int/topics/gender/the-big-picture/gender-and-climate-change

Women, U. N. (2014). Gender Equality and Sustainable Development. Retrieved from https://www.unwomen.org/en/digital-library/publications/2014/1/gender-equality-and-sustainable-development

Women, U. N. (2022). Human Rights, the Environment and Gender Equality. Retrieved from https://www.unwomen.org/en/policy-paper-human-rights-environment-gender-equality

Women, U. N. (2022). Human Rights, the Environment and Gender Equality. Retrieved from https://www.unwomen.org/en/policy-paper-human-rights-environment-gender-equality

KEY TERMS AND DEFINITIONS

Environmental Development: Refers to efforts and processes aimed at promoting sustainable development practices that protect and preserve the environment. It encompasses strategies for managing natural resources, mitigating environmental degradation, and promoting biodiversity conservation while addressing socio-economic needs and promoting human well-being.

Gender: Refers to the socially constructed roles, behaviors, activities, and attributes that a particular society considers appropriate for men, women, and people of diverse gender identities. It encompasses norms, expectations, and identities that individuals and societies attribute to being masculine, feminine, or other gender expressions.

Human Rights: Fundamental rights and freedoms that every person is entitled to, regardless of nationality, ethnicity, gender, or other status. These rights are enshrined in international human rights treaties and customary international law and include civil, political, economic, social, and cultural rights.

Rights-based: Describes approaches, policies, or actions that are grounded in the principles of human rights, including the inherent dignity and equality of all individuals. Rights-based approaches emphasize respecting, protecting, and fulfilling human rights as defined in international legal frameworks and norms.

Women: Refers to adult human females. In discussions of gender and social issues, "women" often refers to individuals who identify as female and/or are recognized as female based on societal norms and biological factors.

Chapter 3
Cells to Society
Let's Talk About Menstruation

Vidhi Verma
Delhi Skill and Entrepreneurship University, India

ABSTRACT

In a world full of technology and modernization, we still live in a society where menstruation and periods are a dirty and negative thing that we hesitate to talk about. Where females are still burdened under the myths and taboos. This research is aimed towards breaking the walls of period related myths and taboos and Introducing the IT involvement in the ground level of this field. We can use technology and create huge machines and algorithms that can detect period related diseases within a second. But what is the use of all these if females will yet not be comfortable in coming out and ask for the technology to test them. It's necessary to start working from the ground level so that females can get everything they need to know about periods without feeling uncomfortable or ashamed. This research is aimed towards understanding and looking at the social and technical aspects of menstruation. This project is a combination of society and technology.

INTRODUCTION

Every month, at a given point of day, approximately 1.8 billion females around the world menstruate. Which is 26% of the world population. According to researchers, around 76% of females have no awareness and knowledge about periods before menarche. These 76% girls get terrified when they get their first period and have no information about how to deal with it. In India, these numbers are super high due to the myths and taboos existing in our minds about periods. These taboos and

DOI: 10.4018/979-8-3693-6069-9.ch003

myths become a barrier when it comes to hygienic menstrual practices. In India, in a group of 10 girls, 3 have no information about the periods. And in places like

Rajasthan, this number is as high as 6 9. Such females with lack of information, hide their periods and perform unsafe practices to do so. Buried under the myths of periods, these girls get caught by serious illnesses and diseases that too they hide as they are afraid of visiting a gynaecologist by having the thought of society. This all happens because of lack of knowledge and the hesitation of asking for advice. This hesitation is generated by the myths and taboos of our society. Out of all, 70% females are suffering from menstrual health issues in India. The reason behind the numbers is that females don't talk about their menstrual health and conditions. They do so with the thought of society and the barriers of myths and taboos. These myths and taboos stop these women from seeking out for help. Technology is growing faster and faster, yet we have very little IT involvement in the field of menstruation and period. Currently, 1.13 billion websites are live over the internet. Can you name 5 websites that are properly dedicated towards periods?

This research is aimed towards using technology at the ground level to solve and clear the myths and taboos generated by this society. Menstrual hygiene and awareness is something that people rarely talk about due to the hesitation and fear generated by myths and taboos. Some people think that menstruation is something very private and should not be talked about publicly. Shame, stigmas and misinformation stops people to take stands and talk about periods. Which ultimately results in poor hygienic practices. The process and methodology I am going to follow is that I will reach out to people of different sectors of the society and communicate with them about their experience of this. The goal is to make everyone aware that PERIODS ARE NORMAL and not a thing to be SHAMED about. The approach is to reach out to females and males, and provide them knowledge and information required for them to have about periods. Providing guidance to them and being a society where periods are normal. We will examine the thinking of people of different age groups and sectors, to know what they think about periods? How much do they know about it? How do they handle it?, and What do they experience? This paper includes some experiments, surveys, interviews, and content analysis for the same.

SOCIETY AND PERIODS

Many people think that menstruation and period is a topic that does not need any open talks and discussions. Their taboos and myths are enough and are all they need to know. These myths and taboos have made periods a term of embarrassment which makes a huge impact on many lives and makes them suffer from many different situations. Such taboos can also be life taking. Menstruation Taboo Claims

Life of 14-year-old Girl in Tamil Nadu during Cyclone Gaja. A girl who was forced to sleep outside her house during a cyclone because she was on her periods and was not allowed inside the house even during a massive cyclone.There are plenty of such incidents out there.

We conducted a survey on 100 people, in which 30 females are above the age of 45 (name as GROUP O), 40 are between the age of 15 to 30 (name as GROUP F). 15 males of different age groups (name as GROUP M), and 15 girls between the age of 9 to 14 (name as GROUP Y). The aim of this survey was to understand and know the current state and more near stories of menstruation.

Figure 1. Research Groups

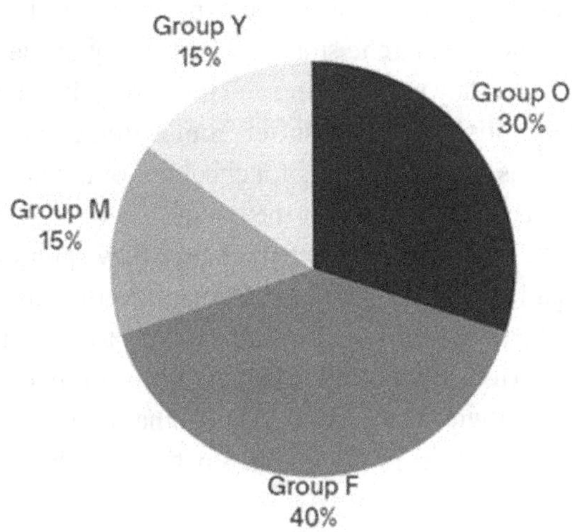

Group O survey showed us the typical mindset and conditions women of the previous generation went through. They said that they used to use clothes during their periods as the pads were really costly and they can't ask for such an amount from their father without telling them the reason. It was really uncomfortable to do day to day activities. 63% of these females are used to stay in a room made up outside their house, during their periods. 37% said that they used to stay in their house during periods but they are not allowed to enter in the kitchen, pooja room, and they are not allowed to touch anyone else's food.

Group Y survey showed us the information girls have before having their menarche. This group involved girls between the age of 9 to 14 who haven't got their first periods. 80% of these girls said that their teacher showed them a video about it but they didn't understand it. They termed the period as "Pain" as they have seen their elder sisters going through pain every month. 87% of them didn't know the correct meaning or periods and what happens during them. 13% of girls had some blurred vision of the correct meaning or menstruation. When we asked them about the correct disposal method of used pads, they mentioned "Secretly" in their explanations. It's in their mind that periods are private and meant to be handled secretly even before their menarche.

Group F, survey showed us the experiences and incidents girls have during their periods. Out of the 40 females, 40% females said a big No for the interview as they didn't want to talk about periods. They felt like we were asking something that should not be asked. From the remaining, 23.3% females agreed for the interview but in an empty room where there should be no male around them. The remaining 36.7% females agreed for the interview without any hesitation. During these interviews, these females shared their stories and experiences with us without any filter or fear. They shared their menarche stories, where many of them thought that they got injured so badly that they are bleeding and they are going to die soon. They also shared that during periods, when they suffer from extreme pain or over blood flow, they use home remedies instead of going for checkups or medical assistance. School and college going girls mentioned that they avoid changing pads in their schools as they try to hide their periods from others, hence they spend the complete 7 to 8 hours in the same pad without changing it. Sometimes they use 2 pads at a time to avoid stains during heavy bleeding. They cancel their tours as they can't visit holy places during their periods. Out of all, 61% females said that they knew about the periods before having their first periods. The Remaining 49% had no idea about periods before their menarche, rather than a repetition of the Results.

Figure 2. Stats of comfortable females

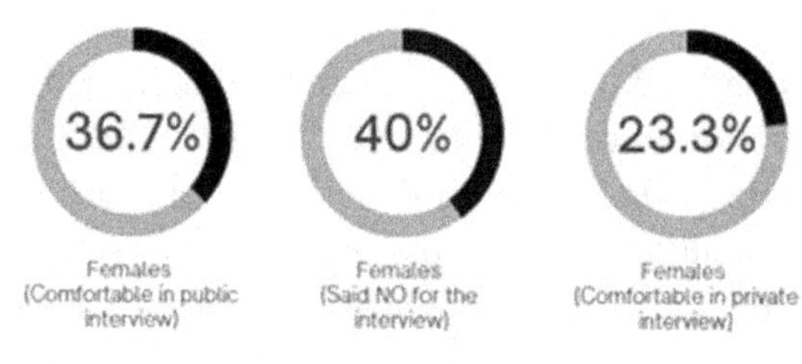

These females shared many things including their weirdest period experience, The craziest thing they have ever heard about periods, What is their take on period and menstrual education, and what do they understand by it. Some of them are: "*It was a strange story for me. I remember I was at a store buying some stuff for my*

younger brother. When I came home, my sister noticed a blood spot on my jeans. I gave the purchased wafers to my brother and went to change.. when i came out of the restroom, i saw the wafers packets were thrown in the dustbin and my mom shouted at me for giving edible things to my brother during periods.."

"*I went to college during my periods. For me it's a very normal thing. I got a stain on my kurta, and a bunch of girls started saying "disgusting" and started laughing at me. Being a girl themself, I think they should understand that it's not disgusting.*`` And many more stories and incidents of period shaming, following the myths and taboos, etc. This survey shows that in 2024 also, such taboos and myths are standing strongly in the society. The urban society girls and females are still facing these taboos in their day to day life. All the females mentioned that menstruating women should avoid touching pickles or fermented foods because their touch can spoil the food. Also many of them mentioned that they are not allowed to talk about their periods in front of their brother or father. Considering the male group, Group M, of our survey, 35.3% males knew the exact meaning of periods and menstruation, where the remaining 64.7% said that menstruation is a process where the body loses old blood and creates new blood. "*I know like it's a monthly cycle which happens to every woman plus it is a normal thing and it's like a filtration where bad or unhealthy blood comes out of the body to generate or make new blood..*"

Figure 3. Males knowledge over periods

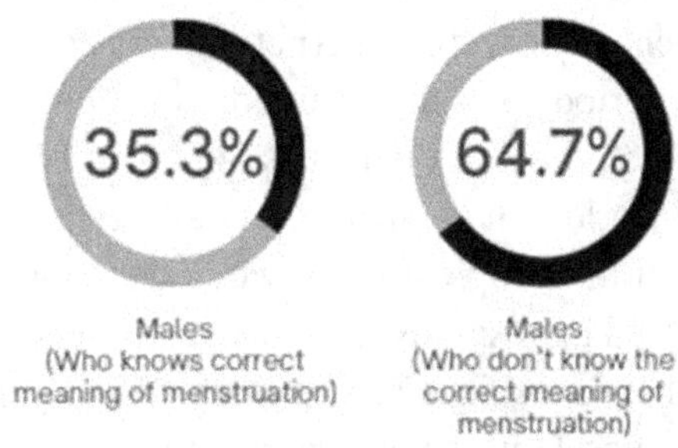

Also few of them said that their teachers used to skip this topic during their science classes which caused them to have a lack of information about periods and menstruation. Also there is a lack of information shared by their own family members and friends. "*not thing, but i remember how our science teacher skipped this topic because it was an all-boys class.*" We asked their views on the importance of menstrual education for boys, and all of them said that it is necessary for males to learn about menstruation and understand it so that they can help their peers and understand what science says, not what society says. On an average of 86% males knew the information like menstrual mood swings, period cramps, and irritability. The Remaining 14% didn't know much about periods as no one told them about it. They said that it's important for parents and schools to give menstrual education to boys as well so that the next generation can understand the truths, not the myths. Education about this is important to break the taboos all around. As a part of our research survey, we also conducted a social experiment on these people. We asked all of them to record a video of themself talking about periods, and send it to us. 98% of them refused and said a big no for that. 2% of people said yes, but we didn't receive a video from them. This experiment gave us the conclusion that people still feel uncomfortable while talking about menstruation. And the discomfort and private treatment results in lack of information and awareness spreading of such crucial topics. Ultimately this results in people using wrong and unhygienic methods for period management. Also because of no information, people believe in myths and taboos because that's all they know about periods.

TECHNOLOGY AND PERIODS

Technology serves an important role in spreading awareness and knowledge all over the world. The most common part of IT that we all encounter every day is websites and apps. If you want to watch a movie, you can find lots of websites that are dedicated towards entertainment. There are 1.13 billion website live currently all over the world. I spent days on the internet looking for a website that is properly dedicated towards periods and menstruation. Where user can get everything they need to know and solutions to their questions. But I found only articles and blogs. In today's smart world, where artificial intelligence and other technologies are so advanced. Most of the people aims towards creating machines and writing algorithms that can be used in hospitals to do fast analysis with more accuracy. But what is the benefit of these machines and technologies if patients are not visiting and using these technologies. During the research, We got the insights that out of all the females, 94% females do not consult any medical expert while facing any kind of menstrual issue. 98% females said that they have never consulted any medical expert or doctor regarding periods.

These females uses home-made remedies without any doctor's prescriptions. While having over bleeding or extreme pain, these females avoid seeking consultancy or check-up. When we asked them why?, the major reason that came out was the society. They are feared that what the society will say or what the other people think. The taboo that periods are meant to be handled privately. So what's the point of creating new machines and technologies if females still continuous to avoid seeking for the medical help because of the fear of society. The technology needs to work on ground level first. To solve an issue which is caused by society, it's very necessary to understand the social aspects first. While doing this research, We understood that females were only comfortable while talking about periods in private. Many of them preferred a written questionnaire rather than a vocal interview. Technology is the most powerful tool today. But very few parts of IT have entered this field. We did another research on 20 IT graduates, and we gave them an option to choose between creating a website dedicated towards menstruation, or a website dedicated towards Entertainment. 94% opted for gaming websites with the reason of profit. They said that menstruation is an underrated topic which no one talks about or looks for. Whereas entertainment is a field that can give them a handsome amount of profit as everyone is keen towards it. The social aspects of menstruation are affecting IT too.

Figure 4. Response graph

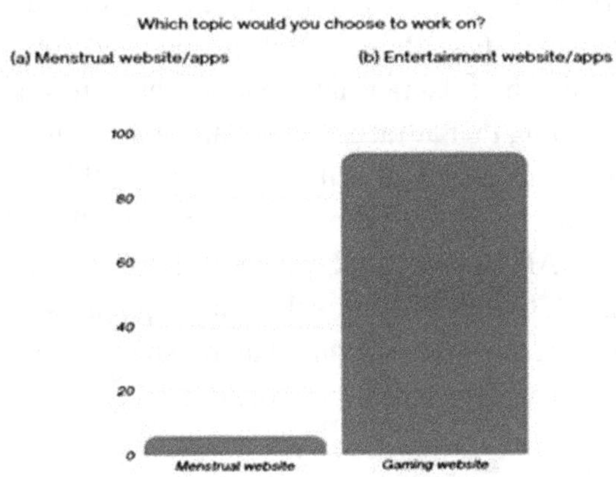

This was a correct choice made by them. Every IT person, or software developer first looks for the scope and future profit of the project. According to

Google trends, In the past 12 months,

"Menstruation" term is searched on google with an average of 10. Whereas "Movies" term is searched with an average of 86. This clearly shows the interest of the public. And to reach max audience and earn profit, it's necessary to work on public interest. This lack of interest is the major reason for such a minimal amount of IT in the field of menstruation.

Figure 5. Search rate of "Period" term over google

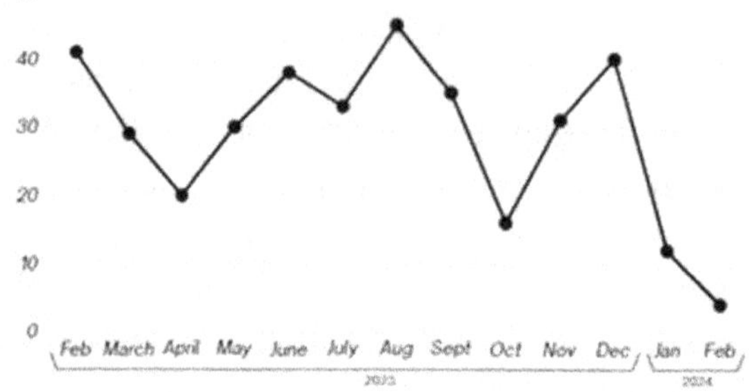

The above chart shows the fall in the search rate of Period term over google, world-wide. (Google Trends) Not only the social taboos and mindset is the reason for this downfall, but also the IT sector. No software/website/app provides what females wants related to periods. As pr the answers we got from our candidates during the research, every site and app have the same information and nothing new or helpful.

They all have some common data and information. There are no updates, innovation, helping factor, or anything that attracts you towards these sites. Not only projects, but there is a lack of research as well in this field. During this research while doing the literature survey, We found only one research paper "DEEP LEARNING FOR POLYCYSTIC OVARIAN SYNDROME CLASSIFICATION USING CONVOLUTIONAL NEURAL NETWORK", jitk, vol. 9, no. 2, pp. 218-226, Feb. 2024" that included IT and menstruation. Even while having smart devices like cramp care heating pads, We still haven't reached the ground level yet.

Figure 6. The impact flow of myths

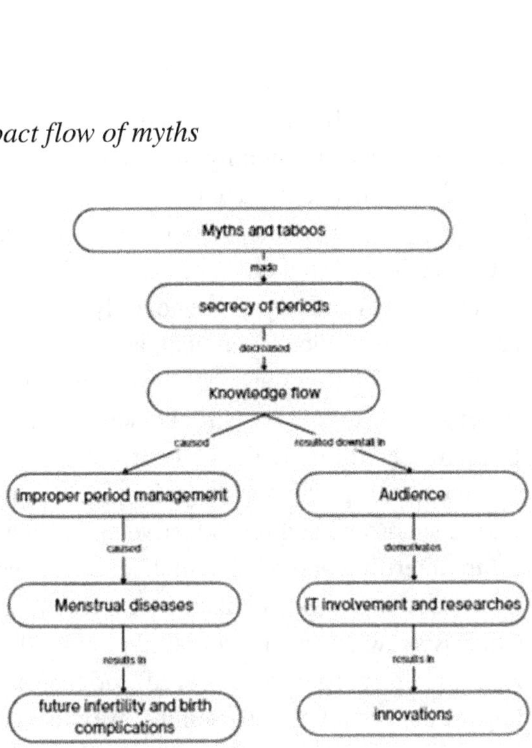

The above flow diagram shows the impact of menstrual myths over the health and the technology sectors. The aim of this diagram is to show the root cause of increasing menstrual diseases and lack of IT involvement in the field of menstruation.

CONCLUSION

After studying the social factors and technical factors of menstruation, myths and taboos, we can say that the social myths and taboos have made a strong barrier for people and for IT too. due to lack of information, people stick to these taboos and myths. Even in 2024, girls still face problems like period shaming. 21st century girls are also not confident enough to talk about periods in front of their father and brother. Such low levels of interest and awareness cause loss to menstrual products creating companies and software developers. Which ultimately makes them switch their field. By our research survey we also got the data that 94% of IT graduates don't want to work in the menstruation field because of such a small audience. Many females said that they find the same articles about menstruation all over the internet, hence they don't find anything new or helpful. Also, they felt shame and hesitation in consulting a doctor regarding menstrual health issues or monthly check-ups. More than 60% of males don't know the correct meaning of menstruation as well as the science behind it. Males above the age of 45 are not even interested in talking about periods. To increase the involvement of IT in this field, it's highly required to do something that can help females in fighting their hesitation, that can help them without making them feel ashamed, and that can show them something new about it. During this research of over 100 people, we can see that on an average of 60% people are not interested in talking about this topic and neither do they want to learn about it. The period related myths and taboos are not only a barrier on the social life of females, but also a barrier on medical, technical, and educational fields. 70% of Indian women face menstrual health issues because of following these myths and taboos and ultimately performing unsafe menstrual practices like using dirty clothes, ashes, leave, using the same pad for more than 6 hours, etc. These unsafe practices made them suffer from menstrual problems like polycystic ovarian syndrome, dysmenorrhea, Premenstrual syndrome and reproductive and urinary tract infections, ultimately causing future infertility and birth complications. Breaking these myths is the most crucial first step that should be taken to resolve all these problems and to improve the growth [as shown in the flow diagram under the heading of "The impact of myths"]. By providing proper menstrual education in schools to both boys and girls and by ending the concept of period shaming, we can create a better society with no myths. By properly using the internet and technology to spread awareness and normalize periods, and also the importance of hygiene during periods, thousands of lives can be saved.

CONFLICT OF INTEREST STATEMENT

This is to certify that there is no conflict of interest in this paper.

REFERENCES

Google Trends. https://trends.google.com/trends/

Chapter 4
Featuring Healthcare, Environment, and Human Rights:
Applications of Artificial Intelligence in the Health Domain

Bhupinder Singh
https://orcid.org/0009-0006-4779-2553
Sharda University, India

Christian Kaunert
https://orcid.org/0000-0002-4493-2235
Dublin City University, Ireland

ABSTRACT

The right to the best possible level of physical and mental health is protected by a number of international legal instruments, such as the International Covenant on Economic, Social, and Cultural Rights. Both liberties and entitlements are included in this right and this includes the protection against torture and against non-consensual medical treatment or experimentation, which is especially important for people with disabilities. The entitlement is the freedom to obtain first-rate medical treatment without encountering prejudice. A human rights-based perspective on health requires nations to set up health systems that respect human rights and are efficient, gender-transforming, integrated, and accountable. It also urges the adoption of public health initiatives targeted at enhancing the fundamental factors that influence health, such access to hygienic conditions and clean water. This chapter deeply dives in the healthcare and human rights perspectives with artificial intelligence applications.

DOI: 10.4018/979-8-3693-6069-9.ch004

1. INTRODUCTION AND BACKGROUND

All the people should be able to get the essential medical treatment whenever and wherever they need it, without having to worry about their finances, according to the concept of universal access to health services (Basulo-Ribeiro & Teixeira, 2024). Nobody should die or become ill because they are too poor to pay for necessary medical care (El Zein et al., 2024; Costa, 2024; Ray et al., 2024; Dinesh Arokia Raj et al., 2024). So, achieving good health is closely related to achieving other fundamental human rights including having access to clean water, proper sanitation, enough food, decent housing, education, and safe working conditions (Yue & Shyu, 2024) (Mithas et al., 2022) (Ivanov et al., 2019). The right to health also includes the freedom of individuals to make decisions about their bodies and health, including having access to services and information related to sexual and reproductive health without facing violence or discrimination (Javaid et al., 2022; Fraga-Lamas et al., 2021). In the modern global context, the interdependence of healthcare, the environment, and human rights creates a complex web of opportunities and difficulties (Felsberger et al., 2022) (Rath et al., 2024). Even while medical progress has led to notable gains in illness prevention, diagnosis, and treatment, human rights abuses and environmental degradation nevertheless have a negative impact on public health outcomes globally (Asadollahi-Yazdi et al., 2020; Meyendorf et al., 2023). Within this framework, the advent of Artificial Intelligence (AI) technology presents unparalleled potential to transform healthcare provision, reduce environmental hazards, and preserve fundamental human rights (Zhong et al., 2017) (Angelopoulos et al., 2020). With utilizing AI-driven technologies, data-driven insights may be used to solve important healthcare concerns and promote fair access to health services that have an influence on the environment and are ecologically sustainable (Lu et al., 2020).

Everyone has the right to privacy, decency, and respect; they shouldn't have to endure forced exams, medical experiments, or treatment without giving their informed permission (Chander et al., 2022). All people possess some inalienable rights only by virtue of being human (Kasowaki & Ahmet, 2024). These rights are independent of nationality, ethnicity, religion, culture, or socioeconomic standing. The only way people can live dignified lives is by exercising our rights without limitations (Sima et al., 2020). Possibly the most vital and related of all our rights is the ability to receive healthcare. Because human life is so fleeting, it is imperative that this right be protected as a matter of public welfare (Tseng et al., 2021). The ensuring that the most marginalized members of society to live dignified lives is contingent upon universal healthcare (Anastasi et al., 2021).

Figure 1. The Landscapes of Introduction Split Sections

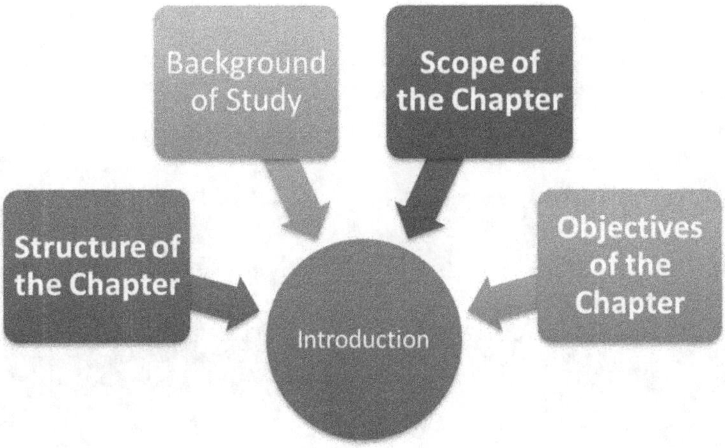

(Source- Original)

1.1 Scope and Objectives of the Chapter

The discourse around healthcare, environmental issues, and human rights has gained prominence in recent times, hence requiring an interdisciplinary approach to achieve practical answers (Dwivedi et al., 2021). This chapter has the following objectives to:

- reconnoiter artificial intelligence (AI) in the healthcare industry, examining its many uses that aim to solve problems in healthcare delivery while advancing environmental sustainability and protecting human rights (Majstorovic & Mitrovic, 2019).
- explain how AI technologies are changing healthcare service delivery, improving environmental monitoring, and supporting human rights in healthcare settings (Badri et al., 2018). It does this by drawing on a variety of literature and case studies (Xu et al., 2018).
- look at the adoption of AI ethical, legal, and social implications (ELSI), highlighting the necessity of appropriate AI governance frameworks (Bhuiyan et al., 2020).
- present a comprehensive view of AI integration that puts equality, transparency, and stability first. It does this by shedding light on the interactions between healthcare, the environment, and human rights (Reier Forradellas & Garay Gallastegui, 2021).

Figure 2. displays the Objectives of the Chapter

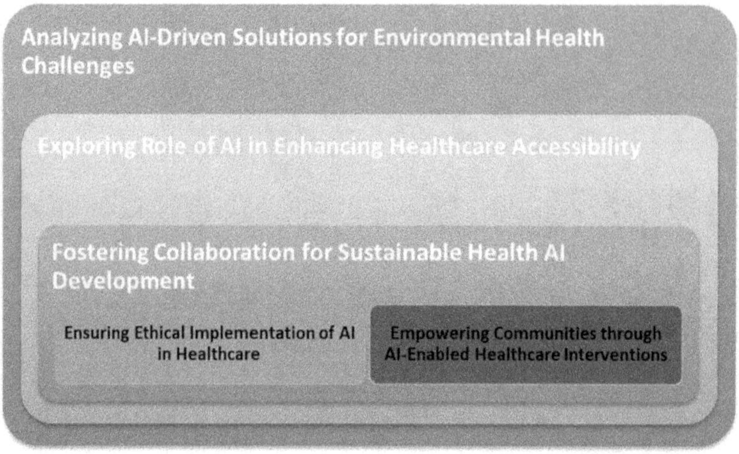

(Source- Original)

1.2 Structure of the Chapter

This chapter comprehensively explores the diverse arena of Featuring Healthcare, Environment and Human Rights: Applications of Artificial Intelligence in Health Domain. Section 2 discusses the Artificial Intelligence in Healthcare. Section 3 elaborates the Environmental Monitoring and Sustainability. Section 4 lays down the Human Rights Implications in Healthcare and Environment Concerns. Section 5 specifies the Case Studies and Practical Applications. Section 6 highlights the Ethical and Governance Considerations. Finally, Section 7 Conclude the Chapter with Future Scope.

Figure 3. Share the Flow of this Chapter

(Source- Original)

2. ARTIFICIAL INTELLIGENCE IN HEALTHCARE

Artificial intelligence (AI) is the capacity of a computer program to carry out operations related to human intellect, including learning and reasoning as well as adaptability, interaction, and sensory perception. As opposed to conventional computer algorithms, which operate according to preset rules, artificial intelligence (AI) systems pick up these rules via training (Singh Rajawat et al., 2021). AI has the power to completely transform the way healthcare is delivered by gleaning insightful information from massive volumes of digital healthcare data (Leng et al., 2020). Algorithms are the focus of the software component of AI systems, which normally consist of both hardware and software components (Javaid et al., 2021). The linked network of neurons seen in the human brain is mimicked by artificial neural networks (ANNs), which provide as a conceptual foundation for the development of AI algorithms. With the use of analytics and these algorithms, machines can find intricate non-linear relationships in large datasets (Hassoun et al., 2023).

Artificial intelligence (AI) has the potential to revolutionize several facets of healthcare delivery, including patient monitoring, tailored medication, and diagnosis and treatment. Specifically, machine learning algorithms have shown to be exceptionally adept at sifting through massive information, finding patterns, and making previously unheard-of diagnoses with surprising precision (Aoun et al., 2021). AI-driven diagnostic systems, for example, may detect illnesses like cancer, cardiovascular conditions, and infectious diseases more quickly, allowing for timely treatments and better patient outcomes. Also, telemedicine systems driven by AI provide distant consultations, improving healthcare accessibility for marginalized communities and mitigating healthcare distribution inequalities (Arden et al., 2021).

The use of novel technology gives rise to apprehensions regarding possible errors and violations of data security. Errors in the vital field of healthcare can have serious repercussions for patients, who frequently seek care at vulnerable times. When AI is used effectively in conjunction with physicians, it may provide evidence-based management and support medical decision-making, among other important benefits (Rane, 2023). Applications of AI in healthcare include diagnostics, drug development, epidemiology, individualized treatment, and operational effectiveness. To prevent possible harm, such as unethical behavior, the incorporation of AI solutions into medical practice requires a strong governance framework (Awan et al., 2021). The Hippocratic Oath is based on the medicalist Hippocrates' teachings, which are the source of ethical standards in medicine. Human rights may be used as a yardstick to gauge how well AI is accepted by society and to identify obstacles and provide recommendations. However, there are gaps to consider because human rights standards existed before artificial intelligence. The fast improvements in AI technologies may raise worries about possible hazards and abuses of human rights (Kumar et al., 2020). The applicability of the human rights framework for assessing AI-related issues, including dangers and benefits, is also called into doubt by this circumstance. With providing individualized health advice, medication reminders, and mental health support, virtual health assistants and chatbots with natural language processing (NLP) skills enhance patient engagement and self-management. In addition, wearable technology and robots powered by AI enable people with impairments to reclaim their freedom and mobility, highlighting the revolutionary potential of AI to improve quality of life (Li et al., 2017).

Figure 4. Key Genesis of Artificial Intelligence in Healthcare

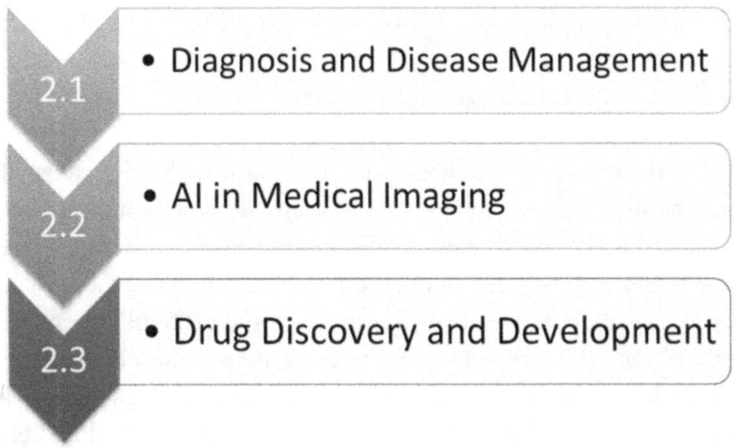

(Source- Original)

2.1 Diagnosis and Disease Management

Artificial Intelligence (AI) represents a swiftly advancing domain within computer science, striving to develop systems capable of executing tasks traditionally associated with human intellect. This encompassing field encompasses methodologies like machine learning (ML), deep learning (DL), and natural language processing (NLP) (Kurniawan et al., 2022). With the advent of digital healthcare technologies like artificial intelligence (AI), 3D printing, robots, nanotechnology, and others, the healthcare industry is undergoing a major upheaval. Numerous potential to reduce human error, improve treatment results, and collect data longitudinally are presented by the digitalization of healthcare (Jakka et al., 2022). Artificial intelligence (AI) approaches, from machine learning to deep learning, are essential in many healthcare fields, such as the creation of innovative clinical systems, the administration of patient data and records, and the treatment of a variety of illnesses (Xing et al., 2021).

2.2 AI in Medical Imaging

Medical imaging refers to a range of technologies that are used to see the human body in order to diagnose, monitor, or treat conditions. These technologies mostly use non-invasive visualization techniques to help medical professionals identify injuries, diagnose illnesses, or treat chronic conditions (Shi et al., 2020). AI analysis in medical imaging helps healthcare practitioners spot subtleties or areas of concern that could go undetected by humans. AI-enhanced medical imaging, for instance, may analyze data points in a medical report to distinguish between healthy and sick tissues and to identify signals over background noise (Finance et al., 2015).

2.3 Drug Discovery and Development

These AI methods are very good at correctly identifying a wide range of illnesses. Artificial intelligence (AI) in healthcare offers previously unheard-of chances to improve patient and clinical team results, cut expenses, and more (Saeed et al., 2023). These models go beyond automation; they allow data generation and suggestion, as well as collaborative analysis, for patients, their families, and healthcare professionals (Kumar & Mallipeddi, 2022). AI can help pinpoint particular populations or geographic areas that are more likely to experience high-risk behaviors or diseases (Bag & Pretorius, 2022). Researchers have shown relationships between the incidence of obesity and the built environment by effectively utilizing deep learning technologies in diagnostic procedures (Morgan et al., 2021).

3. ENVIRONMENTAL MONITORING AND SUSTAINABILITY

Artificial intelligence is essential for environmental monitoring, conservation, and sustainable resource management in addition to its uses in healthcare (Singh et al., 2022). AI systems may more accurately anticipate natural catastrophes, monitor biodiversity, and discover environmental inequities by utilizing data from IoT devices, remote sensing, and predictive analytics (Nica & Stehel, 2021). Artificial intelligence (AI) algorithms enable autonomous drones to monitor threatened species, stop illicit poaching, and carry out environmental assessments in hard-to-reach places, all of which support efforts to conserve animals (Borowski, 2021). Also, by encouraging the integration of renewable energy sources, AI-driven energy management systems minimize carbon emissions, maximize resource consumption, and support global sustainability objectives (Ahmad et al., 2021). AI-based models, for example, provide proactive interventions to prevent environmental degradation and actively maintain ecosystem integrity through real-time monitoring of air quality, water quality assessment, and deforestation surveillance (Wamba-Taguimdje et al., 2020).

The Earth's ecosystem is changing at a rate that might have a significant influence on the seas, agriculture, forests, atmosphere, and water supplies (Popov et al., 2022). The realizing that the Earth is not only ours but also belongs to future generation's means that it has an obligation to leave it a clean and safe place for people to live (Lu et al., 2019). Data from satellite remote sensing is essential for continually monitoring Earth's circumstances both geographically and temporally (Borowski,

2021). The recent technological developments have improved satellite imagery's resolution and coverage, making it possible to monitor polar areas, vegetation, hydrology, air quality, meteorological events, and disasters more thoroughly (Lim et al., 2021) (Vogt, 2021). With utilizing cutting-edge computer power, artificial intelligence (AI) becomes a powerful instrument for producing high-quality satellite photos and conducting global environmental phenomenon analyses (Gadekar et al., 2022). Beyond traditional algorithms, state-of-the-art AI models have the potential to improve satellite data quality for global environmental monitoring (Tao et al., 2021). There are varieties of research areas that use AI techniques in conjunction with remote sensing (Chen et al., 2022).

Figure 5. The Elements of Environmental Monitoring and Sustainability

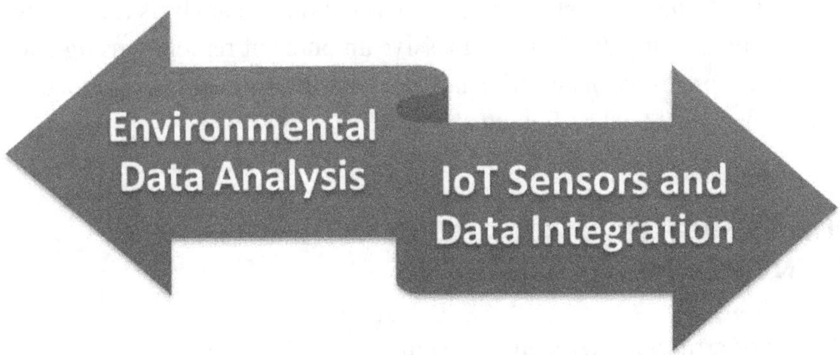

(Source- Original)

3.1 Environmental Data Analysis: Remote Sensing and Satellite Imagery by AI

A technique that makes it possible to collect data without making direct physical touch with the subject is known as remote sensing (Mourtzis et al., 2022). It uses sensors to measure or find various energy types that the item of interest emits, reflects, or scatters such as acoustic waves and electromagnetic radiation (Wan et al., 2018). Many platforms and sensors have been developed over time for the aim of remote sensing (Cioffi et al., 2020). The amount of data generated by remote sensing has increased to previously unheard-of levels because to continuous improvements in sensor technology (Laskurain-Iturbe et al., 2021). The need for effective tools and computational platforms to manage and extract important insights from large datasets is growing due to the increasing quantity and quality of data acquired from remote sensing systems (Brock & Von Wangenheim, 2019). The large-scale observation management, modeling, analysis, and environmental forecasting have all benefited from the use of AI technologies, which have shown to be especially useful for tasks like object detection, data fusion, noise reduction and many other crucial applications (Nguyen et al., 2020) (Hassoun et al., 2022). The collection and archiving of remote sensing data is becoming more and more important as AI technologies advance (Nahavandi, 2019).

3.2 IoT Sensors and Data Integration

Prioritized AI-driven processing systems operate both aboard and, on the ground, taking care of necessary tasks like scaling, filtering, calibration, and data completeness on their own (Zhou et al., 2021). Because these algorithms are good at finding intricate patterns and anomalies, subjectivity and bias in analysis are reduced (de la Pena Zarzuelo et al., 2020). Thus, massive amounts of remote sensing data may be swiftly and precisely managed, analyzed, and interpreted by researchers in an unequaled manner (Sanchez-Sotano et al., 2020).

4. HUMAN RIGHTS IMPLICATIONS IN HEALTHCARE AND ENVIRONMENT CONCERNS

The use of artificial intelligence (AI) in the healthcare sector raises significant human rights issues with nondiscrimination, privacy, and autonomy (Bokhari & Myeong, 2023) (Ali et al., 2022). As AI algorithms significantly rely on vast amounts of personal health data for training and decision-making, protecting patient rights requires maintaining data privacy and confidentiality (Kuzior, 2022). Also, algorithms may perpetuate disparities in healthcare access and treatment results based on demographic criteria like color, ethnicity, or socioeconomic position (Bongomin et al., 2020). This poses a danger of prejudice and discrimination and the strong legal frameworks, transparent algorithmic governance procedures, and proactive steps to improve fairness and impartiality in AI-enabled healthcare systems are needed to address these issues (Pivoto et al., 2021). For people to be empowered to make decisions about their health data and AI-driven interventions, it is also crucial to promote inclusive stakeholder participation and digital literacy (Gupta, 2023). Violations of human rights can have detrimental effects on one's health (Dorfleitner & Braun, 2019). Whether overt or covert, discrimination in the delivery of healthcare is a major barrier to receiving services, lowers the quality of care, and occurs between healthcare personnel and patients (Nassiry, 2018).

Human rights violations not only exacerbate and cause bad health outcomes, but they also put certain marginalized groups at greater risk of encountering abuses of their rights in the hospital context (Singh, 2024) (Chen & Volz, 2021). These include those who are internally displaced, disabled, belong to indigenous groups, are women living with HIV, are sex workers, use drugs, or are transgender or intersex (Yang et al., 2021). Their vulnerability may be increased by forceful or forced treatments and procedures in these abuses (Zhang et al., 2018). Historically, medical law and public health legislation have served as tools for promoting and preserving public health (Migliorelli & Dessertine, 2019). Beyond isolated occurrences, their influence and reach are constrained, and they frequently fall short of effectively addressing chronic maltreatment within the healthcare system (Naderi & Tian, 2022). Using a human rights perspective in patient care is the way to address this deficiency (Singh, 2023). This idea entails redefining constitutional values and international human rights law into a coherent plan that protects and improves both individual and societal health (Thomason 2018). With integrating legal ideas and human rights discourse into the patient care environment, it transcends the limiting focus on healthcare delivery as a transactional consumer good (Liu et al., 2022).

Figure 6. Human Rights Implications in Healthcare and Environment Concerns

(Source- Original)

5. CASE STUDIES AND PRACTICAL APPLICATIONS

5.1 AI-Driven Healthcare Initiatives

In the digital age, healthcare data are especially important and often targeted among the multitude of data kinds (Yan et al., 2022). These data have the potential to greatly improve care delivery and outcomes when used by healthcare systems, providers, and patients (Bayram et al., 2022) (Sachs et al., 2019). This is especially true when the data are combined with cutting-edge analytical technologies like- artificial intelligence (AI) (Campbell-Verduyn, 2023). Both domestic and international courts can impact policy by adopting a human rights approach to patient care (Singh, 2023). The protecting of the rights of disadvantaged and discriminated-against groups who would not have sufficient redress under conventional patient rights frameworks (Schulz & Feist, 2021). With the widespread use of technology like smartphones and the Internet, data has become more and more important in a variety of industries

(Kalaiarasi & Kirubahari, 2023). These data are used for a variety of objectives, such as improving service delivery, corporate strategy guidance, and user behavior comprehension (Chueca Vergara & Ferruz Agudo, 2021). Even though healthcare AI has attracted a lot of attention lately, it is crucial to understand the foundations of these technologies, as well as their benefits and drawbacks, and how they will affect the healthcare industry (Singh, 2022).

5.2 Environmental Conservation Projects: AI Applications in Ecosystem Preservation

The conservation sector is typically viewed as fraught with difficulties and obstacles, making conservation activities look extremely difficult (Sharma & Singh, 2022) (Marke, 2018). But even in the middle of these challenges, there are moments that give rise to optimism instead of despair (Hoang et al., 2022) (Schloesser & Schulz, 2022). This explores the following summary of well-known international conservation programs that are significantly improving the planet which are beneficial environmental efforts are across the world (Singh, 2022) (Harris, 2018). Two new programs are being introduced by the U.S. Agency for International Development (USAID) with the goal of helping Vietnam participate in climate-positive measures to solve the global climate issue (Shih et al., 2023). These programs seek to protect almost 950,000 hectares of Vietnam's wooded lands, which are dispersed over 11 provinces and three national parks (Ozili, 2023) (Bayram et al., 2022). The retaining a strong biodiversity and healthier, denser forests is essential to reducing greenhouse gas emissions (Singh, 2022). Under DAI Global's direction, the Sustainable Forest Management initiative aims to lower carbon emissions from deforestation, forest degradation, and poor management techniques across seven provinces' worth of wood plantations (Micaroni, 2020).

5.3 Community Engagement and Stakeholder Participation

Through the collective participation of many groups, including common people, the media, environmental activists, academics, and scientists, the involvement of stakeholders in environmental decision-making enables citizens to exercise their democratic rights (Dell'Erba, 2021) (Bin Amin et al., 2022). In this context, collaboration refers to a communication strategy that encourages discussion among stakeholders and places an emphasis on problem-solving as opposed to advocacy and dispute (Singh, 2022). The effective cooperation requires open and honest communication, which frequently involves talks about power distribution and establishing a fair atmosphere for all participants (Puaschunder, 2023).

Figure 7. The Case Studies

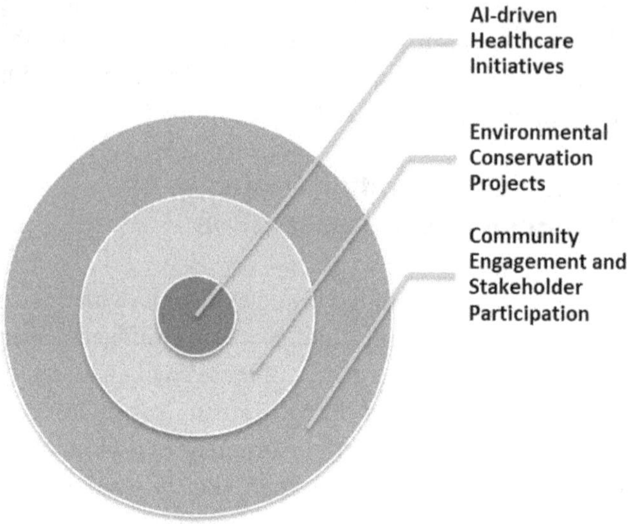

(Source- Original)

6. ETHICAL AND GOVERNANCE CONSIDERATIONS

The use of AI in the healthcare industry raises several ethical concerns and governance gaps, hence it is critical that responsible AI frameworks be developed with a focus on environmental sustainability, human rights, and ethical values (Vikas et al., 2022; Macchiavello, 2023). Transparency in AI decision-making procedures, accountability systems for algorithmic mistakes or biases, and permission systems for data sharing and use are important factors to take into mind (Marke & Sylvester, 2018; Udeagha & Ngepah, 2023). Also, the politicians, healthcare professionals, technologists, and civil society stakeholders must collaborate to navigate the complex ethical, legal, and social implications (ELSI) of AI in healthcare (Hassan et al., 2020). Through the promotion of an inclusive governance and responsible innovation culture, stakeholders may capitalize on AI's revolutionary potential while averting possible hazards and preserving human dignity (Fenwick & Vermeulen, 2020; Singh, 2019).

7. CONCLUSION AND FUTURE SCOPE

The right to enjoy the highest achievable standard of health is an inherent entitlement of every individual, irrespective of race, religion, political affiliation, economic status, or social circumstances (Nenavath & Mishra, 2023). The analysis of human rights, the environment, and healthcare emphasizes how important it is to have a complete viewpoint in order to approach and overcome global health issues (Schulz et al., 2020) (Moro-Visconti et al., 2020). Artificial intelligence is shown itself to be a potent facilitator in the healthcare industry, helping to improve human rights, environmental sustainability, and healthcare delivery (Rizzello & Kabli, 2020). However, to fully realize AI potential, it will be necessary to address ethical, legal, and societal issues as well as promote inclusive governance systems and multidisciplinary collaboration (Jaiwant & Kureethara, 2023). Stakeholders can leverage technology's revolutionary power to build healthcare ecosystems that are more sustainable, human-centered, and respectful of rights by taking a human-centered approach to AI innovation (Bhowmik, 2022) (Radanliev, 2024). These values, which were incorporated into the World Health Organization's Constitution almost seven decades ago, are still incredibly important and relevant (Li et al., 2024). The rights to health include both freedoms and benefits. Freedom includes the right to control one's own health, including the ability to refuse unnecessary medical treatments and experiments (Sreenivasan & Suresh, 2024). The concept of privileges includes the ability to get a full range of healthcare services as well as the essential social determinants of health, all of which are combined to provide fair chances for everyone to attain the best possible quality of health (Anyanwu et al., 2024) (Diro et al., 2024). The idea of the right to health has been central to the WHO's identity and mission since its founding, especially in relation to my main area of interest is the universal health care (Sharma & Dwivedi, 2024) (Singh & Kaunert, 2024).

REFERENCES

Ahmad, T., Zhang, D., Huang, C., Zhang, H., Dai, N., Song, Y., & Chen, H. (2021). Artificial intelligence in sustainable energy industry: Status Quo, challenges and opportunities. *Journal of Cleaner Production*, 289, 125834. DOI: 10.1016/j.jclepro.2021.125834

Ali, A., Septyanto, A. W., Chaudhary, I., Al Hamadi, H., Alzoubi, H. M., & Khan, Z. F. (2022, February). Applied Artificial Intelligence as Event Horizon Of Cyber Security. In *2022 International Conference on Business Analytics for Technology and Security (ICBATS)* (pp. 1-7). IEEE. DOI: 10.1109/ICBATS54253.2022.9759076

Anastasi, S., Madonna, M., & Monica, L. (2021). Implications of embedded artificial intelligence-machine learning on safety of machinery. *Procedia Computer Science*, 180, 338–343. DOI: 10.1016/j.procs.2021.01.171

Angelopoulos, A., Michailidis, E. T., Nomikos, N., Trakadas, P., Hatziefremidis, A., Voliotis, S., & Zahariadis, T. (2019). Tackling faults in the industry 4.0 era—A survey of machine-learning solutions and key aspects. *Sensors (Basel)*, 20(1), 109. DOI: 10.3390/s20010109 PMID: 31878065

Anyanwu, A., Olorunsogo, T., Abrahams, T. O., Akindote, O. J., & Reis, O. (2024). Data confidentiality and integrity: a review of accounting and cybersecurity controls in superannuation organizations. *Computer Science & IT Research Journal*, 5(1), 237–253. DOI: 10.51594/csitrj.v5i1.735

Aoun, A., Ilinca, A., Ghandour, M., & Ibrahim, H. (2021). A review of Industry 4.0 characteristics and challenges, with potential improvements using blockchain technology. *Computers & Industrial Engineering*, 162, 107746. DOI: 10.1016/j.cie.2021.107746

Arden, N. S., Fisher, A. C., Tyner, K., Lawrence, X. Y., Lee, S. L., & Kopcha, M. (2021). Industry 4.0 for pharmaceutical manufacturing: Preparing for the smart factories of the future. *International Journal of Pharmaceutics*, 602, 120554. DOI: 10.1016/j.ijpharm.2021.120554 PMID: 33794326

Asadollahi-Yazdi, E., Couzon, P., Nguyen, N. Q., Ouazene, Y., & Yalaoui, F. (2020). Industry 4.0: Revolution or Evolution? *American Journal of Operations Research*, 10(06), 241–268. DOI: 10.4236/ajor.2020.106014

Awan, U., Sroufe, R., & Shahbaz, M. (2021). Industry 4.0 and the circular economy: A literature review and recommendations for future research. *Business Strategy and the Environment*, 30(4), 2038–2060. DOI: 10.1002/bse.2731

Badri, A., Boudreau-Trudel, B., & Souissi, A. S. (2018). Occupational health and safety in the industry 4.0 era: A cause for major concern? *Safety Science*, 109, 403–411. DOI: 10.1016/j.ssci.2018.06.012

Bag, S., & Pretorius, J. H. C. (2022). Relationships between industry 4.0, sustainable manufacturing and circular economy: Proposal of a research framework. *The International Journal of Organizational Analysis*, 30(4), 864–898. DOI: 10.1108/IJOA-04-2020-2120

Basulo-Ribeiro, J., & Teixeira, L. (2024). The Future of Healthcare with Industry 5.0: Preliminary Interview-Based Qualitative Analysis. *Future Internet*, 16(3), 68. DOI: 10.3390/fi16030068

Bayram, O., Talay, I., & Feridun, M. (2022). Can FinTech promote sustainable finance? Policy lessons from the case of Turkey. *Sustainability (Basel)*, 14(19), 12414. DOI: 10.3390/su141912414

Bhowmik, D. (2022). An Introduction to Climate Fintech. *European Journal of Science. Innovation and Technology*, 2(4), 24–35.

Bhuiyan, A. B., Ali, M. J., Zulkifli, N., & Kumarasamy, M. M. (2020). Industry 4.0: Challenges, opportunities, and strategic solutions for Bangladesh. *International Journal of Business and Management Future*, 4(2), 41–56. DOI: 10.46281/ijbmf.v4i2.832

Bin Amin, S., Taghizadeh-Hesary, F., & Khan, F. (2022). Facilitating green digital finance in Bangladesh: Importance, prospects, and implications for meeting the SDGs. In *Green Digital Finance and Sustainable Development Goals* (pp. 143–165). Springer Nature Singapore. DOI: 10.1007/978-981-19-2662-4_7

Blobel, B. (2020, September). Application of industry 4.0 concept to health care. In *pHealth 2020:Proceedings of the 17th International Conference on Wearable Micro and Nano Technologies for Personalized Health* (Vol. 273, p. 23). IOS Press.

Bokhari, S. A. A., & Myeong, S. (2023). The influence of artificial intelligence on e-Governance and cybersecurity in smart cities: A stakeholder's perspective. *IEEE Access : Practical Innovations, Open Solutions*, 11, 69783–69797. DOI: 10.1109/ACCESS.2023.3293480

Bongomin, O., Gilibrays Ocen, G., Oyondi Nganyi, E., Musinguzi, A., & Omara, T. (2020). Exponential disruptive technologies and the required skills of industry 4.0. *Journal of Engineering*, 2020, 1–17. DOI: 10.1155/2020/8090521

Borowski, P. F. (2021). Digitization, digital twins, blockchain, and industry 4.0 as elements of management process in enterprises in the energy sector. *Energies*, 14(7), 1885. DOI: 10.3390/en14071885

Borowski, P. F. (2021). Innovative processes in managing an enterprise from the energy and food sector in the era of industry 4.0. *Processes (Basel, Switzerland)*, 9(2), 381. DOI: 10.3390/pr9020381

Brock, J. K. U., & Von Wangenheim, F. (2019). Demystifying AI: What digital transformation leaders can teach you about realistic artificial intelligence. *California Management Review*, 61(4), 110–134. DOI: 10.1177/1536504219865226

Campbell-Verduyn, M. (2023). Conjuring a cooler world? Imaginaries of improvement in blockchain climate finance experiments. *Environment and Planning C: Politics and Space*, 23996544231162858.

Chander, B., Pal, S., De, D., & Buyya, R. (2022). Artificial intelligence-based internet of things for industry 5.0. *Artificial intelligence-based internet of things systems*, 3-45.

Chen, Y., Lu, Y., Bulysheva, L., & Kataev, M. Y. (2022). Applications of blockchain in industry 4.0: A review. *Information Systems Frontiers*, , 1–15. DOI: 10.1007/s10796-022-10248-7

Chen, Y., & Volz, U. (2021). Scaling up sustainable investment through blockchain-based project bonds. *ADB-IGF Special Working Paper Series "Fintech to Enable Development, Investment, Financial Inclusion, and Sustainability*.

Chueca Vergara, C., & Ferruz Agudo, L. (2021). Fintech and sustainability: Do they affect each other? *Sustainability (Basel)*, 13(13), 7012. DOI: 10.3390/su13137012

Cioffi, R., Travaglioni, M., Piscitelli, G., Petrillo, A., & De Felice, F. (2020). Artificial intelligence and machine learning applications in smart production: Progress, trends, and directions. *Sustainability (Basel)*, 12(2), 492. DOI: 10.3390/su12020492

Costa, E. (2024). Industry 5.0 and SDG 9: A symbiotic dance towards sustainable transformation. *Sustainable Earth Reviews*, 7(1), 4. DOI: 10.1186/s42055-024-00073-y

de la Peña Zarzuelo, I., Soeane, M. J. F., & Bermúdez, B. L. (2020). Industry 4.0 in the port and maritime industry: A literature review. *Journal of Industrial Information Integration*, 20, 100173. DOI: 10.1016/j.jii.2020.100173

Dell'Erba, M. (2021). Sustainable digital finance and the pursuit of environmental sustainability. *Sustainable Finance in Europe: Corporate Governance, Financial Stability and Financial Markets*, 61-81.

Dinesh Arokia Raj, A., Jha, R. R., Yadav, M., Sam, D., & Jayanthi, K. (2024). Role of Blockchain and Watermarking Toward Cybersecurity. In *Multimedia Watermarking: Latest Developments and Trends* (pp. 103–123). Springer Nature Singapore. DOI: 10.1007/978-981-99-9803-6_6

Diro, A., Zhou, L., Saini, A., Kaisar, S., & Hiep, P. C. (2024). Leveraging zero knowledge proofs for blockchain-based identity sharing: A survey of advancements, challenges and opportunities. *Journal of Information Security and Applications*, 80, 103678. DOI: 10.1016/j.jisa.2023.103678

Dorfleitner, G., & Braun, D. (2019). Fintech, digitalization and blockchain: possible applications for green finance. *The rise of green finance in Europe: opportunities and challenges for issuers, investors and marketplaces*, 207-237.

Dwivedi, Y. K., Hughes, L., Ismagilova, E., Aarts, G., Coombs, C., Crick, T., Duan, Y., Dwivedi, R., Edwards, J., Eirug, A., Galanos, V., Ilavarasan, P. V., Janssen, M., Jones, P., Kar, A. K., Kizgin, H., Kronemann, B., Lal, B., Lucini, B., & Williams, M. D. (2021). Artificial Intelligence (AI): Multidisciplinary perspectives on emerging challenges, opportunities, and agenda for research, practice and policy. *International Journal of Information Management*, 57, 101994. DOI: 10.1016/j.ijinfomgt.2019.08.002

El Zein, B., Elrashidi, A., Dahlan, M., Al Jarwan, A., & Jabbour, G. (2024). Nano and Society 5.0: Advancing the Human-Centric Revolution.

Felsberger, A., Qaiser, F. H., Choudhary, A., & Reiner, G. (2022). The impact of Industry 4.0 on the reconciliation of dynamic capabilities: Evidence from the European manufacturing industries. *Production Planning and Control*, 33(2-3), 277–300. DOI: 10.1080/09537287.2020.1810765

Fenwick, M., & Vermeulen, E. P. (2020). Banking and regulatory responses to FinTech revisited-building the sustainable financial service'ecosystems' of tomorrow. *Singapore Journal of Legal Studies*, (Mar 2020), 165-189.

Finance, A. T. C. C. (2015). *Industry 4.0 Challenges and solutions for the digital transformation and use of exponential technologies. Finance, audit tax consulting corporate*. Swiss.

Fraga-Lamas, P., Lopes, S. I., & Fernández-Caramés, T. M. (2021). Green IoT and edge AI as key technological enablers for a sustainable digital transition towards a smart circular economy: An industry 5.0 use case. *Sensors (Basel)*, 21(17), 5745. DOI: 10.3390/s21175745 PMID: 34502637

Gadekar, R., Sarkar, B., & Gadekar, A. (2022). Key performance indicator based dynamic decision-making framework for sustainable Industry 4.0 implementation risks evaluation: Reference to the Indian manufacturing industries. *Annals of Operations Research*, 318(1), 189–249. DOI: 10.1007/s10479-022-04828-8 PMID: 35910040

Gupta, R. (2023). Industry 4.0 adaption in indian banking Sector—A review and agenda for future research. *Vision (Basel)*, 27(1), 24–32. DOI: 10.1177/0972262921996829 PMID: 36977304

Harris, A. (2018). A conversation with masterminds in blockchain and climate change. In *Transforming climate finance and green investment with blockchains* (pp. 15–22). Academic Press. DOI: 10.1016/B978-0-12-814447-3.00002-1

Hassan, M. K., Rabbani, M. R., & Ali, M. A. M. (2020). Challenges for the Islamic Finance and banking in post COVID era and the role of Fintech. *Journal of Economic Cooperation & Development*, 41(3), 93–116.

Hassoun, A., Aït-Kaddour, A., Abu-Mahfouz, A. M., Rathod, N. B., Bader, F., Barba, F. J., Biancolillo, A., Cropotova, J., Galanakis, C. M., Jambrak, A. R., Lorenzo, J. M., Måge, I., Ozogul, F., & Regenstein, J. (2023). The fourth industrial revolution in the food industry—Part I: Industry 4.0 technologies. *Critical Reviews in Food Science and Nutrition*, 63(23), 6547–6563. DOI: 10.1080/10408398.2022.2034735 PMID: 35114860

Hassoun, A., Prieto, M. A., Carpena, M., Bouzembrak, Y., Marvin, H. J., Pallares, N., Barba, F. J., Punia Bangar, S., Chaudhary, V., Ibrahim, S., & Bono, G. (2022). Exploring the role of green and Industry 4.0 technologies in achieving sustainable development goals in food sectors. *Food Research International*, 162, 112068. DOI: 10.1016/j.foodres.2022.112068 PMID: 36461323

Hoang, T. G., Nguyen, G. N. T., & Le, D. A. (2022). Developments in financial technologies for achieving the Sustainable Development Goals (SDGs): FinTech and SDGs. In *Disruptive technologies and eco-innovation for sustainable development* (pp. 1–19). IGI Global. DOI: 10.4018/978-1-7998-8900-7.ch001

Ivanov, D., Dolgui, A., & Sokolov, B. (2019). The impact of digital technology and Industry 4.0 on the ripple effect and supply chain risk analytics. *International Journal of Production Research*, 57(3), 829–846. DOI: 10.1080/00207543.2018.1488086

Jaiwant, S. V., & Kureethara, J. V. (2023). Green Finance and Fintech: Toward a More Sustainable Financial System. In *Green Finance Instruments, FinTech, and Investment Strategies: Sustainable Portfolio Management in the Post-COVID Era* (pp. 283–300). Springer International Publishing. DOI: 10.1007/978-3-031-29031-2_12

Jakka, G., Yathiraju, N., & Ansari, M. F. (2022). Artificial Intelligence in Terms of Spotting Malware and Delivering Cyber Risk Management. *Journal of Positive School Psychology*, 6(3), 6156–6165.

Javaid, M., Haleem, A., Singh, R. P., Khan, S., & Suman, R. (2021). Blockchain technology applications for Industry 4.0: A literature-based review. *Blockchain: Research and Applications*, 2(4), 100027.

Javaid, M., Haleem, A., Singh, R. P., Suman, R., & Gonzalez, E. S. (2022). Understanding the adoption of Industry 4.0 technologies in improving environmental sustainability. *Sustainable Operations and Computers*, 3, 203–217. DOI: 10.1016/j.susoc.2022.01.008

Kalaiarasi, H., & Kirubahari, S. (2023). Green finance for sustainable development using blockchain technology. In *Green Blockchain Technology for Sustainable Smart Cities* (pp. 167–185). Elsevier. DOI: 10.1016/B978-0-323-95407-5.00003-7

Kasowaki, L., & Ahmet, S. (2024). *Shielding the Virtual Ramparts: Understanding Cybersecurity Essentials* (No. 11700). EasyChair.

Kumar, S., & Mallipeddi, R. R. (2022). Impact of cybersecurity on operations and supply chain management: Emerging trends and future research directions. *Production and Operations Management*, 31(12), 4488–4500. DOI: 10.1111/poms.13859

Kumar, S. H., Talasila, D., Gowrav, M. P., & Gangadharappa, H. V. (2020). Adaptations of Pharma 4.0 from Industry 4.0. *Drug Invention Today*, 14(3).

Kurniawan, T. A., Maiurova, A., Kustikova, M., Bykovskaia, E., Othman, M. H. D., & Goh, H. H. (2022). Accelerating sustainability transition in St. Petersburg (Russia) through digitalization-based circular economy in waste recycling industry: A strategy to promote carbon neutrality in era of Industry 4.0. *Journal of Cleaner Production*, 363, 132452. DOI: 10.1016/j.jclepro.2022.132452

Kuzior, A. (2022). Technological unemployment in the perspective of Industry 4.0. *Virtual Economics*, 5(1), 7–23. DOI: 10.34021/ve.2022.05.01(1)

Laskurain-Iturbe, I., Arana-Landín, G., Landeta-Manzano, B., & Uriarte-Gallastegi, N. (2021). Exploring the influence of industry 4.0 technologies on the circular economy. *Journal of Cleaner Production*, 321, 128944. DOI: 10.1016/j.jclepro.2021.128944

Leng, J., Ye, S., Zhou, M., Zhao, J. L., Liu, Q., Guo, W., Cao, W., & Fu, L. (2020). Blockchain-secured smart manufacturing in industry 4.0: A survey. *IEEE Transactions on Systems, Man, and Cybernetics. Systems*, 51(1), 237–252. DOI: 10.1109/TSMC.2020.3040789

Li, B. H., Hou, B. C., Yu, W. T., Lu, X. B., & Yang, C. W. (2017). Applications of artificial intelligence in intelligent manufacturing: A review. *Frontiers of Information Technology & Electronic Engineering*, 18(1), 86–96. DOI: 10.1631/FITEE.1601885

Li, Z., Liang, X., Wen, Q., & Wan, E. (2024). The Analysis of Financial Network Transaction Risk Control based on Blockchain and Edge Computing Technology. *IEEE Transactions on Engineering Management*, 71, 5669–5690. DOI: 10.1109/TEM.2024.3364832

Lim, C. H., Lim, S., How, B. S., Ng, W. P. Q., Ngan, S. L., Leong, W. D., & Lam, H. L. (2021). A review of industry 4.0 revolution potential in a sustainable and renewable palm oil industry: HAZOP approach. *Renewable & Sustainable Energy Reviews*, 135, 110223. DOI: 10.1016/j.rser.2020.110223

Liu, H., Yao, P., Latif, S., Aslam, S., & Iqbal, N. (2022). Impact of Green financing, FinTech, and financial inclusion on energy efficiency. *Environmental Science and Pollution Research International*, 29(13), 1–12. DOI: 10.1007/s11356-021-16949-x PMID: 34705207

Lu, C., Lyu, J., Zhang, L., Gong, A., Fan, Y., Yan, J., & Li, X. (2020). Nuclear power plants with artificial intelligence in industry 4.0 era: Top-level design and current applications—A systemic review. *IEEE Access : Practical Innovations, Open Solutions*, 8, 194315–194332. DOI: 10.1109/ACCESS.2020.3032529

Lu, Y. (2019). Artificial intelligence: A survey on evolution, models, applications and future trends. *Journal of Management Analytics*, 6(1), 1–29. DOI: 10.1080/23270012.2019.1570365

Macchiavello, E. (2023). Sustainable Finance and Fintech. A Focus on Capital Raising. A Focus on Capital Raising (May 31, 2023). Forthcoming in K. Alexander, M. Gargantini and M. Siri, *The Cambridge Handbook of EU Sustainable Finance-Regulation, Supervision and Governance*. CUP.

Majstorovic, V. D., & Mitrovic, R. (2019). Industry 4.0 programs worldwide. In *Proceedings of the 4th International Conference on the Industry 4.0 Model for Advanced Manufacturing: AMP 2019 4* (pp. 78-99). Springer International Publishing. DOI: 10.1007/978-3-030-18180-2_7

Marke, A. (Ed.). (2018). *Transforming climate finance and green investment with blockchains*. Academic Press.

Marke, A., & Sylvester, B. (2018). Decoding the current global climate finance architecture. In *Transforming climate finance and green investment with blockchains* (pp. 35–59). Academic Press. DOI: 10.1016/B978-0-12-814447-3.00004-5

Meyendorf, N., Ida, N., Singh, R., & Vrana, J. (2023). NDE 4.0: Progress, promise, and its role to industry 4.0. *NDT & E International*, 140, 102957. DOI: 10.1016/j.ndteint.2023.102957

Micaroni, M. (2020). *Sustainable Finance: Addressing the SDGs through Fintech and Digital Finance solutions in EU* (Doctoral dissertation, Politecnico di Torino).

Migliorelli, M., & Dessertine, P. (2019). The rise of green finance in Europe. *Opportunities and challenges for issuers, investors and marketplaces. Cham. Palgrave Macmillan*, 2, 2019.

Mithas, S., Chen, Z. L., Saldanha, T. J., & De Oliveira Silveira, A. (2022). How will artificial intelligence and Industry 4.0 emerging technologies transform operations management? *Production and Operations Management*, 31(12), 4475–4487. DOI: 10.1111/poms.13864

Morgan, J., Halton, M., Qiao, Y., & Breslin, J. G. (2021). Industry 4.0 smart reconfigurable manufacturing machines. *Journal of Manufacturing Systems*, 59, 481–506. DOI: 10.1016/j.jmsy.2021.03.001

Moro-Visconti, R., Cruz Rambaud, S., & López Pascual, J. (2020). Sustainability in FinTechs: An explanation through business model scalability and market valuation. *Sustainability (Basel)*, 12(24), 10316. DOI: 10.3390/su122410316

Mourtzis, D., Angelopoulos, J., & Panopoulos, N. (2022). A Literature Review of the Challenges and Opportunities of the Transition from Industry 4.0 to Society 5.0. *Energies*, 15(17), 6276. DOI: 10.3390/en15176276

Naderi, N., & Tian, Y. (2022). Leveraging Blockchain Technology and Tokenizing Green Assets to Fill the Green Finance Gap. *Energy Research Letters, 3*(3).

Nahavandi, S. (2019). Industry 5.0—A human-centric solution. *Sustainability (Basel)*, 11(16), 4371. DOI: 10.3390/su11164371

Nassiry, D. (2018). *The role of fintech in unlocking green finance: Policy insights for developing countries* (No. 883). ADBI working paper.

Nenavath, S., & Mishra, S. (2023). Impact of green finance and fintech on sustainable economic growth: Empirical evidence from India. *Heliyon*, 9(5), e16301. DOI: 10.1016/j.heliyon.2023.e16301 PMID: 37234625

Nguyen, T., Gosine, R. G., & Warrian, P. (2020). A systematic review of big data analytics for oil and gas industry 4.0. *IEEE Access : Practical Innovations, Open Solutions*, 8, 61183–61201. DOI: 10.1109/ACCESS.2020.2979678

Nica, E., & Stehel, V. (2021). Internet of things sensing networks, artificial intelligence-based decision-making algorithms, and real-time process monitoring in sustainable industry 4.0. *Journal of Self-Governance and Management Economics*, 9(3), 35–47.

Ozili, P. K. (2023). Assessing global interest in decentralized finance, embedded finance, open finance, ocean finance and sustainable finance. *Asian Journal of Economics and Banking*, 7(2), 197–216. DOI: 10.1108/AJEB-03-2022-0029

Pivoto, D. G., de Almeida, L. F., da Rosa Righi, R., Rodrigues, J. J., Lugli, A. B., & Alberti, A. M. (2021). Cyber-physical systems architectures for industrial internet of things applications in Industry 4.0: A literature review. *Journal of Manufacturing Systems*, 58, 176–192. DOI: 10.1016/j.jmsy.2020.11.017

Popov, V. V., Kudryavtseva, E. V., Kumar Katiyar, N., Shishkin, A., Stepanov, S. I., & Goel, S. (2022). Industry 4.0 and digitalisation in healthcare. *Materials (Basel)*, 15(6), 2140. DOI: 10.3390/ma15062140 PMID: 35329592

Puaschunder, J. M. (2023). The Future of Resilient Green Finance. In *The Future of Resilient Finance: Finance Politics in the Age of Sustainable Development* (pp. 185–210). Springer International Publishing. DOI: 10.1007/978-3-031-30138-4_6

Radanliev, P. (2024). Cyber diplomacy: defining the opportunities for cybersecurity and risks from Artificial Intelligence, IoT, Blockchains, and Quantum Computing. *Journal of Cyber Security Technology*, 1-51.

Rane, N. (2023). Transformers in Industry 4.0, Industry 5.0, and Society 5.0: Roles and Challenges.

Rath, K. C., Khang, A., & Roy, D. (2024). The Role of Internet of Things (IoT) Technology in Industry 4.0 Economy. In *Advanced IoT Technologies and Applications in the Industry 4.0 Digital Economy* (pp. 1-28). CRC Press.

Ray, R. K., Chowdhury, F. R., & Hasan, M. R. (2024). Blockchain Applications in Retail Cybersecurity: Enhancing Supply Chain Integrity, Secure Transactions, and Data Protection. *Journal of Business and Management Studies*, 6(1), 206–214. DOI: 10.32996/jbms.2024.6.1.13

Reier Forradellas, R. F., & Garay Gallastegui, L. M. (2021). Digital transformation and artificial intelligence applied to business: Legal regulations, economic impact and perspective. *Laws*, 10(3), 70. DOI: 10.3390/laws10030070

Rizzello, A., & Kabli, A. (2020). Social finance and sustainable development goals: A literature synthesis, current approaches and research agenda. *ACRN Journal of Finance and Risk Perspectives, 9.*

Nikolaou, N. (2018). *The Transition to a Green Global Economy: The Case of Green Finance and Green Bonds.* (Doctoral dissertation), University of Piraeus.

Sachs, J. D., Woo, W. T., Yoshino, N., & Taghizadeh-Hesary, F. (2019). Importance of green finance for achieving sustainable development goals and energy security. *Handbook of green finance: Energy security and sustainable development, 10*, 1-10.

Saeed, S., Altamimi, S. A., Alkayyal, N. A., Alshehri, E., & Alabbad, D. A. (2023). Digital transformation and cybersecurity challenges for businesses resilience: Issues and recommendations. *Sensors (Basel)*, 23(15), 6666. DOI: 10.3390/s23156666 PMID: 37571451

Sánchez-Sotano, A., Cerezo-Narváez, A., Abad-Fraga, F., Pastor-Fernández, A., & Salguero-Gómez, J. (2020). Trends of digital transformation in the shipbuilding sector. In *New Trends in the Use of Artificial Intelligence for the Industry 4.0.* IntechOpen. DOI: 10.5772/intechopen.91164

Schloesser, T., & Schulz, K. (2022). Distributed ledger technology and climate finance. In *Green Digital Finance and Sustainable Development Goals* (pp. 265–286). Springer Nature Singapore. DOI: 10.1007/978-981-19-2662-4_13

Schoenmaker, D., & Volz, U. (2022). *Scaling up sustainable finance and investment in the Global South.* CEPR Press.

Schulz, K., & Feist, M. (2021). Leveraging blockchain technology for innovative climate finance under the Green Climate Fund. *Earth System Governance*, 7, 100084. DOI: 10.1016/j.esg.2020.100084

Schulz, K. A., Gstrein, O. J., & Zwitter, A. J. (2020). Exploring the governance and implementation of sustainable development initiatives through blockchain technology. *Futures*, 122, 102611. DOI: 10.1016/j.futures.2020.102611

Sharma, A., & Singh, B. (2022). Measuring Impact of E-commerce on Small Scale Business: A Systematic Review. *Journal of Corporate Governance and International Business Law*, 5(1).

Sharma, S., & Dwivedi, R. (2024). A survey on blockchain deployment for biometric systems. *IET Blockchain*, 4(2), 124–151. DOI: 10.1049/blc2.12063

Shi, Z., Xie, Y., Xue, W., Chen, Y., Fu, L., & Xu, X. (2020). Smart factory in Industry 4.0. *Systems Research and Behavioral Science*, 37(4), 607–617. DOI: 10.1002/sres.2704

Shih, C., Gwizdalski, A., & Deng, X. (2023). *Building a Sustainable Future: Exploring Green Finance*. Regenerative Finance, and Green Financial Technology.

Sima, V., Gheorghe, I. G., Subić, J., & Nancu, D. (2020). Influences of the industry 4.0 revolution on the human capital development and consumer behavior: A systematic review. *Sustainability (Basel)*, 12(10), 4035. DOI: 10.3390/su12104035

Singh, B. (2019). Profiling Public Healthcare: A Comparative Analysis Based on the Multidimensional Healthcare Management and Legal Approach. *Indian Journal of Health and Medical Law*, 2(2), 1–5.

Singh, B. (2020). Global science and jurisprudential approach concerning healthcare and illness. *Indian Journal of Health and Medical Law*, 3(1), 7–13.

Singh, B. (2022). COVID-19 Pandemic and Public Healthcare: Endless Downward Spiral or Solution via Rapid Legal and Health Services Implementation with Patient Monitoring Program. *Justice and Law Bulletin*, 1(1), 1–7.

Singh, B. (2022). Relevance of Agriculture-Nutrition Linkage for Human Healthcare: A Conceptual Legal Framework of Implication and Pathways. *Justice and Law Bulletin*, 1(1), 44–49.

Singh, B. (2022). Understanding Legal Frameworks Concerning Transgender Healthcare in the Age of Dynamism. *Electronic Journal of Social and Strategic Studies*, 3(1), 56–65. DOI: 10.47362/EJSSS.2022.3104

Singh, B. (2023). Blockchain Technology in Renovating Healthcare: Legal and Future Perspectives. In *Revolutionizing Healthcare Through Artificial Intelligence and Internet of Things Applications* (pp. 177-186). IGI Global.

Singh, B. (2023). Federated Learning for Envision Future Trajectory Smart Transport System for Climate Preservation and Smart Green Planet: Insights into Global Governance and SDG-9 (Industry, Innovation and Infrastructure). *National Journal of Environmental Law*, 6(2), 6–17.

Singh, B. (2024). Legal Dynamics Lensing Metaverse Crafted for Videogame Industry and E-Sports: Phenomenological Exploration Catalyst Complexity and Future. *Journal of Intellectual Property Rights Law*, 7(1), 8–14.

Singh, B., & Kaunert, C. (2024). Future of Digital Marketing: Hyper-Personalized Customer Dynamic Experience with AI-Based Predictive Models. In Revolutionizing the AI-Digital Landscape (pp. 189-203). Productivity Press.

Singh, S. K., Sharma, S. K., Singla, D., & Gill, S. S. (2022). Evolving requirements and application of SDN and IoT in the context of industry 4.0, blockchain and artificial intelligence. *Software Defined Networks: Architecture and Applications*, 427-496.

Singh, V. K. (2022). Regulatory and Legal Framework for Promoting Green Digital Finance. In *Green Digital Finance and Sustainable Development Goals* (pp. 3–27). Springer Nature Singapore. DOI: 10.1007/978-981-19-2662-4_1

Singh Rajawat, A., Bedi, P., Goyal, S. B., Shukla, P. K., Zaguia, A., Jain, A., & Monirujjaman Khan, M. (2021). Reformist framework for improving human security for mobile robots in industry 4.0. *Mobile Information Systems*, 2021, 1–10. DOI: 10.1155/2021/4744220

Sreenivasan, A., & Suresh, M. (2024). Start-up sustainability: Does blockchain adoption drives sustainability in start-ups? A systematic literature reviews. *Management Research Review*, 47(3), 390–405. DOI: 10.1108/MRR-07-2022-0519

Tao, F., Akhtar, M. S., & Jiayuan, Z. (2021). The future of artificial intelligence in cybersecurity: A comprehensive survey. *EAI Endorsed Transactions on Creative Technologies*, 8(28), e3–e3. DOI: 10.4108/eai.7-7-2021.170285

Thomason, J., Ahmad, M., Bronder, P., Hoyt, E., Pocock, S., Bouteloupe, J., & Shrier, D. (2018). Blockchain—powering and empowering the poor in developing countries. In *Transforming climate finance and green investment with blockchains* (pp. 137–152). Academic Press. DOI: 10.1016/B978-0-12-814447-3.00010-0

Tseng, M. L., Tran, T. P. T., Ha, H. M., Bui, T. D., & Lim, M. K. (2021). Sustainable industrial and operation engineering trends and challenges Toward Industry 4.0: A data driven analysis. *Journal of Industrial and Production Engineering*, 38(8), 581–598. DOI: 10.1080/21681015.2021.1950227

Udeagha, M. C., & Ngepah, N. (2023). The drivers of environmental sustainability in BRICS economies: Do green finance and fintech matter? *World Development Sustainability*, 3, 100096. DOI: 10.1016/j.wds.2023.100096

Vikas, N., Venegas, P., & Aiyer, S. (2022). Role of Banks and Other Financial Institutions in Enhancing Green Digital Finance. In *Green Digital Finance and Sustainable Development Goals* (pp. 329–352). Springer Nature Singapore. DOI: 10.1007/978-981-19-2662-4_16

Vogt, J. (2021). Where is the human got to go? Artificial intelligence, machine learning, big data, digitalisation, and human–robot interaction in Industry 4.0 and 5.0: Review Comment on: Bauer, M.(2020). Preise kalkulieren mit KI-gestützter Onlineplattform BAM GmbH, Weiden, Bavaria, Germany. *AI & Society*, 36(3), 1083–1087. DOI: 10.1007/s00146-020-01123-7

Wamba-Taguimdje, S. L., Fosso Wamba, S., Kala Kamdjoug, J. R., & Tchatchouang Wanko, C. E. (2020). Influence of artificial intelligence (AI) on firm performance: The business value of AI-based transformation projects. *Business Process Management Journal*, 26(7), 1893–1924. DOI: 10.1108/BPMJ-10-2019-0411

Wan, J., Yang, J., Wang, Z., & Hua, Q. (2018). Artificial intelligence for cloud-assisted smart factory. *IEEE Access : Practical Innovations, Open Solutions*, 6, 55419–55430. DOI: 10.1109/ACCESS.2018.2871724

Xing, K., Cropley, D. H., Oppert, M. L., & Singh, C. (2021). Readiness for digital innovation and industry 4.0 transformation: studies on manufacturing industries in the city of salisbury. *Business Innovation with New ICT in the Asia-Pacific: Case Studies*, 155-176.

Xu, L. D., Xu, E. L., & Li, L. (2018). Industry 4.0: State of the art and future trends. *International Journal of Production Research*, 56(8), 2941–2962. DOI: 10.1080/00207543.2018.1444806

Yan, C., Siddik, A. B., Yong, L., Dong, Q., Zheng, G. W., & Rahman, M. N. (2022). A two-staged SEM-artificial neural network approach to analyze the impact of FinTech adoption on the sustainability performance of banking firms: The mediating effect of green finance and innovation. *Systems*, 10(5), 148. DOI: 10.3390/systems10050148

Yang, Y., Su, X., & Yao, S. (2021). Nexus between green finance, fintech, and high-quality economic development: Empirical evidence from China. *Resources Policy*, 74, 102445. DOI: 10.1016/j.resourpol.2021.102445

Yue, Y., & Shyu, J. Z. (2024). A paradigm shift in crisis management: The nexus of AGI-driven intelligence fusion networks and blockchain trustworthiness. *Journal of Contingencies and Crisis Management*, 32(1), e12541. DOI: 10.1111/1468-5973.12541

Zhang, X., Aranguiz, M., Xu, D., Zhang, X., & Xu, X. (2018). Utilizing blockchain for better enforcement of green finance law and regulations. In *Transforming climate finance and green investment with blockchains* (pp. 289–301). Academic Press. DOI: 10.1016/B978-0-12-814447-3.00021-5

Zhong, R. Y., Xu, X., Klotz, E., & Newman, S. T. (2017). Intelligent manufacturing in the context of industry 4.0: A review. *Engineering (Beijing)*, 3(5), 616–630. DOI: 10.1016/J.ENG.2017.05.015

Zhou, J., Zhang, S., Lu, Q., Dai, W., Chen, M., Liu, X., . . . Herrera-Viedma, E. (2021). A survey on federated learning and its applications for accelerating industrial internet of things. *arXiv preprint arXiv:2104.10501*.

Chapter 5
Decoding Women's Reproductive Autonomy: A Critical Examination of Abortion Laws in India

Barkha Dodai
Ramaiah College of Law, Bengaluru, India

Abhilash Arun Sapre
https://orcid.org/0000-0002-7680-0894
Gujarat National Law University, Gandhinagar, India

ABSTRACT

Abortions have been an age-old practice and prevalent since the ancient times. Different religions have had different connotations, always finding a way to control women's bodies and their choices. India had been the hub of unsafe abortions with an extremely high maternal mortality ratio (MMR). The Medical Termination of Pregnancy Act 1971 was India's first attempt marking a positive step, decriminalizing abortions and giving a higher degree of bodily autonomy to the women. Consequently, the government during various phases came up with policies to provide safe and accessible abortions to the women of the country.This paper makes a thorough analysis of the historical view and statistics on abortion, the functioning of the MTP Act, government schemes, and abortions from a global perspective. An attempt has been made to look into the other statutes which have a direct impact on the MTP Act – the statute which governs abortion in India. A critical analysis has also been provided concerning the outcomes of the Act – affirmative and non-affirmative.

DOI: 10.4018/979-8-3693-6069-9.ch005

INTRODUCTION

Abortion in plain words can be addressed as *"the deliberate termination of human pregnancy"*. Abortions reflect a woman's reproductive choices, which can be exercised to procreate and abstain from procreating. In matters where the condition of the fetus or the mother's health is at stake, abortions become necessary. Historically looking, abortion is a practice as old as pregnancy. Abortion in India have been defined individually by different religions. In the Vedic texts which are central to Hinduism, abortions have been equated to the killing of one's parents (Sathya Narayanan,2016). Though they are allowed to be performed if the women's life has a direct threat due to the presence of the fetus. In Islam, India's second most practiced religion, the holy book Quran leaves the interpretation to the individual states where some schools of thought forbid abortions at any point and others have reservations over the same (Al-Matary,2014). Abortions in India were a serious offense under Section 312 of the Indian Penal Code till 1971. Abortions were legalized only after the implementation of the Medical Terminations of Pregnancy (MTP) Act, 1971.

ABORTIONS – TYPES AND STATISTICS

There are mainly two types of abortions:

1) Medical abortions - The medical abortions are usually the ones which need no invasive procedure or surgery. They are performed by taking drugs, most of the times; it is a combination of two – mifepristone followed by misoprostol. Misoprostol alone can be used if mifepristone is not available (Laura Collins,2022).
2) Surgical abortions - Surgical abortions are performed involves suction techniques by inserting small suction tube in the uterus. The suction removes fetus and the other pregnancy related material from the uterus.

The lancet global health journal in one of its reports (Guttmacher,2017) published that a total of 15.6 million abortions took place in India in 2015, against the figure of 7 lakhs, published by the Centre for the last 15 years. It was found that the vast majority of abortions (81%) were achieved using medical abortions referred to as medical methods of abortion (MMA) that were obtained either from a health facility or other source. The study also estimated that out of 48.1 million pregnancies in 2015, half of them were unintended and abortions were undertaken in one-third of these cases. The study further stated that India's abortion rate is 47 per 1000 almost similar to that of Pakistan and Bangladesh at 50 and 39 respectively. The public sector which is the primary source of healthcare for the poor accounted for only a quarter of total abortions (Guttmacher,2017).

Despite having legislative measures through MTP Act, Abortions were accounted to be the third major cause of female mortality in the country. The use of medicines has brought it down to 8% from an earlier 12%, a figure which is nonetheless huge according to the United Nations Population Fund (UNFP) (State World Population Report of 2022). One in seven unintended pregnancies in the world takes place in India. The report stated that 67% of abortions in India between 2007-2011 were unsafe causing nearly 8 deaths in a day. Some of the States have a high maternal mortality rate in India which includes Uttar Pradesh, Rajasthan, Madhya Pradesh, Chhattisgarh, Odisha, Bihar and Assam (Susheela Singh et al,2018). The Health Management Information System (HMIS) by the Ministry of Health and Family Welfare (MoHFW) states the total number of unsafe abortions at around 9lacs in the years 2014-15 and 2015-16 and around 9.7 lacs for the year 2016-17(Ministry of family health and welfare,2022).

AN OVERVIEW OF SEXISM AND GENDER BASED OPPRESSION WITHIN MEDICINE BOTH IN THE US AND IN INDIA

The medical realm is an open and blooming ground for sexism and oppression against gender to be seen. This shows up differently—from partial cases where access to medical care is much investigated to lopsided research studies and discriminating against both patients and healthcare professionals. In this light, this overview identifies these systemic problems within the US and India and outlines the uniformities and differences found in how gender bias and sexism are ingrained within the medical systems.

This is due to the fact that sexism in medicine is quite a well-documented problem in the United States, with quite a few studies totally bringing to light the fact that, women do receive poor quality healthcare compared to men. Research shows that these three problems—procrastination, failing to reconcile with others, and perfectionism—were responsible for the largest proportion of failures within the first few days of making changes. Studies show that women with heart disease are less likely to receive aggressive treatment for their heart disease, and they are more often involved in having their reports on pain dismissed and not taken seriously compared to men.

They have also shown the gender gap through underrepresentation in higher medical specialties and holding leadership. Though these numbers are changing with more women joining the medical profession, for women in this sector, many glass ceilings still remain, and they are at higher risk of getting impacted by such issues as the gender pay gap and sexual harassment. In fact, women at academic medical centers were getting paid much less than men even after you controlled for specialty, age, and hours worked, according to a 2016 study in the "Journal of the American Medical Association.".

This case emanates from India—a country that predisposes one to cultural, and in general, systemic forms of gender oppression. With reference to the attitudes in culture on gender roles in society, such ideas seem to weigh down both access to healthcare. Women seem to take less decisional autonomy when compared to men in making choices about health. This is worsened by economic constraints and levels of literacy, where especially in the rural setup, women may not even seek medical help due to societal norms and a lack of economic liberation.

Gender bias in medical education and practice also finds its way into patient care in India. This is according to research that indicates a dearth of sensitive gender-specific data and insensitivity to issues facing women's health by the system. For example, the definition equates reproductive health with women's health and, in so doing, obliterates other equally critical areas, such. Women in this profession have to take a lot of heat in the form of sexual harassment and a career glass ceiling where promotions are hardly visible.

CONSTITUTIONAL ASPECTS OF ABORTIONS and MTP ACT, 1971

"A woman's freedom of choice whether to bear a pregnancy or abort it are areas which fall under the realm of privacy" (Soli Sorabjee,2018)

In the landmark decision of (***Suchitra Srivastava and Anr. v. Chandigarh Administration*,2009**) the Supreme Court of India observed that reproductive rights of a pregnant woman are covered well within the ambit of Article 21 of Indian Constitution. The above-mentioned decision reflected the reproductive choices of woman, which can be exercised to procreate as well as to refrain from procreating. Ignorance of the same may deprive a woman's bodily sovereignty in abortion matters. It further states the right to abortion can be read from the right to privacy which is implicit in Article 21. The Supreme Court in the (***KS Puttaswamy v. Union of India,2017***) established that reproductive choice is facet of personal liberty which is guaranteed under Article 21 of the Indian Constitution. In the Indian Penal Code 1860, abortion laws find mentioned between Section 312 to Section 316.

The liberalization of abortion laws began in 1964 as recommended by Shantilal Committee. The Government of India appointed Shah Committee in 1966 to examine and review various facets of abortion in India. The report of this committee recommended the legalization of abortion on compassionate and medical grounds. Based on the recommendations, the Medical Termination of Pregnancy Bill, 1967 was passed by the parliament followed by the assent of the president in 1971, and the MTP Act 1971 was enforced from April 1st, 1972 in India. The Act is based on the model of The Abortion act of 1967 of England. The Act was amended thrice, first in the year 1975 in 2002, and of late in the year 2021 to make the procedural requirements easier and to make the services readily available. It will be beneficial to have an overview of the recent amendment in the MTP Act and its functioning under the new set of clauses.

OBJECTIVES OF THE MEDICAL TERMINATION OF PREGNANCY ACT, 1971

The Medical Termination of Pregnancy Act 1971 was the first concrete attempt to regulate abortions in India. Its main objectives were:

- It legalizes and liberalizes the termination of pregnancies.
- It seeks to raise the maternal health of women and limit the high mortality rate of women caused as a result of unsafe and illegal abortion.
- It renders protection to registered medical practitioners who would otherwise be liable under the Indian Penal Code (Section 315-316).
- It creates rules for monitoring and regulating the activities of medical practitioners relating to abortion services.

Key Elements Of Medical Termination Of Pregnancy Amendment Act 2021

It came into force on the 24th of September and it alters the MTP 1971. Some of the key provisions of this Act are –

1) **Upper Gestation Limit**- It raises the gestational age limit from 20 weeks to 24 weeks for special category of women which is specifically mentioned in the Act. This would include rape survivors, incest victims, and other vulnerable women like minors and women with mental illness.

2) **Opinion required before termination of pregnancy-** abortions performed before 20 weeks require ascent of one medical practitioner, whereas between 20-24 weeks requires the ascent of two registered medical practitioner (RMP). For special categories of women, abortions performed between 20-24 weeks requires the opinion of two medical practitioners.
3) **Medical termination of pregnancy beyond 24 weeks** – permission would be granted by a medical board constituted by the respective government at the approved facility and will be performed by two medical practitioners based on the decision of the medical board. The state government will appoint a gynecologist, pediatrician, radiologist/oncologist, and other members of the board. This process would only be undertaken after due consideration ensuring that it would not be safe for the woman and the fetus would be incompatible with life or born with serious deformities.
4) **Marital status not required** – contraceptive failures suffice for a valid reason for abortions. There is no need to disclose the marital status for the same.
5) **Confidentiality regarding the identity of the woman-** A clause of confidentiality has been reiterated, with only the medical practitioner having the authority to disclose the name to the person authorized by law. Violation would be punishable with imprisonment of up to a year, or fine, or both.

JUDICIAL RECOGNITION OF RIGHT TO ABORTION BEYOND 24 WEEKS

After the amendment of 2021 in the MTP Act there have been instances where the courts have recognized abortion even after the 24 weeks pregnancy ceiling.

Recently in (***Smt. Nivedita Basu v.The State of West Bengal & Ors.,2022***)– The Calcutta High Court allowed medical termination of pregnancy at 34 weeks and 6 days old pregnant lady after considering the, medical report which stated the remote chance of the baby's survival or even if the baby is born it would have severe abnormalities and there was also health risks to the petitioner.

In (***Ms. X (minor) through her father v. State of Uttarakhand & Ors,2022)–*** The Uttarakhand High Court held that there is right to terminate pregnancy in the situation of rape. The rape survivor has a choice whether to carry the baby or not subject to the conditions mentioned the MTP Act. The court also stated that not allowing her to terminate the pregnancy would cause grave agony to her mental and physical health and compelling her to continue without abortion would also violate her right to live with human dignity under article 21 of the Constitution.

***In (Sangeeta Thapa v. Government of NCT of Delhi & Ors.*,2021)** - The Delhi High Court while allowing 28 weeks pregnant woman to terminate the pregnancy due to severe fetal abnormalities and mental distress to the mother observed that "*reproductive choice is a one of the facet of the reproductive rights of the woman and dimension of her personal liberty.*"

***(XXXX v. State of Telangana*,2022)** - The Telangana High Court allowed pregnant rape survivor to terminate her 26 weeks old fetus and held that '*right of the unborn cannot be placed on a higher pedestal than the life of the pregnant woman.*" The court reiterated Article 21 of the constitution which includes right to dignity, self - respect and healthy living. Further the court stated that the Woman has the right not to carry the pregnancy subject to the reasonable restrictions imposed by the statues.

JUDICIAL POSITION BEFORE THE AMENDMENT OF 2021.

Courts have had different opinions with regard to termination of pregnancy post 24 weeks before the Amendment in the MTP Act, 1971.

In (***Z v. The State of Bihar and Ors.*,2017**) the petition was filed by a rape survivor to terminate the pregnancy which happened as a result of rape. She approached the authorities she was 13 weeks pregnant and due to delays she was not given the facility of abortion till 20 weeks. Therefore she was forced to approach to the High Court who denied the petition which led her file a appeal to the Supreme Court. The Court stated that the manner in which is treated is violative of her fundamental rights. The option of MTP at this stage posed a risk to petitioner's health so the court was forced to deny the permission for abortion. She was granted compensation for the "*incalculable harm and irreversible injury giving rise to emotional trauma.*"

***In (Nandini Tushar Rawool v. State*,2018)-** The petitioner was 25 weeks pregnant when she approached the court for the permission of termination of pregnancy. On the basis of the opinion of the medical board who stated that the baby could be managed, the court denied the permission to abort. The Court stated that mental anguish is a part of life which is different from the anguish caused by sexual assault. Also, if the pregnancy is terminated at this stage the child would be born alive and the doctors would face ethical dilemma with respect to potential non-resuscitation.

In (***Priyanka Yadav v. State*,2017**) - the Madhya Pradesh High Court based on the medical board's report refused to allow abortion at 29 weeks of pregnancy to the petitioner who pregnant as a result of rape. The board opined that the risks related the termination of pregnancy and chances of the fetus born alive are high.

A writ petition was filed before the Supreme Court in (***Mrs. X and Mrs. Y v. Union of India and Ors.,2017***) In this case, Mrs. X, on her 26th week of pregnancy received her sonography report on August 2012, it was found that the fetus has Anencephaly, which means the fetus was developed but the brain and head were not. It was medically possible to terminate her pregnancy at that point but not legally. She was forced to continue the pregnancy in spite of knowing it could adversely affect her body and the fetus has no chance of survival. After painful delivery, the child died after 3 hours.

DEVELOPMENT OF ABORTION LAWS AT THE GLOBAL FRONT LINE:

India is a developing country which often gets influenced by the changing dimensions of laws and related aspects of other developed countries.

On a global frontline, the development of reproductive rights for the woman started in Ancient Rome and Greece, but only in the works of a jurist. The Greece philosopher, Aristotle discussed the question of abortion in his book VII of the politics, where he seems to allow it for population control, he stated,

"As for the exposure of nurture of infants, let there be a law for nourishing those that are deformed, but if exposing offspring because the number of children one has is prohibited by the customary rule, then a numerical limit must be set upon procreation. But if children are conceived by some of those who have intercourse in violation of this, an abortion must be induced before the onset of sensation and life. For what is holy will be distinguished from what is not through a sensation of life."(Aristotle,1997)

As per the above-mentioned context, we can consider that those abortions that are performed "before the onset of life" are permissible. It can be regarded that Aristotle's views on abortion would yield a pro-life conclusion.

United Kingdom

In 1803 English Law considered abortion after quickening a criminal act, which was punishable by the death penalty if abortion is performed before "quickening" i.e. before the fetus moves in the womb between 18-20 weeks into the pregnancy. Until this time people consider a fetus as a part of the woman which gave no rights to the unborn, hence its destruction posed no ethical dilemma. After some time, in 1837 the English law abolished the rule of quickening & grievous punishment of the death penalty, the abortion was partially allowed to save the life of a woman. In 1938 court changed the obsolete law in the landmark case of **(*Rex V. Bour*ne,1938)** and held that the mother's mental suffering was sufficient reason for an abortion.

United States of America

Laws against abortion were specifically prevailing in the 1800s, and by 1900 abortion were declared illegal everywhere in the USA, except if conducted to save the life of the mother. But the law doesn't always remain stringent, it has to develop, grow and change with society. Hence, the US Supreme Court in the landmark of (***Roe V. Wade,1973***) held that a set of Texas statutes that make criminal all abortions except when medically advised for saving the life of the mother are an unconstitutional invasion of privacy.

The controversial ruling came in June 2022. American Supreme Court overturned the *Roe V. Wade* ruling declaring that the right to abortion was unconstitutional, rendering abortions a crime in large swathes of the country. The decision has called for backlash from various civil society groups, women organizations, and other countries of the world. The magnitude of the decision is such that even traditional allies of the United States have minced no words in criticizing the medieval era dictum by the Supreme Court of the United States

Gonzales v. Carhart (2007): The case lay on whether the Partial-Birth Abortion Ban Act of 2003 was legal and constitutional. In doing so, the Supreme Court pointedly signaled that there would be no departure from the relentless sociocultural disputes that have marked issues of reproductive rights within the nation, by supporting the national ban on this procedure. This is a decision, in essence, that reflects great social and cultural challenges to women in relation to their right to decide over their health and access to some specific health services. The ruling has been quite controversial; to many, this is seen as a big blow to women's repro rights and points further to the struggle that still remains for women to have equal access to medical care without gender discrimination.

ABORTION: PERSPECTIVE OF AN INDIAN WOMAN

"No woman can call herself free who does not control her own body. No woman can call herself free until she can choose consciously whether she will or not be a mother."- Margaret Sanger

Woman's Right To Abortion And Bodily Sovereignty

The MTP Act, of 1971 provided an opportunity for terminating their pregnancy, but simultaneously it forces the women to justify the reason for abortion. Whether such reason is justifiable or not and whether such abortion should be allowed or not is decided by the doctor concerned. The amendment of the Act in 2021 also didn't serves this purpose. A woman has no right to Abortion unless supported by any justifiable reason. There are instances where a woman's decision to abort is based on various factors like her financial capacity to bear the expenses of the child, career goals, and complications in a matrimonial relationship which may further lead to divorce. The Act lacks in considering these circumstances as justifiable one. All these factors which reflect the 'right of choice' is necessarily avoided by the MTP Act. Hence in this context, a woman is usually deprived of her right to choose and decide on abortion matters. The broad scope of Article 21 must include it under its ambit; ignorance of the same may deprive a woman of her bodily sovereignty in abortion matters.

Need To Prioritize Women's Rights Over The Unborn Child

There are two contrasting views on prioritizing woman's right to life, liberty, and privacy over the life of the unborn. One view is that the unborn has a right to life. Ancient Hinduism believed that the unborn has a life and killing an unborn is a sin (Jayaram,2018). It is believed that life begins at the conception and abortion is thus considered similar to committing murder of a human life. The other view is that the life of the unborn begins at birth and the ultimate right to decide whether to give birth or not should lie with the mother. Article 1 of the Universal Declaration of Human Rights states that *"all human beings are born free and equal in dignity and rights"*. And it is also specified that *"According to international human rights law a person is vested with human rights only at birth and unborn is not an entity with human rights"* (High Court on its Own Motion v. State of Maharashtra, Suo Moto P.I.L. No. 1 of 2016).As per the Indian laws, it is an offense to forcefully abort a child but to make a pregnant woman carry the child against her will is a greater act of cruelty violating the mother's right to life which is ignored by the Act. Thus, there is a dire need that a mother's rights should be prioritized over the interests of the unborn.

GOVERNMENT PROGRAMMES REGARDING ABORTIONS

The MTP Act has been passed by the government which regulates abortions in the country. Abortions are fully covered under the government's public national health insurance funds – Ayushman Bharat and Employees State Insurance scheme both for surgical as well as medical abortions.

1. The National Family Planning Program (NFPP) was launched in 1952 with an emphasis on population control through sterilization but it later evolved to encompass maternal and child health services, including abortion. Under NFPP the government trains healthcare providers in abortion techniques and offers abortion services at government healthcare facilities.
2. The implementation of the Medical Termination of Pregnancy (MTP) Act in 1971 legalized abortion under circumstances to reduce unsafe abortion practices and maternal mortality. Government initiatives are focused on ensuring access to abortion services by effectively implementing the provisions of the MTP Act especially targeting vulnerable populations.
3. Janani Suraksha Yojana (JSY) introduced in 2005 aims to decrease neonatal mortality rates by encouraging institutional deliveries and providing cash incentives to pregnant women delivering in healthcare facilities. While not directly linked to abortion JSY indirectly influences abortion services by promoting care where information on safe abortion options can be shared.
4. Introduced in 2013 the Reproductive, Maternal, Newborn, Child, and Adolescent Health (RMNCH+A) Strategy concentrates on enhancing access, to reproductive healthcare services that include family planning and safe abortion. Government initiatives within the RMNCH+A framework strive to blend abortion services with reproductive health initiatives guaranteeing a seamless provision of care, for women.
5. Comprehensive Abortion Care (CAC) -Indian Government under the National Health Mission formulated the component of CAC and it is an integral part of the objective of bringing down the maternal mortality ratio. CAC is a training package which comes with a trainers and providers manual and is established by the maternal wing of MOHFW. Concurrently, National Health Systems Resource Centre (NHSRC) have worked alongside MOHFW to train Accredited Social Health Activists (ASHA) to provide them with information regarding abortions at community level. The health management information system is an initiative by MOHFW, a health portal for all the types of information on all indicators for public sector. Abortion services are also recorded on the HMIS portal. The government in 2014 also launched the first national mass media campaign over safe abortions. The drugs for use in India for the medical methods

of abortion (MMA) is approved by Drugs Controller General of India (DCGI) under MOHFW. A coalition of individuals and organizations – the Pratigya campaign has also done considerable work in safe abortion care.

ABORTIONS AND RELATION WITH OTHER STATUTES

Abortions in India are directly linked with other factors which have to be taken into consideration. One of them is the age of abortion in young girls is detrimental to their mental and physical health. For all practical purposes, this directly affects maternal mortality rates. The recent amendment in the Prohibition of Child Marriage Act, 2006 stated that the minimum number of years of girls to marry would be increased to 21. Similarly, Pre conception and Pre natal Diagnostic Techniques Act, (PCPNDT) 1994 regulates the misuse of diagnostic techniques for the determination of the sex of the fetus. The strict implementation of this act has also put the medical practitioners into a fix as abortions if proved on the sex determination count would be met with stringent legal actions. This incites reluctance on abortions. The Protection of Children from the Sexual Offenses (POCSO) Act, 2012 states that sex with a woman who is a minor is to be termed rape. If there is a case of minor pregnancy, the hospital ultimately has to report it to the police. This disturbs the objective of safe abortions as the POCSO Act nowhere mentions that if the sex is consensual, what is the course of action that will be followed. Treating every act of sex with a minor as an offense is detrimental to the awareness and accessibility related to abortions in the country. Any alternative discourse undertaken to escape this route on the part of the individual will pose a huge barrier to her sexual and reproductive health and overall well-being.

ANALYSIS

The laws on abortions have to be made more accommodative in the country. The number of unsafe abortions is extremely high in the country. Women in rural areas have typically been the major sufferers. Social stigma and lack of access to proper healthcare facilities are the main reasons. One of the major aspects on which these laws have to be redefined has to be the consent of the woman as is the case with the POCSO Act. These provisions cater to the situation of none as there is no role of the individual, who if at all decides to exercise her consent. The mere cover of a minor cannot strip an individual of her life and privacy which are enshrined as the basic tenet of the Constitution. Similarly, the government should come up with robust plans for abortions in rural areas. There has been a widespread call on training grassroots health workers to keep women away from quacks. The ASHA workers should be trained to carry out safe abortions in rural places which have been not allowed till now.

There is also an urgent need to make people and corporations aware of the emotional as well as the physical aspects of abortions. The provisions should be added for paid leaves under the Maternity Benefit Act. This should cover miscarriages as well as willful abortions. A robust checking on the implementation is required. The MTP Act abstains to mention abortions concerning transgender and queers which during the current times is nothing short of dastardly. It is almost like excluding a section of the population from the benefits of rights which are guaranteed in the Indian Constitution. MTP Act also doesn't provide any remedy for rape survivors seeking abortions beyond the period of 24 weeks and the victims have to approach the Courts for seeking permission to terminate the pregnancy. This is devastating considering the physical, mental, and social stigma that a rape victim goes through. Policy adjustments are required in these regards. A section of AYUSH doctors and paramedics should also be trained to carry out procedures like abortions. The Indian medical association has earlier objected to the same, but the data which comes up is grim in regards to safe abortions in the country. With proper training and supervision, the scope of medical practitioners who can perform abortions would increase, which is the need of the hour in a nation like India, where the rural population is huge and the availability of basic health facilities is far and few.

CONCLUSION AND SUGGESTIONS

The examination of the system in India regarding a women's right to abortion shows a complex situation influenced by various legal, social and cultural factors. India's laws on abortion have changed significantly over time aiming to balance women's rights with concerns about protecting fetuses and public health. The Medical Termination of Pregnancy Act from 1971 is central to abortion regulations in India offering guidelines for abortion and partially safeguarding women's reproductive rights. Despite these measures challenges persist in ensuring universal access to safe and legal abortion services throughout India. Social stigma, lack of awareness, insufficient healthcare infrastructure and bureaucratic obstacles often hinder women from exercising their reproductive rights. Moreover, inconsistencies within the framework create uncertainty and impede effective implementation particularly in cases involving pregnancies resulting from sexual violence or fetal abnormalities. Additionally, the issue of abortions based on sex and gender discrimination remains a significant concern. This reflects seated patriarchal beliefs and underscores the necessity for comprehensive legal and social actions to combat gender inequality and encourage healthcare practices sensitive, to gender issues.

Given these obstacles there are recommendations for future studies and governmental actions to support women's reproductive rights and enhance the availability of safe abortion services, in India. Abortion is one of the oldest practices being performed in the Indian ancient era. Since the beginning abortion was considered dangerous and abortionists tended to kill a lot of women because of a lack of training as abortion was an illegal practice. The coming of the MTP Act in 1971 and its further amendments in 2021 has been a welcome change in the Indian context. United Nations Human Rights Committee has already made it clear that access to abortion and prevention of maternal mortality are a part of Human Rights. The recent development in the US reversing the judgments given in *Roe v. Wade* has come like a shocker to the world during the times when equality and individual rights are of utmost paramount than they have ever been in the history of civilized mankind. The world looks to the United States as the beacon of individual rights and freedom, the decision in that context is nothing less than a Mongol diktat of the medieval ages. India has seen great improvement in the aspects of female freedom, the positive amendment in the MTP Act is one such glaring example in recent times. India cannot be complacent and the necessary tweaks required should be catered to immediately. Women compromise almost half the total population of the country. The earlier social norms, policies, and obsolete cultural practices have failed women as a group. By giving them necessary freedom only, India can realize its dreams of becoming a "Vishwa-guru". As Dr. Jaydeep Tank from FIGO Safe Abortion Committee has quoted – *"any one woman or girl dying or becoming disabled from an unsafe abortion is one too many"*.

Suggestions-

1. The Influence of Societal Stigma and Economic Factors;
    - Delve into the societal perceptions surrounding abortion and how they influence women's choices and their ability to access services.
    - Examine the economic challenges, such as poverty, limited education and inadequate healthcare facilities, that particularly affect marginalized women seeking safe abortion.
    - Study how community engagement and advocacy initiatives can help reduce stigma and improve access to services.
2. Challenges in Implementing the MTP Act;
    - Conduct a thorough examination of how the MTP Act is put into practice in various states, highlighting regional variations and obstacles faced by healthcare providers.

- Evaluate the effectiveness of current mechanisms for addressing grievances of women who are denied legal abortion services.
- Consider potential modifications to the MTP Act to tackle barriers to access, including discussions on mandatory counseling and spousal consent requirements.

3. Broadening the Discussion; Looking Beyond Women's Rights
   - Assess the legal and ethical aspects related to abortion access for transgender individuals and those who do not conform to traditional gender norms, as addressed in recent Supreme Court rulings.
   - Investigate how telemedicine and digital platforms can play a role in enhancing access to abortion services, especially in remote areas.
   - Explore the legal framework concerning fetal viability and its implications on accessing abortions during later stages of pregnancy.

REFERENCES

Al-Matary, A., & Ali, J. (2014). Controversies and considerations regarding the termination of pregnancy for Foetal Anomalies in Islam. *BMC Medical Ethics*, 15(1), 10. DOI: 10.1186/1472-6939-15-10 PMID: 24499356

Aristotle. (1997). *Politics Book VII & VIII, translated by commentary by Richard Kraut*. Oxford Univ. Press.

Collins, L. *The complicated life of the abortion pill,* New Yorker (Last updated on July 5, 2022), https://www.newyorker.com/science/annals-of-medicine/emile-baulieu-the-complicated-life-of-the-abortion-pill/amp

ghdx.healthdata.org. https://ghdx.healthdata.org/record/india-hmis-standard-report-data-itemwise-monthly-2016-2017 (Last updated on Feb 28, 2022).

guttmacher.org. https://www.guttmacher.org/news-release/2017/national-estimate-abortion-india-released (Last updated on Dec.11, 2017).

High Court on its Own Motion v. State of Maharashtra, Suo Moto P.I.L. No. 1 of 2016.

Jayaram, V. *Hinduism and Abortions*, Hindu Website (Last updated on Sept 01, 2018, 11:35 AM), https://www.hinduwebsite.com/hinduism/h_abortions.asp

KS Puttaswamy v. Union of India AIR 2017 SC 4161.

Mrs X and Mrs Y v. Union of India, W.P. 81 of 2017.

Nandini Tushar Rawool v. State – W.P No.8313 of 2018.

Narayanan, S. *What Hinduism says about abortion Article*, speakingtree.in (Last updated on Oct 24, 2016, 1:15 PM), https://www.speadkingtree.in/allslides/what-hinduism-says-about-abortion-article

Priyanka Yadav v. State W.P. No. 23307 of 2017.

Rex v. Bourne, (1938)3 All E.R. 615.

Roe v. Wade, (1973) 410 U.S. 113.

Sangeeta Thapa v. Government of NCT of Delhi & Ors. W.P.(C) 15241/2021.

Singh, S., Hussain, R., Shekhar, C., Acharya, R., Moore, A. M., Stillman, M., ... & Ball, H. (2018). Abortion and unintended pregnancy in six Indian states: findings and implications for policies and programs.

Smt. Nivedita Basu v. The State of West Bengal & Ors., W.P. 2513 of 2022.

Sorabjee, S. (2018). *Abortion: Its every women's right to choose*, Hindustan Times (Last updated on Jan 8, 2018 4:39 PM).https://www.hindustantimes.com/opinion/abortion-it-s-every-woman-s-right-to-choose/story-52qv723N3yXESybsvyHR0J.html

Suchita Shrivastav and Anr. v. Chandigarh Administrator, SLP (C) 17985 of 2009.

unfpa.org. (2022). *State of World Population Report.* (Last updated on Apr. 1, 2022), https://india.unfpa.org/sites/default/files/pub-pdf/1_key_insights_india_final_web.pdf

X (minor) through her father v. State of Uttarakhand & Ors. W.P No. 201 of 2022.

XXXX v. State of Telangana, W.P. 24850 OF 2021.

Z v. The State of Bihar and Ors. MANU/SC/1011/2017.

Chapter 6
Traditional Knowledge System of the Cold Desert of Leh–Ladakh (India) for the Environment Management

Mahesh Kumar Gaur
https://orcid.org/0000-0003-4649-0554
ICAR-Central Arid Zone Research Institute, India

ABSTRACT

Situated in the northernmost reaches of India lies the Cold Desert of Leh-Ladakh, the temperatures fluctuate dramatically from -30°C to 30°C. This region is inhabited by indigenous communities whose traditional ecological knowledge (TEK) is deeply intertwined with their environment, aiding them in adapting and thriving in this harsh landscape. Over centuries, the Ladakhi, Tibetan, and Changpa peoples have developed intricate understandings of the area's ecology, climate, and biodiversity, passed down through oral traditions. TEK plays a crucial role in natural resource management, bolstering locals' resilience to climate change impacts. However, as modernization and globalization encroach upon their lifestyles, there's a need to preserve their age-old practices. The rich cultural heritage of the region faces a challenge as younger generations display dwindling interest in these traditions, emphasizing the necessity of documenting traditional knowledge before it fades away.

DOI: 10.4018/979-8-3693-6069-9.ch006

INTRODUCTION

Traditional Ecological Knowledge (TEK) emerges from generations of experience, providing profound insights into specific ecosystems. It encompasses indigenous knowledge passed down through oral tradition and cultural practices such as arts, crafts, and ceremonies, as well as activities like traditional food preparation and cultivation (Finn et al., 2017).

Though cultivated by indigenous communities worldwide for millennia, TEK gained recognition in the 1980s when tribal elders introduced it as a conceptual framework to elucidate the interconnectedness between humans and the natural world (Bureau of Indian Affairs, 2016). It has since expanded to inform various fields such as resource management, law, mental health, ethnobotany, and, more recently, environmental health and climate change research. The definition of TEK varies across literature, reflecting its diverse applications (Alcorn, 1989; Tsosie, 1996; McGregor, 2009; Flint et al., 2011; Gone, 2012; Maldonado et al., 2015; Moorehead et al., 2015). Integrating indigenous knowledge, ethics, values, and cultural identity into land and wildlife stewardship offers promise for improving resource management and mitigating human health risks (Houde, 2007).

Berkes (2001) characterizes TEK as a cumulative reservoir of knowledge, practices, and beliefs that evolve through adaptive processes and are transmitted across generations through cultural means. It encompasses the interrelations between living beings, including humans, and their environment, blending observation, experimentation, and religious traditions within societies. TEK is both accumulative and dynamic, drawing from experience and adapting to changing circumstances (Gaur and Gaur, 2004). It typically manifests in societies with historical continuity in resource utilization within specific territories, often found in non-industrial or less technologically advanced societies, many of which are indigenous or tribal. While TEK is a subset of indigenous knowledge, the latter is generally defined as local knowledge unique to a particular culture or society (Gaur, 2006).

Indigenous peoples of the Cold Desert hold a wealth of knowledge acquired through generations of living closely with nature. Their practices, including traditional agriculture, pastoralism, water management, medicinal plant use, and biodiversity conservation, are deeply rooted in observation, experimentation, and a profound understanding of local ecosystems, weather patterns, and natural resources. This traditional knowledge not only sustains livelihoods but also fosters resilience in the face of environmental challenges like extreme cold, water scarcity, and unpredictable weather events.

INDIGENOUS COMMUNITIES OF LADAKH

According to the National Commission for Scheduled Tribes (NCST), the Union Territory of Ladakh is predominantly a tribal region in the country. The Scheduled Tribe population represent 66.8 percent in Leh, 73.35 percent in Nubra, 97.05 percent in Khalsti, 83.49 per cent in Kargil, 89.96 per cent in Sanku and 99.16 per cent in Zanskar areas of the Ladakh region. So, the total tribal population in Ladakh region is more than 97 percent. The Bhots, Ladakhi Bodh, Ladakshpa or Bodriksha are the dominant communities of Ladakh and socially stratified into *gyalpo* (rulers), *kahlen* (bureaucrats), *mangrikas* (peasants), *rengon* (artisans), *serger* (goldsmith) etc. Mann (2002) has included twelve tribes, encompassing Muslims, Argons, Christians, Hindus, and Sikhs, as an integral part. However, as per the NCST, the primary scheduled tribes (STs) in Ladakh are Balti, Beda (Musician-drum player), Bot (or Boto), Brokpa (or Drokpa, Dard, Shin), Changpa (Pastoralists of highland), Garra (Ironsmith), Mon (Masculine flute player) and Purigpa. Three main tribes are found in Ladakh which can be distinctively recognized due to their peculiar traditions. These are: the Mons of Indian Origin, nomads of Tibetan origin and the people of coming from the central Asia, a pure Aryan breed, the *Dards*.

Table 1. The Buddhist tribes in Ladakh

Buddhist tribes	Religious affiliations	Language groups	Occupation
Ladakhi Bodh	Geluka, Dropka and Naima	Bhotia (Tibeto-Burman)	Agriculture
Broq-pa	Geluka	Brok, Shad, Dard (Indo-Aryan)	Agriculture
Gara	Geluka, Dropka	Tibeto-Burman	Blacksmith
Mon	Geluka, Dropka	Tibeto-Burman	Muscians
Beda	Geluka	Tibeto-Burman	Musicians
Changpa	Drokpa	Changkyet	Animal husbandry and cattle farming

Source: https://www.undv.org/vesak2011/panel3/11AnandSinghFINAL.pdf

THE LADAKH: AN OVERVIEW

Eastern Ladakh is characterized by the high-altitude plateau of Changthang, with settlements scattered between 2600 m and 4500 m. Known for its rugged terrain, Ladakh's high mountain region boasts an average elevation exceeding 3000 meters

(Figure 1). Administrative division in 1979 led to Leh and Kargil districts, further divided into tehsils and subdivisions with separate headquarters.

Figure 1. Location map of Ladakh

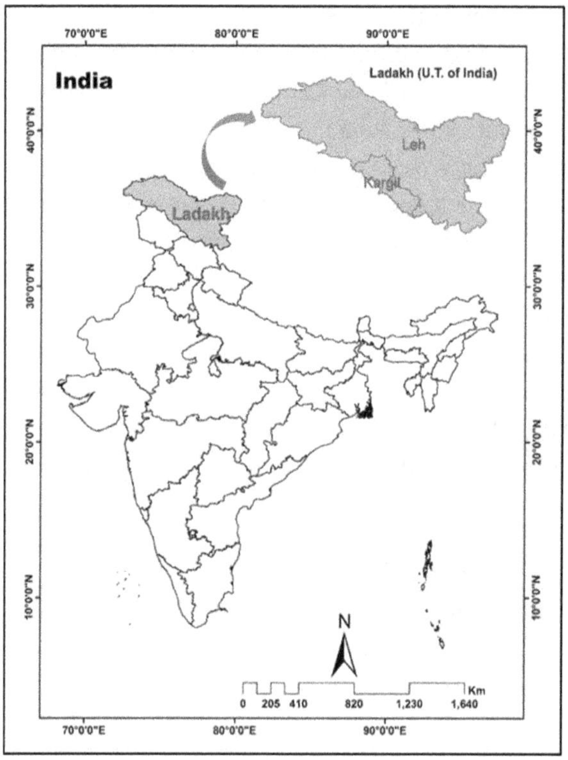

The region experiences significant seasonal temperature fluctuations, with average temperatures ranging from approximately -15.6°C in Dras to -5°C in Leh during January and February, and rising considerably in July and August, from 17.1°C in Dras to 23.7°C in Kargil (Figure 2). Agriculture relies on irrigation channels diverting meltwater from glaciers, snowfields, or main rivers to settlements and cultivated areas along river courses or in inter-valleys.

Figure 2. Weather chart of Ladakh

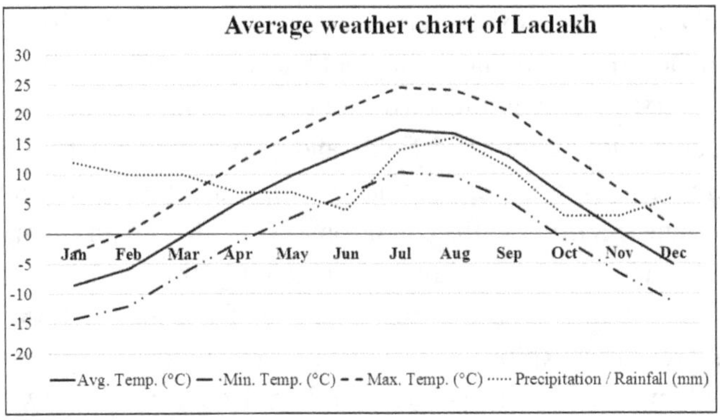

The 2011 Census reported 274,289 inhabitants in Ladakh, with 133,487 in Leh district and 140,802 in Kargil district. The Buddhist population in Leh is 88,635, while in Kargil, the Muslim population is 108,239. In 1971, the total population was 154,422, with 44,556 Buddhists and 6,333 Muslims in Leh, and 42,798 Muslims and 10,009 Buddhists in Kargil. By 1981, Ladakh's total population was 134,372, with 51,407 Muslims and 12,862 Buddhists in Kargil, and 55,514 Buddhists and 10,475 Muslims in Leh (Census of India, 2011).

The agricultural window spans from May to September, with barley and wheat as primary staple crops on irrigated terraces. Vegetables like turnips, cabbage, potatoes, onions, carrots, and leafy greens are cultivated in kitchen gardens. Single-cropping prevails due to altitude constraints, with double-cropping feasible below approximately 3000 meters. Wheat thrives between altitudes of 3600 to 3800 meters, while barley grows above 4000 to 4400 meters. Sparse forest cover consists mainly of willow and poplar trees along water channels and cultivated land edges, providing resources for heating, construction, tools, and animal feed. Fruit trees such as apples and apricots, along with wild herb collection, contribute to diverse land use. The agricultural system revolves around the symbiotic relationship between crop cultivation and animal husbandry.

ADAPTATION STRATEGIES

Adaptation is pivotal in mitigating vulnerability to climate change-induced global warming, aiming to reduce adverse effects on both social and biological systems. However, adaptation capacity varies among human populations due to factors such as poverty, inadequate policy frameworks, limited access to advanced technology, and weak institutional structures. The ability of a country to adapt is significantly influenced by its level of social and economic development, as the costs associated with climate change adaptation tend to be substantial (Kumar, 2014).

In the Cold Desert, traditional ecological knowledge (TEK) offers insights into various adaptation strategies employed by indigenous communities. These strategies, ranging from constructing terraced fields and canals for irrigation to developing artificial glaciers and cultivating cold-hardy crops, are continually adapted to suit the region's unique environmental conditions. Nomadic pastoralism, practiced by communities like the Changpas, involves managing grazing lands and migration routes to ensure sustainable pasture resource use. Traditional taboos and customs regulate natural resource exploitation, contributing to biodiversity conservation and ecosystem resilience. The village-level traditional governing body, led by Goba, plays a crucial role in conserving and managing local flora and fauna, promoting cultural values, and ensuring sustainable natural resource utilization.

In Ladakh's frigid, arid region, integrating traditional ecological wisdom is essential for sustainable development. Key adaptation strategies leveraging traditional knowledge include:

a. *Community Engagement and Participation:* Local communities are required to actively participate in decision-making processes regarding natural resource management. This involvement ensures the incorporation of traditional insights into development initiatives, fostering sustainable practices promoted by village-level governing bodies led by *Goba*.
b. *Agricultural and Irrigation Systems:* Agricultural and irrigation practices in Ladakh align with the zodiac's movements, influenced by altitude and locally crafted calendars. Rituals conducted by village Lamas invoke blessings for a fruitful harvest. These practices should be preserved and adapted to changing climatic conditions.
c. *Combining Traditional and Scientific Approaches:* Collaboration between traditional knowledge holders and scientific experts is essential to innovate solutions for environmental challenges, maintaining production equilibrium and sustainable systems amidst limited resources.
d. *Promoting Agroecology and Indigenous Farming Practices:* Supporting traditional farming methods such as terrace farming, crop rotation, and using local seeds suited to Ladakh's harsh climate is crucial. These methods are resilient to climate variability and require fewer external inputs.
e. *Water Harvesting and Management Techniques:* Traditional water harvesting and management methods, such as building artificial glaciers and maintaining ancient irrigation channels, should be utilized to conserve water resources and mitigate water scarcity impacts.

Glacial water, vital for sustaining life in the region, is channeled to fields via *kuhls*, or water channels. Participatory management ensures equitable water distribution. *Kuhls* are strategically constructed along hill gradients to maintain proper gravity for irrigation and running watermills. Flooding with glacial water serves a dual purpose: removing toxins and acting as a disinfectant against pathogens.

In this area, irrigation relies mainly on *kuhls* drawn from streams or nallas. Most hamlets, situated on plateaus beside the main river, receive water from streams cascading down cliffs onto these plateaus. Alternatively, spring water is collected into small reservoirs on high uplands, with water drawn as needed. Only a small portion of land is irrigated directly from rivers due to steep banks. Fields are located on naturally level plateaus or terraced on hillsides, ensuring safety from avalanches or falling rocks.

Some villagers maintain reservoirs above cultivated lands, using the field to cover snow before early summer raking. Somewhere they rely on perennial torrents for irrigation. Irrigation channels are lengthy, requiring significant annual labor for repair post-snowmelt. Typically, all beneficiaries cooperate for repair work, with substitutes provided for absent individuals. Repairing these channels often resembles constructing new ones, as many are destroyed by avalanches. This longstanding community practice ensures efficient irrigation in otherwise dry, porous soils through the meticulous construction and maintenance of *kuhls*.

The intricate system of water management begins with the collection of *Kangs-chhu*, or melted snow water, from various rivulets. These streams converge to form a *togpo*, flowing through valleys and touching numerous villages. The primary conduit for this water is the *ma-yur*, the main channel constructed along the mountainside, acting as a retaining wall or dyke. From the *ma-yur*, water is diverted into smaller canals called *yu-ra*, which in turn irrigate the fields. The point of diversion of togpo water into the *ma-yur*, and from the *ma-yur* into the *yu-ra*, is termed *yurgo*. The *ska* marks the point where water from the yu-ra is redirected to individual fields. Channels known as *snang* guide the water throughout the fields, ensuring even distribution. Oversight of water distribution for irrigation falls under the responsibility of a *chhur-pon*, an official selected by the villagers. The *chhur-pon* plays a crucial role in managing water supply, making them a pivotal figure in the community's agricultural activities (Angchok, *et al*., 2010).

Artificial glaciers are located closure to the villages and are at lower elevations. They melt earlier than natural glaciers, channeling the resulting water into small reservoirs called *'Zings'*. Despite their size, these reservoirs offer considerable potential for crop irrigation. During November and December, channels are constructed to redirect runoff water to the shaded slopes of mountains, where it can freeze gradually. Retaining walls are erected along the terrain to slow down the flow and promote freezing. Moreover, the water beneath the frozen snow is captured and allowed to freeze. By April, the melted snow water serves as the initial irrigation for barley fields, a practice known as *Thachus*. The specific combination of topography and climate contributes to the small size and high altitude of glaciers, almost all of which terminate above 5200 m. While the glaciated area of central Ladakh totals 997 km2 with more than 1800 glaciers, 79% of them are smaller than 0.75 km2. A general glacier decrease is evident for the time between 1969 and 2016, with a high variability across different watersheds, with glaciated area reduction ranging from 0.2 to 0.9% year^{-1} (Schmidt and Nüsser 2017)

f. *Agropastoralism and Grazing Land Management:* Traditional agropastoral practices, guided by community institutions, promote sustainable resource management. These practices, rooted in a deep understanding of local ecology, should be preserved amidst changing social structures and increasing external pressures.

During the agricultural season from April to May, villagers engage in three primary tasks. Firstly, they prepare water channels and fields for farming. Secondly, they send various animals such as cows, yaks, dzos, sheep, and goats to pastures located on the higher altitudes for grazing, both daily and seasonally. Thirdly, they establish the institution of lorapa, with households taking turns under the supervision of the Goba, to monitor grazing livestock in open cropped areas near the village and impose penalties on violators. Seasonal migration to these pastures serves several purposes: meeting food and fodder needs, reducing risks associated with subsistence agriculture, promoting livestock growth and development with access to diverse forage species, providing respite for agricultural activities without interference from free-grazing animals, and generating additional income through trade items like wool, hair, and skin.

g. *Cultural Revitalization:* Efforts to revitalize Ladakhi culture and traditional ecological knowledge are crucial amidst the rise of cash-driven tourism and changing dietary patterns. Recognizing the cultural significance of traditional practices fosters pride in local heritage and strengthens intergenerational transmission of knowledge. In remote regions, where the contributions of indigenous peoples to ecosystem preservation and management have been recognized, the cultural and social factors of ecological restoration hold even greater significance. Ladakhis have engaged in strong discourse around constructing and strengthening a Ladakhi cultural identity. However, with the influx of a large of national and international tourist, Ladakhi culture is going global and even new diets are gradually becoming integral part of their habit.

h. *Adaptive Livelihood Diversification:* Diversifying livelihood options based on traditional practices, such as pastoralism, handicrafts, and eco-tourism, reduces dependence on fragile natural resources and enhances community resilience to external shocks.
i. *Policy support and institutional collaboration:* TEK encompasses the accumulated knowledge, practices, and beliefs passed down through generations within a community, relating to the use and management of natural resources and ecosystems. Policy support and institutional collaboration for Traditional Ecological Knowledge (TEK) in Ladakh are crucial for preserving indigenous knowledge systems, promoting sustainable resource management, and fostering community resilience. This would foster collaboration between government agencies, and local communities to implement these policies effectively.

Therefore, integrating traditional ecological knowledge into adaptation strategies is essential for sustainable development in the Cold Desert region for ensuring resilience to climate change impacts while preserving cultural heritage and ecological integrity.

CHALLENGES AND OPPORTUNITIES

Traditional ecological knowledge (TEK) confronts challenges amidst rapid socio-economic changes, urbanization, and the erosion of cultural traditions. Often marginalized within Western scientific paradigms, TEK offers nuanced insights deeply entrenched in local contexts. However, the allure of modern lifestyles among younger generations threatens the transmission of TEK to future generations, necessitating urgent documentation and preservation efforts. Despite these challenges, opportunities exist to rejuvenate indigenous knowledge systems through community-driven initiatives, educational programs, and collaborations with researchers and conservation organizations. By merging traditional wisdom with modern science and technology, indigenous communities can enhance their resilience to environmental shifts while safeguarding their cultural heritage.

CONCLUSION

Policy support and institutional collaboration are vital for the preservation of TEK in Ladakh, promoting sustainable development and bolstering community resilience. Integrating TEK into governance structures and decision-making processes can harness the wisdom of traditional knowledge systems, paving the way for a more resilient and sustainable future. As indigenous communities navigate the complexities of modernization and environmental change, the preservation and revitalization of TEK emerge as imperative. Recognizing the significance of indigenous knowledge systems and supporting initiatives aimed at documenting, transmitting, and integrating this wisdom into sustainable development strategies are crucial steps towards ensuring the welfare of both indigenous populations and the ecosystems they inhabit. Amidst uncertainties posed by climate change, TEK serves as a valuable complement to conventional scientific research, bridging interdisciplinary gaps and offering unconventional approaches to climate change research, information dissemination, analysis, and policymaking. Thus, TEK plays a multifaceted role in offering understanding and guidance in our rapidly evolving world. Situated in the northernmost reaches of India lies the Cold Desert of Leh-Ladakh, where temperatures fluctuate dramatically from -30°C to 30°C. This region is inhabited by indigenous communities whose traditional ecological knowledge (TEK) is deeply intertwined with their environment, enabling them to adapt and thrive in this harsh landscape. Over centuries, the Ladakhi, Tibetan, and Changpa peoples have developed intricate understandings of the area's ecology, climate, and biodiversity, passed down through oral traditions. TEK plays a crucial role in natural resource management, bolstering locals' resilience to climate change impacts. However, as modernization and globalization encroach upon their lifestyles, preserving their age-old practices becomes imperative. The rich cultural heritage of the region faces a challenge as younger generations show dwindling interest in these traditions, underscoring the necessity of documenting traditional knowledge before it fades away.

REFERENCES

Alcorn, J. B. (1989). Process as resource: The traditional agricultural ideology of Bora and Huastec resource management and its implications for research. *Adv Econ Botany*, 7, 63–77.

Berkes, F. 2001. Religious Traditions and Biodiversity, appeared in *Encyclopedia of Biodiversity*, Vol. 5, pp. 109-120.

Census of India, 2011, A-11 Individual Scheduled Tribe Primary Census Abstract Data and its Appendix, Last modified July 7, 2016.

Finn, S., Herne, M., & Castille, D. (2017, August 29). The Value of Traditional Ecological Knowledge for the Environmental Health Sciences and Biomedical Research. *Environmental Health Perspectives*, 125(8), 085006. DOI: 10.1289/EHP858 PMID: 28858824

Flint, C. G., Robinson, E. S., Kellogg, J., Ferguson, G., BouFajreldin, L., Dolan, M., Raskin, I., & Lila, M. A. (2011). Promoting wellness in Alaskan villages: Integrating traditional knowledge and science of wild berries. *EcoHealth*, 8(2), 199–209. DOI: 10.1007/s10393-011-0707-9 PMID: 21915737

Gaur, M. K. 2006. Traditional Environmental Knowledge: Barriers in Documenting Conservation Practices in Thar Desert. *Nat. Conf.: Environment Planning & Conservation by Traditional Methods.* Dec. 22-23, 2006. Govt. College, Ajmer. pp. 125-134.

Gaur, M. K., & Gaur, H. (2004). *Combating Desertification: Building on Traditional Knowledge System of the Thar Desert Communities. Internat. J Environmental Monitoring and Assessment*. Kluwer Academic Publishers.

Gone, J. P. (2012). Indigenous traditional knowledge and substance abuse treatment outcomes: The problem of efficacy evaluation. *The American Journal of Drug and Alcohol Abuse*, 38(5), 493–497. DOI: 10.3109/00952990.2012.694528 PMID: 22931084

Houde, N. (2007). The six faces of traditional ecological knowledge: Challenges and opportunities for Canadian co-management arrangements. *Ecology and Society*, 12(2), 34. DOI: 10.5751/ES-02270-120234

Kumar, V. (2014). Role of Indigenous Knowledge in Climate Change Adaptation Strategies: A Study with Special Reference to North- Western India. *J Geogr Nat Disast*, 4, 131.

Maldonado, J., Bull Bennett, T. M., Chief, K., Cochran, P., Cozzetto, K., & Gough, B.. (2015). Engagement with indigenous peoples and honoring traditional knowledge systems. *Climatic Change*, 135(1), 111–126. DOI: 10.1007/s10584-015-1535-7 PMID: 26346998

Moorehead, V. D.Jr, Gone, J. P., & Damia, D. (2015). A gathering of native american healers: Exploring the interface of indigenous tradition and professional practice. *American Journal of Community Psychology*, 56(3-4), 383–394. DOI: 10.1007/s10464-015-9747-6 PMID: 26351006

Tsosie, R. (1996). Tribal environmental policy in an era of self-determination: The role of ethics, economics, and traditional ecological knowledge. *Vermont Law Review*, 21, 225.

Chapter 7
Human Rights of Incarcerated Women in India: Empty Rhetoric or Evolutionary Progress?

Himanshi Bhatia
School of Law, Graphic Era Hill University, Dehradun, India

Karun Sanjaya
Symbiosis Law School, Nagpur, Symbiosis International University (Deemed), Pune, India

ABSTRACT

The plight of female prisoners is not unknown globally; the UN Rules for the Treatment of Women Prisoners and Non-custodial Measures for Women Offenders, popularly known as "Bangkok Rules", is the primary international instrument dedicated to incarcerated women. The Indian National Prison Manual, 2016 is based upon the principles of these rules along with another UN-led "Nelson Mandela Rules". However, the Ministry of Women & Children Development led an exclusive study in 2018 titled "Women in Prisons" that recorded the major issues & human rights violations of such women behind bars due to various reasons like inadequate female staff, accommodation, medical access along with basic right of healthy food. This paper hypothesized that according to the historical & current scenario, prisons in the country specifically still do not cater to the necessities for the subsistence of the incarcerated women. The study intends to analyze major concerns like obligations of the Indian Prison Administration vis-à-vis the progress made till now, specifically concerning Female Prisoners.

DOI: 10.4018/979-8-3693-6069-9.ch007

1. INCARCERATED WOMEN *'OUT OF SIGHT, OUT OF MIND'*: AN INTRODUCTION

According to the Prison Statistics India Report of 2020, female prisoners constitute a minority population of 4% out of the total population of prisoners in India. However, the fact that Indian Judiciary even though being a staunch believer in rehabilitation & reformation of the incarcerated, often restricts itself to the prisoners in general ignoring the special needs of women prisoners. Although imprisonment limits, if not eliminates, many freedoms, it does not deprive people of their most fundamental constitutional and human rights. The unprecedented rise of human rights violation complaints in the recent year from the prisoners or in favor of prisoners to the National Human Rights Commission (NHRC) & State Human Rights Commission (SHRC) against the prison staff predicts the sad state of affairs. The category of female prisoners is worst hit at first instance by being incarcerated in a man's world.

Though India has signed agreements & commitments on the reformation of the women living in imprisonment. Yet there are just 29 dedicated female prisons in India, which cater to only 17% of the female inmates while the rest still adjust in female enclosures within the male prisons (Ministry of Women & Child Development, 2018). Thus, such an arrangement can be said as "Prison within the Prison" which is exclusively for one particular gender. Unfortunately, the so-called "minuscule prison population" is still on the backfoot after countless prison reforms committee & hon'ble supreme court & high courts judgments. Prisons being the state subject (The Constitution of India, 1950) also creates many differences in the progress of the states according to the respective budgetary allocation & administration mechanism.

This paper attempts to deal with the history & struggles of the women in prisons with access to a bare minimum. The study will try to the argument that it can be said that the criminal justice system has just ignored the existence of a minority population of general inmates in the prisons. Hence, the continuance violation of human rights is not a concerning issue for executives as of now.

This leads to the primary questions like, ***Whether the Indian Criminal Justice system adheres to the minimum standards of protection & access for the incarcerated women in the prisons?*** Basic entitlements like sanitation, adequate female prison staff, nutrition, child & pregnancy, menstrual health, and protection from sexual harassment or violence are some of the woman prisoners-oriented needs that the criminal justice system shall provide to the prison inmates without any discrimination. However, the menace of violation of human rights & unfair treatment raises the above-stated question concerning the Indian Prisons.

Secondly, ***Do the incarcerated women have any mechanism or procedure to reiterate their right to live with dignity behind the bars?***

Amongst the most noticeable ways to incorporate human rights into legal rights is through the unique application of constitutional provisions as they are superior legal rules that influence conventional legal norms (Bhat, 2017). Similarly, the Indian Constitution does not recognize the rights of incarcerated women in prisons directly but the core principles of the constitution favor the upliftment of human rights of the convicted in the form of Articles 20 & 21. However, judicial activism & evolving constitutionalism has led to the creation of prison reform committees & guidelines for incarcerated woman in the last few decades.

2. EXISTING FRAMEWORK FOR INCARCERATED WOMEN: NATIONAL & INTERNATIONAL PERSPECTIVES

From the first national committee on women prisoners (1986) under the guidance of Justice Krishna Iyer to the path-breaking judgments (*Annexure II, Ministry of Home Affairs*, 2009) on up-gradation & protection of incarcerated women, the ground reality has not changed much. As stated above the long historical commitments of India about international standards or judicial pronouncements are still unable to predict a positive report card. Some of these instruments are enumerated below:

2.1 International Initiatives

The significance of human rights was recognized post-second world war atrocities with the enactment of the ***Universal Declaration of Human Rights, 1948*** (*The Universal Declaration of Human Rights (UDHR),* 1948). With the 30 articles forming the edifice of the human rights regime, the instrument also subtly recognizes the prisoner's rights with the principle of no torture or cruelty to any person (*UDHR*, art. 5) along with the recognition of "person" before the court of law (*UDHR,* art. 6) which recognizes the prisoners as the person which deserves at least humane treatment even in jails.

Subsequently, the ***International Covenant on Civil & Political Rights, 1966*** (ICCPR, GA Res 2200A, GAOR, UN Doc A/ Res/ 2200/A, 1966) was adopted to substantiate the former declaration. With the foundation based on enjoyment of civil & political rights, it took care of the discrimination of rights enjoyment between men & women (ICCPR, art. 3). On the issue of human rights of prisoners, the rudiments of covenants prescribed an exhaustive code of conduct to maintain the liberty & humane conditions for the incarcerated persons as per the crimes & age irrespective of gender (ICCPR, art. 10).

The foremost dedicated instrument towards the realization of women's human rights is the ***Convention against Elimination of All Forms of Discrimination against Women, 1970 (CEDAW).*** With the aim to attain equality & non-discriminatory practices for women in all spheres of life like law, family, profession, or stereotypical roles. However, originally the female prisoners were not within the ambit of the convention. With the advocacy of female prisoners' human rights violations, instrumental committees were set up to look after the complaints of harassed female prisoners globally (Rope, 2013).

To commemorate the legacy of the Late President of South Africa, Nelson Mandela who spent approximately 27 years of his life in prison as an activist against discrimination & human rights violations, the United Nations General Assembly endorsed the ***"Mandela Rules***(U.N. General Assembly, 2016)" in 2015. Though the objective of this resolution was to maintain minimum standards for prisoners in general it includes various women prisoners' specific rules (Mandela Rules, 2016) as well. From the separate detention space for women to women's right to claim pre & post-natal care in the prison all have been covered under the rules as minimum standards for the detained female prisoners.

The most crucial & only gender-specific resolution by the United Nations for the treatment of female prisoners & their children is popularly known as the ***"Bangkok Rules"***(U.N. General Assembly, 2011). Considering the minutiae of women offenders & prisoners the rules are divided into four parts to cover the different facets of the women-oriented criminal justice system. It includes prison research, planning, re-integration steps, non-custodial measures, or sanctions to cater to the special needs of female convicted be it pregnancy or indigenous status. The peculiar feature of the instrument is evident from its first of all initiative to raise gender-based discrimination & issues for the marginalized community of society i.e., women prisoners.

2.2 Indian Initiatives

The colonial history of India gave birth to the present penal laws & penitentiary system. The notorious detention laws by the authorities to the rigorous imprisonment culture in the form of the infamous Cellular Jail in Andaman, the sub-continent has witnessed a long chain of human rights violations in the colonial period.

However, Independent India ushered in the same archaic laws & detention policies. The development & narrative of the sanctity of human rights via constitutional provisions & ideology led to the various initiatives & amendments to protect the prisoners from gross human rights violations. Some of the major reforms are as follows:

On the issue of congestion of under-trial prisoners in jails, the ***79th Law Commission*** (*79th Report on Congestion of Under-Trial Prisoners in India*, 1979) was set up. The inordinate delay in investigation & confinement of under-trial persons in the prisons was essentially taken up as a violation of human rights by the committee. To solve the menace of delayed disposal of cases, the committee suggested segregation of prisoners as per their probable bail criterion and so on & so forth. The common issue of under-trial prisoners' detention with convicted prisoners was also taken up. Though the focus of the commission was not upon women prisoners these reforms were considered in general for all types of prisoners.

Popularly known as the ***Mulla Committee*** (Ministry of Home Affairs, 1983) under the chairmanship of Justice A.N. Mulla, is considered one of the most exhaustive reports on prison reforms. The extremity of the national policy on prison reform was suggested by it. Along with it, the study suggested various progressive guidelines towards handling women prisoners like – handling of women prisoners by the police & jail staff with due dignity, accessibility of easy bail, pre-natal & post-natal medical care, allowance of incarcerated children till the age of five to stay with their mothers, proper working & rehabilitation arrangements, etc.

The Pioneer Committee on Women Prisoners was formed under the Chairmanship of Justice V.R. Krishna Iyer titled ***"The National Committee on Women Prisoners (1986)".*** The creation of more women prisons & up-gradation of dilapidated central jails was recommended to maintain the minimum standards of living for the vulnerable population. Special provisions like furlough, commutation & parole were put up for the women prisoners whose dependent older children can not live with them in the prisons to fulfill their respective maternity obligations. The path-breaking recommendations by the committee were the bedrock for the creation of the Bureau of Police Research & Development (BPR&D).

The role of BPR&D as a correctional administration agency was realized with the preparation of the ***Model Prison Manual (2003)*** by it. It dedicated a chapter (Bureau of Police Research and Development, 2003a) on women prisoners on their management by the prison authorities. It released the mandate on the subject of diet, pregnancy, medical aid, child care facility, exclusion of male staff & separate enclosures security, vocational training, rehabilitation & reintegration schemes. The distinct characteristic of this manual was its mandatory stance rather than the advisory guidelines or recommendations which were floated by the above-stated expert committee & reports.

The latest regulatory document on the prison reforms is the ***Model Prison Manual, 2016*** which succeeded the above-stated 2003 manual. The objective of the manual is to bridge the gap in terms of prison administration across the country. It also included a chapter (Bureau of Police Research and Development, Chap. 26) on women prisoners which began with the establishment of more women prisoners &

safety of the enclosures within the general jail. Few progressive recommendations which were discussed in the manual were as follows-

Mental Health Consultation for female prisoners (Bureau of Police Research and Development, Chap. 26.19), ***Gynaecology access to the Pregnant Prisoners*** (Bureau of Police Research and Development, Chap. 26.28), ***Arrangement for child-birth on temporary release from the prison, and Special sessions to understand the female criminality & behaviour to be conducted by the prison authorities.***

The above-stated rules are just a few amongst many other advisories, parliamentary debates, reports, state laws & judgments covering the different regions of the country. Thus, we have plenty of human rights instruments, constitutional provisions & committee reports dedicated solely to curbing the menace of violation of the human rights of prisoners. However, does it fulfilled the purpose can be analysed by the present statistics & the Apex Court Judgements on the same?

3. ROLE OF JUDICIARY: 'BARE MINIMUM EXECUTION' FOR THE MINIMUM STANDARDS

India is a signatory to several instruments on unalienable rights of the prisoners some of them have been discussed above as well. Therefore, the Indian State realizes its commitments to the constitutional principles as the backdrop of all the enactments & concerned manuals. Though the Indian Constitution does not expressly entertain the rights of the prisoners or women prisoners in specific there are various provisions that the judicial organ of the country has interpreted for the upliftment of this section of the society.

The recent Special Leave Petition to the Supreme Court on the issue of release of detained prisoners titled ***The Home Secretary (Prison) & Ors. Vs. H. Nilofer Nisha*** 73 SC SCC OnLine 2020., the court essentially held that it is the duty of the court that if it can curtail the liberty of the person by incarcerating them for the offense as per the law then it should also ensure the maintenance of his/her entitled fundamental rights of the incarcerated behind the bars.

If we analyse, these decisions in the light of the current status or statistics under different heads for the incarcerated women. It can be said that the difference between the ideal & the real outcome is quite poles apart. The most common parameters discussed by the hon'ble apex court for the living standard of female prisoners can be examined concerning the present data as follows:

3.1 Occupancy of Women in Prisons

To cater to the ever-increasing population (National Crime Records Bureau, 2021) of women prisoners over the past few years, there are only 29 dedicated women jails that can take in a total of 6179 inmates that too in just 14 UTs/States (National Crime Records Bureau, 2021). According to the data, the occupancy rate of these jails is 50.1% as on the nation's aggregate (National Crime Records Bureau, 2021 at 12) which looks like a rosy depiction in terms of overcrowding of prisons. However, there are three states (National Crime Records Bureau, 2021 for Bihar, West Bengal & Maharashtra).with more than 100% occupancy rate in their women jails.

Thus, the fallacy of maintenance of occupancy can be broken with the data that these women jails serve only 15% of the total incarcerated women i.e., 20,046 at the national level (National Crime Records Bureau, 2021). Henceforth, approximately 86% (16,962) of incarcerated women are imprisoned in general prisons. The situation is much worse here as a large chunk of states still does not have even a single women's prison. Consequently, there are 6 states with occupancy rates of more than 100% of female inmates in other jails (National Crime Records Bureau, 2021).

This issue of overcrowding of female prisoners in the jails was also put up in the latest government-led study titled "Women in Prison". As per the report, overcrowding in women's prions or enclosures can have severe ramifications on the hygiene conditions & psychological disorders of the inmates. These issues were discussed decades back by the Supreme Court in a judgment of ***R.D. Upadhyaya Vs. State of A.P.*** AIR 1999 SC 2183 concerning sufficient care & protection to incarcerated women & their children. Wherein, prison is the state subject the court-ordered affidavits from all the states & UTs on the status of living & accommodation conditions due to lack of hygiene.

3.2 Undertrial-Convict Proportion

Often the expert committee & judicial organ emphasis on speedy trial, bail & segregation of the undertrial & convicted prisoners to preserve the sanity of the undertrials from the trial & prison trauma. However, the data reveals that the majority of the population in the prisons are undertrial than convicts (National Crime Records Bureau, 2021). If we analyse women's data, then in the year 2020, approximately 15167 were undertrials while just 4713 were convicts in different jails (Ministry of Women & Child Development, 2018). Even the women's jails recorded 62.6% of inmates as undertrial prisoners (Ministry of Women & Child Development, 2018).

The outbreak of the ongoing pandemic introduced administrative bodies to the repercussions of overcrowding in the prisons due to the huge number of undertrial inmates. In the Suo moto order, (*Contagion of Covid 19 Virus in Prisons, Suo*

Motu Writ Petition(C) No.1/2020, 2021) the supreme court ordered the high courts to look for de-congestion avenues to manage the spread of disease. Subsequently, the incarcerated women within the enclosures of general jails had to suffer from various struggles, restrictions, violence & discrimination (Baxi, Singh, 2021). The landmark case of ***Re-inhuman Conditions in 1382 Prisons*** (WRIT PETITION (CIVIL) NO. 406 OF 2013) briefly discussed the inability of the criminal justice administration to implement Sec 436A of the CrPC which essentially contributes to the overcrowding in the prisons. Thus, if we analyze the data majority of the female prisons have been arrested or charge-sheeted for compoundable or bailable offenses. Even in the case of non-bailable offenses, the ***National Commission for Women issued an advisory*** (National Commission for Women, 2020) on the release of women inmates under the proviso of Sec 437(1) CrPC to facilitate humane conditions for the minority population of the prisons (women here). The harsh reality is that there is a huge proportion of undertrial women still languishing in jails some even with their children waiting for the end of their trial.

3.3 Inadequate Female Staff

The premise of the recent criminal amendments is set on the grounds of the protection of the integrity of the women victims as well as offenders. These amendments create the need for more criminal justice administration workforce as female officers & administrators. Even the Prison Manual, 2016 prescribes the creation of the post of Lady Superintendent, Assistant Superintendent, Female chief warden & head warden for the convicted women (Ministry of Home Affairs, Government of India, 2016). It even mandated one post for the lady DIG, to the Prison's Headquarter which will look after the complaints pertaining to women staff, prisoners & administration.

Notwithstanding, the immense shortage & above-stated recommendations there are only 8,380 women staff in India across different jails. Whilst, as per the statistics, the largest state Uttar Pradesh covers approximately 14 inmates per staff which depict the sorry state of affairs (National Crime Records Bureau, 2021). The situation is worst in the jails of small district jails wherein the acute shortage of female staff just adds misery to the lives of incarcerated women.

Consequently, the dearth of female staff transpires to the circumstances wherein mostly male staff has been given the duty for the female prisoners. This is disregarding the sanctity of the gender-specific requirements of the inmates. The non-observance of the sanctioned strength of female staff just aggravates the vulnerability of these inmates to discrimination, sexual assault, harassment, torture & even gender-based violence.

These issues are not unfamiliar, the landmark judgment on prison reforms ***Rama Murthy Vs. State of Karnataka*** AIR 1997 SC 1739 which the three-judge bench analyzed that the performance & recruitment of female staff & doctors is not satisfactory in the majority of jails. Thus, the Hon'ble Supreme Court ordered the increase in prison staff strength along with the training & education schemes for the new recruitment to be aware of the constitutional rights of the prison inmates as well.

However, the above-quoted judgment which was issued years ago still holds relevance in present India. The unfortunate progress in the placement of prison staff as per the population of women inmates can be considered a failure on the part of the Government.

3.4 Infrastructural Lapse for Women Prisoners

With the vision to smoothen the criminal justice system, the Government have had initiated various programs & policies. One such mechanism is budgetary allocation towards up-gradation & alignment of the functioning of the various organs including prisons.

The Union Budget 2022-23 (Ministry of Finance, 2022) allocated more than 500 crores rupees for the inter-operability of the criminal justice system. Along with it, another 500 crore rupees were set aside for the modernisation of the prisons. Interestingly, there was no allocation under the head modernization of prisons in the financial year 2021-22. As the aim of modernization of prisons was under the cold sack since 2009 which is being revamped with the current union budget (Suresh, 2022). Even though prison comes under the state list, the Union's financial allocation & schemes facilitate the prison administration across the country.

Howbeit, the budgetary allocation by various states at all Indian levels has been reported by the Prison Statistics Report to be declined by 2.9% in the financial year 2020-21 from the preceding year (National Crime Records Bureau, 2021). Furthermore, just a fraction of it is spent on prison inmates, as it was just 34.7% of the total budget in the year 2020-21 at the national level (National Crime Records Bureau, 2021).

3.5 Criminal Justice for Female Prisoners

Another loophole is the accessibility of legal aid to women prisoners. This issue was raised by the journalist named Sheela Barse's writ petition 1983 AIR 378, wherein lack of legal aid & custodial violence of the women prisoners was considered by the hon'ble court. Thus, this case made it mandatory for the state authorities to prepare a list and aware prison inmates of their legal rights & DLSA assistance. Similarly, on the issue of "Protective Custody," the women undertrials were held in illegal detention by the court being violative of article 21 of the Constitution (*Hussainara Khatoon (iv) & Ors. v. Home Secretary. Bihar*, 1979).

Similarly, the arrest of the women accused by the police officials in the absence of any female officer was also a kind of vulnerability for women convicts. The no arrest policy of women accused before sunrise & after sunset in the absence of a woman constable was ordered in lieu of the rising instance of harassment (*Christian Community Welfare Council of India v. Govt of Maharashtra,* 1996).

4. CONCLUSION: 'A RAY OF HOPE

The legislative & judicial organs endeavor towards the upliftment of human rights & rehabilitation of women prisoners. Howbeit, the actual fruition of these efforts still awaits a stronger mechanism & justice system. The ultimate object of incarceration is reformation which is generally ignored in the process of disciplining the inmates. The implementation of the international standards & recent model prison manual, 2016 is the need of an hour. The ministry of women & child development recommendations for the women prisoners can be taken into consideration like the establishment of a national commission of prisons, the setting up of robust grievance redressal, removal of social stigma attached to the detainment of the women accused & their pregnancy in the prisons (Sanjaya & Praveen, 2020). The fact as discussed in the paper shows that the largely male-centric prison system still awaits major reforms for the rehabilitation & human rights entitlements for female prisoners in the long run. The administrative & policy of the state will play an active role in the realization of such rights.

REFERENCES

Government of India. (2009). *Annexure II, Ministry of Home Affairs.*

Law Commission of India. (1979). *79th Report on Congestion of Under-Trial Prisoners in India.*

Baxi, S. P., Navsharan. (2021, April 4). *A Case for the Release of All Women Undertrial Prisoners – The Leaflet.* Theleaflet.in. https://theleaflet.in/a-case-for-the-release-of-all-women-undertrial-prisoners/

Bhat, P. I. (2017). "Women's Rights As Human Rights": A Dynamic Approach Towards Gender Justice. *CMR University Journal for Contemporary Legal Affairs*, 1(1), 67.

Bureau of Police Research and Development. (2003a). *Model Prison Manual for the Superintendence and Management of Prisons in India* (p. Chapter 3). Ministry of Home Affairs.

Bureau of Police Research and Development. (2003b). *Model Prison Manual for the Superintendence and Management of Prisons in India* (p. Chapter 26).

Christian Community Welfare Council of India v. Govt of Maharashtra, (Supreme Court of India 1996).

Contagion of Covid 19 Virus in Prisons, Suo Motu Writ Petition(C) No.1/2020, (Supreme Court of India, May 2021).

Hussainara Khatoon (iv) & Ors. v. Home Secretary. Bihar, (Supreme Court of India, 1979).

International Covenant on Civil and Political Rights (ICCPR), Article 3 (1966).

Ministry of Finance. (2022). *Union Budget 2022-23.* Government of India.

Ministry of Home Affairs. (1983). *Report of the All-India Committee on Jail Reforms 1980-83.* Government of India.

Ministry of Home Affairs, Government of India. (2016). *Model Prison Manual, 2016, table 26.157.*

Ministry of Women & Child Development. (2018). *Report on Women in Prisons.* Government of India.

National Commission for Women. (2020, April 22). *Press Note* .

National Crime Records Bureau. (2021). *Prison Statistics India 2020*. Ministry of Home Affairs.

Rope, O. (Ed.). (2013, August 19). *Protecting the rights of women offenders: A job for the CEDAW Committee? (Olivia Rope – Penal Reform International)*. Optional Protocol to CEDAW. https://opcedaw.wordpress.com/2013/08/19/protecting-the-rights-of-women-offenders/

Mandela Rules, Rule. 11, 28, 45, 48, 58 & 81. (2016).

Sanjaya, K., & Praveen, D. (2020). Ethics in the Criminal Justice System. In Deb, S., & Subhalakshmi, G. (Eds.), *Delivering Justice* (p. 223). Routledge., DOI: 10.4324/9780429324062-20

Suresh, N. (2022, February 22). *Explained: How The Union Budget Funds India's Justice System*. Www.indiaspend.com. https://www.indiaspend.com/explainers/how-the-union-budget-funds-indias-justice-system-804821#:~:text=It%20used%20the%20allocated%20Rs

The Constitution of India, List II, Entry 4. (1950).

The Universal Declaration of Human Rights, Article 5 (1948). International Covenant on Civil and Political Rights (ICCPR), GA Res 2200A, GAOR, UN Doc A/Res/ 2200/A, (1966).

The Universal Declaration of Human Rights (UDHR),GA Res 217 A, GAOR, UN Doc A/Res/ 217/A. (1948). U.N. General Assembly.

U.N. General Assembly. (2011, March 16). *United Nations Rules for the Treatment of Women Prisoners and Non-custodial Measures for Women Offenders (the Bangkok Rules), GA Res 65/229, UN Doc A/RES/65/229.*

U.N. General Assembly. (2016, January 8). *United Nations Standard Minimum Rules for the Treatment of Prisoners (the Nelson Mandela Rules), GA Res A/RES/70/175 .*

Writ Petition (Civil) No. 406 OF 2013, (Supreme Court of India, 2013).

Chapter 8
Sacred Landscapes:
The Intersection of Indigenous Women's Spirituality and Environmental Protection in Canada

Swati Chakraborty
https://orcid.org/0000-0003-0799-1954
GLA University, India

ABSTRACT

This research paper explores the profound connection between Indigenous women's spirituality and their roles in environmental protection in Canada. By examining traditional ecological knowledge, spiritual practices, and contemporary environmental activism, the study aims to highlight the integral role of Indigenous women in safeguarding their ancestral lands. The paper will delve into case studies of specific Indigenous communities, analyze the impact of colonization and industrialization, and offer insights into how Indigenous women's spiritual perspectives can inform broader environmental conservation efforts.

INTRODUCTION

For centuries, Indigenous women have served as the stewards of their communities' spiritual and physical realms in these lands that we now call Canada. Part of their role involves a spiritual tie to the land, seen as sentient and not merely something existing for our use. This article examines the nexus between Indigenous women's spirituality and environmental activism, illustrating how their spiritual practices uphold their conservation efforts. This understanding is of the essence, so that we

DOI: 10.4018/979-8-3693-6069-9.ch008

continue to value and recognize what Indigenous women contribute in an important way to the conservation of nature and henceforth start viewing necessary policies being designed around their leadership.

A crisis of global proportions, compelling the world to respond with immediate and inclusive action. In Canada, Indigenous women have long been at the forefront environmental stewardship by leveraging centuries of traditional ecological knowledge. However Indigenous women in Canada have long been the custodians of both the spiritual and physical realms of their communities. Their roles encompass a deep-seated connection to the land, which is seen as a living entity rather than a resource to be exploited. This paper explores the intersection of Indigenous women's spirituality and environmental protection, highlighting how their spiritual beliefs and practices underpin their environmental stewardship. By understanding this relationship, we can appreciate the vital contributions of Indigenous women to environmental conservation and recognize the need for policies that support and amplify their leadership.

Indigenous women in Canada play a pivotal role in environmental stewardship and advocacy, deeply rooted in their historical and spiritual connections to the land. Historically, Indigenous women have been integral to their communities, serving as keepers of traditional ecological knowledge and leading sustainable practices that respect the environment (Kwaymullina, 2018). Despite the challenges posed by colonial policies, which disrupted their roles and marginalized their voices, Indigenous women have demonstrated remarkable resilience and leadership in environmental movements (Hoover et al., 2012).

The spiritual connection to the land is fundamental to Indigenous women's identities, where the environment is seen as a living entity deserving of respect and care (Wilson, 2006). This perspective is evident in practices such as the Haudenosaunee Thanksgiving Address, which acknowledges the interdependence of all life forms and the importance of gratitude towards the natural world (Beverley, 2020). Women often lead these ceremonies, emphasizing their role as cultural and spiritual leaders within their communities (Campbell, 2017).

Furthermore, Indigenous women actively confront environmental racism and advocate for justice within their territories. Their leadership is critical in addressing issues such as land dispossession and the impacts of resource extraction, as seen in movements like the Clayoquot Sound protests (Tindall & Robinson, 2017). By integrating their cultural values and ecological practices, Indigenous women not only work towards environmental sustainability but also strive for the recognition of their rights and the protection of their communities (Smart Prosperity Institute, n.d.; Wain, 2021).

In summary, Indigenous women's leadership in Canada exemplifies a unique intersection of environmental advocacy, cultural resilience, and human rights, highlighting the need for inclusive approaches that honor their contributions and address the systemic challenges they face (Fury for the Sound, n.d.; Eco-Feminist Appropriations, n.d.). Their ongoing efforts are essential for fostering a sustainable future that respects both the land and Indigenous sovereignty.

The environmental crisis poses unprecedented challenges globally, necessitating urgent and inclusive solutions. Indigenous women in Canada play a pivotal role in environmental stewardship, drawing on centuries of traditional ecological knowledge. However, their contributions are often overlooked in mainstream environmental discourse. This paper aims to highlight the indispensable role of Indigenous women's leadership in environmental advocacy and how their unique perspectives foster a human rights-based approach to sustainability.

Indigenous women have historically been the caretakers of the land, with deep-rooted connections to their natural surroundings. This stewardship is embedded in their cultural practices, oral histories, and community roles. Despite the historical marginalization and systemic barriers faced, Indigenous women have continued to lead environmental movements, leveraging their knowledge to combat climate change, protect biodiversity, and advocate for sustainable land use.

HISTORICAL CONTEXT AND SPIRITUAL CONNECTION TO THE LAND

Indigenous cultures in Canada are diverse, yet a common thread among them is the spiritual connection to the land. This connection is often rooted in creation stories, traditional practices, and community rituals. The land is seen as a sacred entity that sustains life and embodies the spirits of ancestors and deities. For Indigenous women, this spiritual bond is particularly significant, as they are traditionally seen as the nurturers and protectors of life.

Historically, Indigenous women's roles have been crucial in maintaining the health and balance of their ecosystems. They have been responsible for gathering medicinal plants, ensuring sustainable agricultural practices, and conducting rituals that honor the land. This deep ecological knowledge, passed down through generations, is intertwined with their spiritual beliefs.

The understanding of historical context, and the spiritual ties to the land also serve as a basis for comprehension surrounding Indigenous women in Canada taking on environmental stewardship. Natural World: Indigenous cultures range from the Arctic to tropical rainforests. But all have a deep and intrinsic natural world for many Indigenous peoples this connection is not just utilitarian, but inseparable from their spiritual beliefs and cultural identity.

Historical Context

Before the Colonization: Before European settlers reached Canada, Indigenous peoples in every region of present-day country lived on and with their lands - according to traditional ecological knowledge that guided them through thousands of years. This was in the form of total knowledge, which did not simply address practical tools for physical survival but also spiritual and cultural functions. For examples, in all the centuries of European habitation in North America before 8000 BP (Before Present), indigenous societies developed sustainable practices that maintained their ecosystems.

Colonial Disruption: The arrival of European settlers brought profound disruptions to Indigenous ways of life. Colonial policies aimed at assimilating Indigenous peoples into European ways of life included the appropriation of land, the imposition of foreign governance structures, and the marginalization of Indigenous spiritual practices. The Indian Act of 1876 and the establishment of residential schools were particularly detrimental, as they sought to erase Indigenous cultures and identities. These schools forcibly removed children from their families, stripping them of their language, traditions, and spiritual beliefs. The dispossession of land and resources disrupted the traditional roles of Indigenous women and their connection to the land.

Resilience and Revival: Despite these hardships, Indigenous women have shown remarkable resilience. Over the past several decades, there has been a revival of Indigenous cultures, languages, and spiritual practices. Indigenous communities have been actively reclaiming their rights and restoring their traditional relationships with the land. Women have been at the forefront of this movement, revitalizing traditional ecological knowledge and leading efforts to protect and preserve their environments.

Spiritual Connection to the Land

Land as a Living Entity: For many Indigenous cultures, the land is not an inanimate object but a living entity imbued with spirit. It is a source of life, sustenance, and identity. This spiritual connection is reflected in the belief that all elements of the natural world – animals, plants, rivers, mountains – possess a spirit and are interconnected. This worldview fosters a deep respect for the environment and a commitment to its stewardship.

Role of Women in Spiritual Practices: Indigenous women often hold key roles in maintaining the spiritual health of their communities. They are the keepers of ceremonies, rituals, and traditional practices that honor the land. These practices are integral to the community's spiritual well-being and environmental health. For instance, the practice of giving thanks to the earth for its bounty, known as "thanksgiving" or "giving thanks," is a common ritual that underscores the reciprocal relationship between humans and nature.

Intergenerational Knowledge Transmission: The transmission of traditional ecological knowledge is often facilitated by women through storytelling, ceremonies, and hands-on teaching. This knowledge encompasses understanding the life cycles of plants and animals, seasonal patterns, and sustainable harvesting techniques. It is deeply intertwined with spiritual beliefs and practices, ensuring that environmental stewardship is guided by a holistic understanding of the ecosystem.

Cultural and Spiritual Resurgence: In recent years, there has been a cultural and spiritual resurgence among Indigenous communities in Canada. Women are leading efforts to revive traditional practices and reconnect with the land. This resurgence is not only about preserving cultural heritage but also about asserting sovereignty and reclaiming the right to manage and protect their ancestral territories.

THE ROLE OF INDIGENOUS WOMEN IN ENVIRONMENTAL STEWARDSHIP

Indigenous women's spirituality informs their approach to environmental protection. Their spiritual practices often involve ceremonies and rituals that emphasize the interconnectedness of all life forms and the need to maintain harmony with nature. For example, many Indigenous communities have ceremonies that mark the changing seasons, which are not only spiritual events but also opportunities to impart ecological knowledge and sustainable practices.

In contemporary times, Indigenous women have become prominent leaders in environmental advocacy. Their efforts are often rooted in their spiritual obligation to protect the land. For instance, the Idle No More movement, which began in Canada, has seen significant involvement from Indigenous women. This grassroots movement advocates for the protection of land and water against industrial projects that threaten the environment and Indigenous ways of life.

CASE STUDIES

The Haida Nation and Gwaii Haanas: The Haida Nation's management of Gwaii Haanas National Park Reserve and Haida Heritage Site is a prime example of Indigenous women's leadership in environmental conservation. Haida women play vital roles in the governance of the park, which integrates traditional ecological knowledge with modern conservation practices. This approach not only preserves the natural environment but also maintains the spiritual and cultural heritage of the Haida people.

The Akwesasne Women's Traditional Knowledge Project: The Akwesasne Women's Traditional Knowledge Project focuses on the conservation of the St. Lawrence River ecosystem. This initiative, led by Mohawk women, emphasizes the use of traditional knowledge and spiritual practices in environmental stewardship. The project addresses issues such as water pollution and the loss of biodiversity, while also educating younger generations about the spiritual significance of the river.

The Tsleil-Waututh Nation's Opposition to the Trans Mountain Pipeline: The Tsleil-Waututh Nation's opposition to the Trans Mountain Pipeline expansion is driven by a commitment to protect their ancestral lands and waters. Indigenous women have been at the forefront of this resistance, invoking their spiritual duty to safeguard the environment. Their leadership has brought national and international attention to the environmental and cultural risks posed by the pipeline project.

BARRIERS TO INDIGENOUS WOMEN'S ENVIRONMENTAL LEADERSHIP

Despite their critical roles, Indigenous women face numerous barriers in environmental leadership. These include systemic discrimination, lack of access to resources, and political marginalization. Furthermore, their spiritual and ecological knowledge is often undervalued in mainstream environmental discourse, which tends to prioritize scientific and economic perspectives.

Colonial policies have historically disrupted Indigenous ways of life and governance structures, undermining the roles of Indigenous women. Land dispossession, cultural assimilation programs, and the imposition of patriarchal systems have all contributed to the erosion of Indigenous women's authority and influence.

INTEGRATING INDIGENOUS KNOWLEDGE SYSTEMS WITH MODERN ENVIRONMENTAL PRACTICES

Indigenous knowledge systems are holistic, encompassing spiritual, cultural, and practical wisdom about the environment. This knowledge is invaluable in addressing contemporary environmental issues. Indigenous women's leadership in environmental initiatives often involves integrating these traditional practices with modern scientific approaches. For example, in forest management, Indigenous techniques of controlled burns and plant-based soil enrichment are being recognized for their effectiveness in maintaining ecosystem health and resilience.

IMPACT OF CANADIAN POLICIES ON INDIGENOUS ENVIRONMENTAL RIGHTS AND LEADERSHIP

Canadian policies have historically marginalized Indigenous communities, often leading to environmental degradation and loss of traditional lands. However, recent legislative changes, such as the recognition of Indigenous rights in the United Nations Declaration on the Rights of Indigenous Peoples (UNDRIP), offer a framework for better inclusion of Indigenous voices in environmental governance. Despite these advances, significant barriers remain, including limited access to decision-making processes and inadequate implementation of protective measures.

BARRIERS TO INDIGENOUS WOMEN'S PARTICIPATION IN ENVIRONMENTAL GOVERNANCE

Indigenous women face multiple barriers in environmental governance, including gender discrimination, lack of access to resources, and political marginalization. These challenges are compounded by broader social and economic inequalities. Addressing these barriers requires targeted policies that empower Indigenous women, ensuring their active participation in environmental decision-making processes.

OPPORTUNITIES FOR ENHANCING INDIGENOUS WOMEN'S LEADERSHIP

To enhance Indigenous women's leadership in environmental governance, it is crucial to provide platforms for their voices and integrate their perspectives into policy frameworks. Educational programs that highlight Indigenous knowledge, community-led research initiatives, and partnerships between Indigenous organizations and governmental bodies can facilitate this inclusion. Additionally, supporting Indigenous women's networks and advocacy groups can amplify their influence in environmental dialogues.

Recommendations for Policymakers, Environmental Organizations, and Communities

1. **Policy Development:** Adopt a human rights-based approach to environmental policy that recognizes and incorporates Indigenous knowledge and leadership. Ensure that Indigenous women have a seat at the table in environmental decision-making bodies.
2. **Capacity Building:** Invest in education and training programs for Indigenous women to enhance their leadership skills and knowledge of environmental governance.
3. **Community Engagement:** Foster collaborations between Indigenous communities and environmental organizations to co-create sustainable solutions.
4. **Legal Frameworks:** Strengthen legal protections for Indigenous lands and resources, ensuring that policies are enforced, and Indigenous rights are upheld.
5. **Research and Documentation:** Support research initiatives that document and validate Indigenous women's environmental knowledge and contributions.

To support and enhance Indigenous women's leadership in environmental protection, several policy measures can be undertaken:

- **Recognizing Indigenous Land Rights:** Governments should recognize and enforce Indigenous land rights, ensuring that Indigenous communities have control over their territories. This includes respecting land treaties and supporting land restitution initiatives.
- **Integrating Traditional Knowledge:** Environmental policies should integrate traditional ecological knowledge and practices. This can be achieved through collaborative governance models that involve Indigenous women as key decision-makers.
- **Supporting Indigenous Women's Organizations:** Providing funding and resources to Indigenous women's organizations can enhance their capacity to lead environmental initiatives. This includes supporting educational programs that focus on traditional knowledge and leadership skills.
- **Promoting Inclusive Environmental Governance:** Environmental governance structures should be inclusive, ensuring that Indigenous women's voices are heard and respected. This involves creating platforms for dialogue and consultation between Indigenous communities and policymakers.

CONCLUSION

Indigenous women's spirituality is deeply intertwined with their environmental stewardship. Their leadership is rooted in a profound spiritual connection to the land, which drives their commitment to protecting and sustaining their ecosystems. Recognizing and supporting the contributions of Indigenous women is essential for achieving sustainable and equitable environmental practices. By valuing their spiritual and ecological knowledge, we can foster a more inclusive and comprehensive approach to environmental protection that honors the interconnectedness of all life.

Indigenous women's leadership is indispensable for advancing environmental dialogue and promoting a human rights-based approach to sustainability in Canada. Their unique perspectives and traditional knowledge offer valuable insights and solutions to contemporary environmental challenges. Recognizing and supporting their contributions is not only a matter of justice but also a crucial step towards achieving sustainable and equitable environmental practices. Policymakers, environmental organizations, and communities must work together to ensure that Indigenous women's voices are central to the environmental dialogue, fostering a future where human rights and environmental stewardship go hand in hand.

The historical and spiritual connection to the land is central to understanding the environmental stewardship of Indigenous women in Canada. Despite the disruptions caused by colonialism, Indigenous women have maintained and revitalized their spiritual practices and ecological knowledge. Their deep spiritual connection to the land guides their environmental advocacy and stewardship, underscoring the importance of integrating Indigenous perspectives into contemporary environmental policies and practices. Recognizing and supporting Indigenous women's roles as spiritual and environmental leaders is crucial for fostering sustainable and equitable approaches to environmental protection.

REFERENCES

APA PsycNet. (n.d.). Retrieved April 16, 2023, from https://psycnet.apa.org/

Bantjes, R., & Trussler, T. (1999). Feminism and the Grass Roots: Women and Environmentalism in Nova Scotia, 1980–1983. *Canadian Review of Sociology/Revue Canadienne de Sociologie, 36*(2), 179–198. DOI: 10.1111/j.1755-618X.1999.tb01274.x

Beverley, A. (2020). Uranium Mining, Interdisciplinarity, and Ecofeminism in Donna Smyth's Subversive Elements. *Canadian Literature*, 143, 40–59.

Books, A. (n.d.). There's Something In The Water. Retrieved April 17, 2023, from https://atlanticbooks.ca/theres-something-in-the-water

Braided Way Magazine. (2018, November 3). *Women Are the First Environment.* Interview with Mohawk elder Katsi Cook. Retrieved April 1, 2023, from https://braidedway.org/women-are-the-first-environment-interview-with-mohawk-elder-katsi-cook/S

Campbell, J. (2017). Gendered Environments in Canada: An Analysis of Women and Environments Magazine, from 1976 to 1997 (Master's thesis, University of Saskatchewan). Retrieved from https://harvest.usask.ca/handle/10388/8521

Eco-Feminist Appropriations of Indigenous Feminisms and Environmental Violence. (n.d.). The Feminist Wire. Retrieved April 16, 2023, from https://thefeministwire.com/2018/10/eco-feminist-appropriations-of-indigenous-feminisms-and-environmental-violence/

Environmental racism and Black environmentalism in Canada. (n.d.). Smart Prosperity Institute. Retrieved April 1, 2023, from https://institute.smartprosperity.ca/library/research/environmental-racism-and-black-environmentalism-canada

Fury for the Sound: The Women at Clayoquot. (n.d.). Moving Images Distribution. Retrieved April 1, 2023, from https://www.movingimages.ca/store/products.php?pid=1312

Global Affairs Canada. (2017). Canada's Feminist International Assistance Policy. Retrieved April 1, 2023, from https://www.international.gc.ca/world-monde/issues_development-enjeux_developpement/priorities-priorites/policy-politique.aspx?lang=eng

Hofman, K. (2021). A defense of ecofeminism: Re-examining the Clayoquot Sound peace camp. Archived from the original on April 1, 2023. Retrieved April 1, 2023, from https://kaylahofman.com/defense-of-ecofeminism-clayoquot-sound

Hoover, E., Cook, K., Plain, R., Sanchez, K., Waghiyi, V., Miller, P., Dufault, R., Sislin, C., & Carpenter, D. O. (2012). Indigenous Peoples of North America: Environmental Exposures and Reproductive Justice. *Environmental Health Perspectives*, 120(12), 1645–1649. DOI: 10.1289/ehp.1205422 PMID: 22899635

Kwaymullina, A. (2018). In Stevens, L., Tait, P., & Varney, D. (Eds.), *You Are on Indigenous Land: Ecofeminism, Indigenous Peoples and Land Justice. In Feminist Ecologies: Changing Environments in the Anthropocene* (pp. 193–208). Springer International Publishing., DOI: 10.1007/978-3-319-64385-4_11

Moore, N. (2011). Eco/feminism and rewriting the ending of feminism: From the Chipko movement to Clayoquot Sound. *Feminist Theory*, 12(1), 3–21. DOI: 10.1177/1464700110390592

OHCHR. (n.d.). *Special Rapporteur on toxics and human rights*. Retrieved April 1, 2023, from https://www.ohchr.org/EN/Issues/ToxicsAndHumanRights/Pages/SRToxicsAndHumanRightsIndex.aspx

Rowland, L. (2015). Speculative Solutions: The Development of Environmental and Ecofeminist Discourse in Margaret Atwood's MaddAddam. Studies in Canadian Literature. ISSN 1718-7850.

Smith College Finding Aids. (n.d.). *Mother's Milk Project*. Retrieved April 1, 2023, from https://findingaids.smith.edu/repositories/4/resources/146

There's Something In The Water. (n.d.). Fernwood Publishing. Retrieved April 17, 2023, from https://fernwoodpublishing.ca/book/theres-something-in-the-water

Tindall, D. B., & Robinson, J. L. (2017). Collective action to save the ancient temperate rainforest: Social networks and environmental activism in Clayoquot Sound. *Ecology and Society*, 22(1), art40. Advance online publication. DOI: 10.5751/ES-09042-220140

Uranium. (n.d.). The Mining Association of Nova Scotia. Retrieved April 1, 2023, from https://tmans.ca/uranium

Wain, D. (2021). An Ecofeminist Treatment of Nourishment and Feeding in Margaret Atwood's MaddAddam Trilogy. In Dystopias and Utopias on Earth and Beyond (pp. 25–36). DOI: 10.4324/9781003152989-4

Wilson, K. (2006). Ecofeminism and First Nations Peoples in Canada: Linking culture, gender and nature. *Gender, Place and Culture*, 12(3), 333–355. DOI: 10.1080/09663690500202574

KEY TERMS AND DEFINITIONS

Activism: The efforts to promote, impede, or direct social, political, economic, or environmental change, often through campaigns and advocacy.

Collective Action: Cooperative efforts by a group to achieve a common goal, often in the context of social movements or environmental advocacy.

Cultural Resilience: The ability of a community to maintain and adapt its cultural practices in the face of external pressures and changes.

Ecofeminism: A philosophical and social movement that combines ecological concerns with feminist perspectives, advocating for both environmental and gender justice.

Environmental Racism: The disproportionate impact of environmental hazards on marginalized communities, particularly Indigenous and racialized groups.

Environmental Stewardship: The responsible management and care of the environment and natural resources.

Gender Equality: The state of equal access to rights, responsibilities, and opportunities regardless of gender.

Human Rights Framework: A set of international standards that protect the rights and dignity of individuals, including the rights of Indigenous peoples and women.

Indigenous Peoples: Original inhabitants of a region, often possessing distinct cultures, languages, and spiritual beliefs.

Intersectionality: A framework for understanding how various forms of social stratification, such as race, gender, and class, intersect and contribute to unique experiences of oppression.

Land Sovereignty: The right of Indigenous peoples to govern their lands according to their traditions and laws, distinct from colonial or state governance.

Resource Extraction: The process of removing natural resources from the environment, often associated with significant ecological and social impacts.

Spiritual Connection to the Land: The belief that the land is not just a physical space but a living entity that is integral to identity, culture, and spiritual practices.

Sustainable Development: Development that meets the needs of the present without compromising the ability of future generations to meet their own needs.

Traditional Ecological Knowledge (TEK): Indigenous knowledge systems that encompass the understanding of ecological processes and relationships based on generations of cultural practices.

Chapter 9
The Overlooked Narratives of Women in Rural India and Africa, as Depicted in Contemporary Literature

Anwesha Biswas
Syamaprasad College, Kolkata, India

ABSTRACT

This comparative study of Mariama Bâ's "Une Si Longue Lettre" and Ashapurna Devi's "Pratham Pratishruti (The First Promise)" explores the feminist consciousness of these two pioneering authors. Despite being from different continents, eras, and cultures, their novels highlight similar struggles of women within patriarchal societies. During times when global focus was on wars and colonization, these authors brought attention to the overlooked domestic struggles of women. The study demonstrates how Bâ and Devi's protagonists break free from the confines of patriarchy, finding their paths to freedom. The research aims to uncover the overlooked narratives of women in rural India and Africa, emphasizing the domestic realms depicted in these contemporary literary works.

INTRODUCTION

Once Swami Vivekananda said that the best thermometer of a nation's progress is the treatment it reserves for its women and that there is no chance for the welfare of the world if the condition of women is not improved. Therefore, the place of women

DOI: 10.4018/979-8-3693-6069-9.ch009

in society is a critical factor in judging the success of any civilization. All around the world, various social norms, in one form or another, deprive women of their right to education, healthcare, economic opportunities, and political participation. Literary writers and filmmakers, through their works, have attempted to demonstrate the roles and identities of women in the family and society.

Literature is prudently a periphery where human beings have the freedom to express their misery, exultation, protests, moans, and laughter without distinction of race, gender, and religion. Simply put, through literature, every individual gains the power to express their own point of view. Literature gives voice to the oppressed and the defeated. Therefore, it is very important for women to have access to the world of literature where they can express their opinions with the help of ink. If one looks through the windows of history, one encounters many women writers who have used their pens as weapons to fight against domination.

The narratives of women in rural India and Africa have often been overshadowed by broader socio-political discourses, leaving their everyday struggles and resilience largely unexplored. Contemporary literature has sought to illuminate these hidden stories, with writers like Mariama Bâ and Ashapurna Devi playing pivotal roles in this endeavor. Through their novels "So Long a Letter" (Bâ, 1981) and "Pratham Pratishruti" (Devi, 1964), they delve into the intricacies of domestic realms and offer profound insights into the lives of women in their societies.

This paper, based on the comparative study of two novels from different eras and continents, pays tribute to two pioneers of modern feminism, Mariama Bâ and Ashapurna Devi, and shows how their feminist consciousnesses intertwine across time, culture, and language. Both feminists illustrate how writing is often a means to escape distress, and how, despite personal sufferings, one can still succeed in saying something meaningful about life and society.

Mariama Bâ, winner of the Noma Prize in 1981, uses the epistolary format in "So Long a Letter" to depict a society where patriarchy prevails to such an extent that a fifty-year-old Muslim woman who bore a man twelve children could be abandoned at the age of fifty-three for her eldest daughter's beautiful friend (Bâ, 1981). After the sudden death of her husband Modou, Ramatoulaye writes a letter to her friend Aissatou in America about her life, which becomes a scathing critique of a patriarchal society that consciously neglects the oppression of women. Through this novel, Ramatoulaye circumvents the forty-day silence mandate imposed by Islam on a widow, making the letter a powerful symbol of a woman seizing the pen.

Ashapurna Devi, despite having no formal education, has made significant contributions to feminism through her numerous novels and short stories, shedding light on the condition of Bengali women (Devi, 1964). Devi's writings, particularly her trilogy comprising "Pratham Pratishruti" [The First Promise] (1964), "Subarnalata" (1967), and "Bakulkatha" [The Story of Bakul] (1974), articulate the struggles of women during the colonial and post-colonial periods in Bengal. The trilogy expresses the lives of three generations of the same family—Satyabati, Subarnalata (Satyabati's daughter), and Bakul (Subarnalata's daughter). Through the character of Bakul, Devi ultimately establishes a victory over women's emancipation.

This paper emphasizes "Pratham Pratishruti" [The First Promise], where the protagonist Satyabati embodies a true feminist spirit by protesting against gender discrimination and the irrelevant norms of patriarchal society (Devi, 1964). Devi portrays Satyabati as fearless and rebellious, questioning everything and breaking the chains of preconceived notions of patriarchal society. Satyabati is not like other women and children; she is different. Her legacy of rebellion continues until "Bakulkatha," where Bakul achieves true victory over women's emancipation (Devi, 1974).

Thus, these two novels, "So Long a Letter" and "Pratham Pratishruti" [The First Promise], explore the struggles and battles of women confined within the four walls to make their voices heard in a patriarchal society. Bâ and Devi's protagonists show us that silence was forced upon them, but they were never silent. Through these novels, these feminist voices come out of the claustrophobic four walls and begin to resonate in the fresh air of freedom.

RELIGION: A MECHANISM OF PATRIARCHAL CONTROL

Since antiquity, women have been subjugated by prevailing patriarchal conditions in society, and this subjugation originates within the four walls of the domestic world. The first weapon used by patriarchy is religious norms that depreciate women's rights. Religion is the most practiced weapon of the patriarchal mentality to dominate women in the domestic world, and this domination within the four walls is the true representation of the larger world. No one can dominate women as religion can.

The objective of this chapter is to highlight the disempowering effects of any religion, particularly concerning women in the traditional African Muslim context of Senegal and the traditional Hindu context of Bengal before independence, which are well reflected in these two novels.

Firstly, the focus is on the epistolary novel *So Long a Letter* by Mariama Bâ (1979), which illustrates the marginalization of Muslim women. The women in this novel can be considered "victims" of the patriarchal society dominated by a particular religion. Consequently, this chapter analyzes the battle of the sexes in the novel by examining the treatment, exploitation, and manipulation of female characters established based on religious and traditional principles.

In this novel, Bâ (1979) describes the effects of Islam and its tradition on Senegalese women. At the beginning, we learn that the protagonist, Ramatoulaye, has just lost her husband. She decides to write a letter to her friend Aissatou to cope with the ritual of Iddah, where a woman must observe silence while confined within four walls after her husband's death or after a divorce, during which she cannot marry another man. She writes about her emotional struggle to regain her life shortly after learning of her husband's second marriage at the age of fifty. Although the laws of Islam are compatible with the actions of her husband Modou, it is considered a brutal rejection of their thirty years of life together. It is also considered a betrayal of the wife's trust, encouraged by religion since Islamic law allows polygamy, where a Muslim man can be married to four women at the same time (Bâ, 1979). Religion treats women as inferior to men in society and condemns them to be second-class citizens. Even Aissatou, Ramatoulaye's childhood friend, faced the same heart-breaking ordeal before Ramatoulaye when her husband Mowdo decided to take another wife. But unlike Ramatoulaye, despite her Islamic background, Aissatou boldly leaves her husband and her problems behind, moves to France, and starts her life alone. She breaks the old habits that perpetuate women's subjugation and oppression (Bâ, 1979).

Men take women for granted according to their convenience, just as Modou and Mowdo take this assertion as their own. Although Mowdo loves Aissatou, the impact of African Muslim culture and traditions makes Mowdo accept his cousin Nabou as his second wife, a decision likely influenced by his mother's wishes, which she later regrets. The influence of religion is such that neither he nor Modou considers their wives before deciding on a second marriage.

Religion plays a significant role in the domination and oppression of women, as depicted in this long epistolary letter by Mariama Bâ, where the lives of women in African Muslim society are confined within religious boundaries. Religious customs like "Polygamy" and "Iddah" have made their lives nothing short of hell. These customs of the Muslim religion have denied all rights to women in society (Bâ, 1979). A similar image is beautifully projected in the film Escape from Taliban, based on the story *A Kabuliwala's Bengali Wife* by Sushmita Banerjee (Banerjee, 1997), who experienced the same sufferings while being trapped in the customs of the Muslim religion in her in-laws' house in Afghanistan.

It is the religious tradition that does not even allow women to study, as expressed by Bâ (1979) through the character of Aunt Nobou Mawdo's mother, who teaches her niece that "a woman doesn't need too much education" (Bâ, 1979, p.30). Religion denies women the right to freedom and education. However, in this novel, both protagonists succeed in enlightening themselves with Western education and culture after their primary schooling at the Koranic school.

On the other hand, a similar scenario is depicted in Ashapurna Devi's (1964) masterpiece, *Pratham Pratishruti (The First Promise)*, although it belongs to a different century, a different culture, and a different religion. The novel is set in a village in Bengal before partition. Its theme focuses on a social structure based on religious superstition, prejudice, and injustice toward women.

In this masterpiece by Devi (1964), there is an account of almost all kinds of religious and social practices, norms, and systems prevalent in contemporary rural Bengal. First, if we look at the widows in this novel, we can notice that they are the worst victims of irrelevant religious practices. There is an enormous representation of the widow system in this novel. Like the Iddah ritual in Muslim culture, in Hinduism too, a woman must suffer a lot after her husband's death. The lives of widows are equally miserable in contemporary Hindu society in Bengal. In this novel, the five widows of Ramakali's family—Dinatarini, Kashiswari, Shankari, Shibjaya, and Mokshada—lead a strict life by their own choice. There was a separate place for the widows to cook, and not everyone could enter. They are responsible for all the practical problems of managing a household from dawn to dusk. They are required to strictly adhere to the rules governing widowhood, rules that they reinforce by demanding that other female members of the family learn to observe them in a male-dominated society (Devi, 1964).

Another case where these two novels intertwine is polygamy, nurtured by both religions. In Pratham Pratishruti, the prevalence of polygamy is a normal occurrence. It is the "belief system" that has legalized the polygamy system, and women have accepted it without protest, showing the patriarchal process of socializing the girl child. In this novel, one of Ramakali's nephews, Rashbehari, is forced to contract a second marriage due to obligations imposed on a Kulin Brahmin, but at the same time, a woman is not allowed to start a new life even after her husband's death. One of the five widows in Ramakali's house, Shankari, elopes with the man who courts her, Nagen, bringing disgrace to the entire family (Devi, 1964).

But unlike the character of Ramatoulaye in *So Long a Letter*, Rashbehari's first wife, Sarada, vigorously protests her husband's second marriage by threatening to commit suicide. Although Sarada is not very educated, her protest against polygamy reminds us of the character of Aissatou in Bâ's novel *So Long a Letter* (Bâ, 1979; Devi, 1964).

In great similarity to Bâ's *So Long a Letter*, Devi's novel *Pratham Pratishruti* (1964) also shows that the Hindu religion of that period denies women the right to education and approves child marriage. But like the protagonists of So Long a Letter, Ramatoulaye and Aissatou, Satyabati, Devi's protagonist, succeeds in enlightening herself with education.

Although these two novels belong to different centuries, different cultures, and different religions, the religious domination over women forms a strong link between them. These novels focus on the negative effects that Islam, Hinduism, and polygamy have on women. They emphasize the eternal victimization of African and Indian women by their religion and culture. They also focus on child marriages, which were prevalent in Bengal and Senegal. In these two novels, the writers depict the Senegalese and Bengali societies of religious and patriarchal influences, where families exert much pressure and influence on the young to enforce traditional, irrelevant taboos, norms, and privileges (Bâ, 1979; Devi, 1964).

ELDERLY WOMEN OF THOSE CENTURIES: UNWAVERING PROPONENTS OF PATRIARCHY

Both Bâ and Devi use their narratives to highlight the intersection of gender, culture, and tradition in shaping the lives of women. Their works illustrate that the domestic realm is not a static, passive space but a dynamic site of struggle and negotiation. By focusing on the everyday lives of women, they challenge the marginalization of these experiences in mainstream historical and literary discourses.

The domestic realm in both novels is depicted as a space where power dynamics are constantly at play. In *So Long a Letter*, Ramatoulaye's reflection on her life and relationships reveals the subtle ways in which women exert influence and navigate their positions within the family (Bâ, 1979). Similarly, in *Pratham Pratishruti*, Satyabati's interactions with her family members highlight the complexities of female agency within a rigidly structured society (Devi, 1964). Both authors underscore the importance of education and self-awareness as tools for empowerment, suggesting that these are crucial for women to challenge and transform their circumstances. Moreover, Bâ and Devi's works emphasize the need for intergenerational solidarity among women. They portray how older women, who are often seen as custodians of tradition, can either perpetuate or challenge oppressive practices. This dual role reflects the broader societal tensions and the potential for transformative change.

Generally, feminist research attributes the domination of patriarchal ideology in society to the activities of men while considering women as innocent victims of patriarchal authority. However, a careful reading of texts by certain women writers like Mariama Bâ and Ashapurna Devi critiques this view and examines the direct and indirect roles played by some women in perpetuating patriarchal oppression. The focus is on older women in the domestic world who are often so ignorant, selfish, and manipulative that they make life difficult for other women. Thus, this study differs from conventional feminist research, which generally emphasizes "men-women" oppression relations as an ideal means of understanding male domination and female subordination. It focuses on the role of older women in the family to understand the complexities of patriarchal oppression and contributes to the debate on the nature of patriarchy and its role in the oppression of women, particularly in reference to women in the traditional African society of Senegal and the traditional Bengali society before independence, as well reflected in these two novels.

First, I will shed light on the novel *So Long a Letter* by Bâ, where I will analyze the roles of certain women in the patriarchal system in the context of the domestic world of African society. The common belief among most feminist critics of African literature today is that men are the worst enemies of women. These feminists have denounced men, accusing them of being the major source of women's unhappiness, especially within the family. They claim that men oppress, mistreat, and exploit women by inhibiting and restricting their self-realization (Frank, 1987). Critics like Katherine Frank argue that it is neither possible nor necessary to negotiate and compromise with men, the enemies. She therefore suggests a complete rejection of patriarchy in all its forms to create a world without men, where women would be safe, sane, and supportive of each other (Frank, 1987).

However, the issue is more complex, and one wonders if women are not the aggressors of other women and partially responsible for their own marital unhappiness. In this novel, it can be well noted that Mawdo and his wife Aïssatou were happy and satisfied in their marriage until Mawdo began to face external pressures from his mother. Mawdo's mother, Aunt Nabou, cannot accept that her son chose a wife from a lower caste. She then educates and trains her niece Nabou to be obedient, docile, sober, and well-informed in the traditional roles of a woman. Aunt Nabou later lies to her son Mawdo, saying, "My brother Farha gave you young Nabou as a wife, to thank me for the dignified manner in which I raised her" (Bâ, 1979, Ch.12). She further warns that her refusal to obey her will and marry her cousin could kill her: "I will never recover if you do not take her as your wife. Shame kills faster than disease" (Bâ, 1979, Ch.12). Whether it is the stratagem of a mother or not, she succeeds and obtains the double joy of breaking her son's previous marriage and cementing the family's aristocracy. Clearly, Bâ criticizes Aunt Nabou for her central role in the collapse of Aissatou's marriage. Aunt Nabou is also criticized for destroying her niece's future. Instead of helping young Nabou emancipate, she wants her to remain ignorant, perfectly docile, and as silent as possible. She teaches her that "a woman does not need too much education" (Bâ, 1979, p.30). Aunt Nabou is an example of women whose actions are detrimental to the modern interests of women in their society. She is as selfish and manipulative as the men in her culture.

Traditionally, the value of an African woman is measured by the number of sons she has and the success of these sons. Thus, Modou's mother, Ramatoulaye's mother-in-law, brings a constant stream of visitors to parade through Ramatoulaye's well-maintained house. She infringes the rules of hospitality with the intention of showcasing her son's social success, and therefore her own. Being proud of her son's social success, Modou's mother considers herself the main architect of his success and wants to fully enjoy it in the presence of her friends. Ramatoulaye invariably interrupts her busy schedule to respond to her mother-in-law's wish, and there is always a neatly folded banknote in the mother-in-law's hand when she leaves Ramatoulaye's house (Bâ, 1979).

Another character of this kind is Binetou's mother, whom the author ironically calls Dame Belle-Mère. She was more concerned with putting food on the table than with her daughter's education, thus encouraging Binetou to interrupt her studies and marry Modou, who is older than her and of her father's age. In the process, this Dame Belle-Mère destroys another woman's home and subsequently compromises her daughter's chances of becoming financially independent. Without the necessary education to find a job, Binetou becomes economically dependent on Modou. Binetou's mother considers this marriage as a way to end her misery and to be propelled to the category of women with heavy bracelets, sung by the griots (Bâ, 1979). By removing Binetou from school to "auction her off" to the wealthy but old Modou, adults like Dame Belle-Mère stifle the aspirations of children like Binetou.

Ultimately, the mothers-in-law of Aïssatou and Ramatoulaye and Binetou's mother are the real accomplices of polygamous marriages, destroying not only the happiness of Aïssatou and Ramatoulaye but also the future of the two young wives, Nabou and Binetou. These older women are not only the servants of patriarchal structures but also their guardians. Their complicity is strongly condemned by Ramatoulaye's daughter, Daba, as she says: "How can a woman undermine another woman's happiness?" (Bâ, 1979, Ch.21). Daba also advises young girls like Binetou and many others not to sacrifice their youth, studies, and dreams to satisfy their mother's greed.

On the other hand, if we analyze Ashapurna Devi's novel *The First Promise*, we can notice that the older widows in Ramakali's family dominate and establish the patriarchal tradition more than a man (Devi, 1964). Devi observed the gargantuan gulf between masculine and feminine values. She never saw men as the enemy of women but also pointed out that if women could rise above such pettiness and inhumanity and consider with an open mind what is good and desirable for them, some of the problems could be solved. In this novel, most of the "mothers" we find are timid by nature; they are extremely loving toward their sons but at the same time very cruel to their daughters-in-law. But at the same time, there are mothers like Bhubaneshwar and Saudamini, who are conformist, devoted, and ardent to both their children and in-laws (Devi, 1964). At the beginning of this novel, we see the character of Jashoda (Ramakali's great-aunt) complaining to Jaykali (Ramakali's father) that Ramakali mocks God and does not respect religious rituals, and that this act of Ramakali must not be tolerated as neglecting religious rituals is nothing but a sin according to Jashoda (Devi, 1964). The character of this old lady in the family clearly indicates that older women in the domestic world of that time were more vigorous and conscientious about following and maintaining strict religious activities and they oblige other young members of the household to perform and follow these irrelevant traditional rituals.

Then we see that the five widows in Ramakali's family—Dinatarini, Kashiswari, Shankari, Shibjaya, and Mokshada—lead a strict life by their own choice. They are responsible from dawn to dusk for taking care of all practical problems of running a household. They are required to strictly follow the rules governing widowhood, rules they reinforce by demanding that other female family members learn to observe them in a male-dominated society (Devi, 1964).

Another incident that clearly indicates that women also play a role in the suffering of other women is when one of Ramkali's nephews, Rashbehari, due to the obligations imposed on a kulin Brahmin, is forced into a second marriage, which his first wife, Sarada, vigorously protests by threatening to commit suicide. As a result, her husband refrains from sleeping with the second wife. The other jealous women in the house cannot accept Sarada's success in blackmailing her husband and manage to persuade Rashbehari to sleep with the second wife. And then, when one of the five widows, Shankari, elopes with the man courting her, Nagen, bringing disgrace to the whole family, these other widows vehemently oppose and criticize Shankari. It is evident that out of jealousy, these women are incapable of accepting the success or independence of other women and thus become the cause of other women's suffering (Devi, 1964). These older women in the family, that is, Satyabati's grandmothers, try hard to oppress Satyabati, but she always confronts them with her logic and rebellious nature. So, Satyabati's grandmother starts blaming Satyabati's mother for Satyabati's unconventional activities. At the very beginning of the novel, we find her debating her desire to catch fish with a fishing rod because, according to her grandmother, it is forbidden for a girl to catch fish with a fishing rod (Devi, 1964).

All women are victims of these dominant traditions of society. Some never realize their victimization and tend to blindly follow these societal norms, and due to their tendency to blindly follow these customs, other women who intend to achieve freedom become victims. In these two novels, characters like Ramatoulaye's mother, her mother-in-law, Aïssatou's mother-in-law, and the second wives of their husbands and Binetou's mother, and the widows, i.e., Satyabati's grandmothers who support the polygamous system, are the greatest examples of such women who blindly believe in patriarchy more than a man and desperately want to continue this tradition (Bâ, 1979; Devi, 1964).

This chapter aims to show that although men are often accused of oppressing women, it is quite evident that women also participate in the oppression of women. These ignorant, selfish, manipulative, and insensitive older women are a real obstacle to women's emancipation. Therefore, there is an absolute need for cordial relationships, fraternity, and, above all, solidarity among women, whether married or single, to save themselves from male oppression and domination.

WOMEN'S EDUCATION: A KEY TO FREEDOM

In the very first sentence of the introduction to this study, I quote a saying from the great Indian monk, Swami Vivekananda, who stated that the best thermometer of a nation's progress is the treatment of its women and that there is no chance for the welfare of the world if the condition of women is not improved. He identified that the ignorance of women was the main obstacle to the progress of any nation. He insisted that women should be placed in positions of power to solve their own problems in their own way, and this cannot be possible without education.

In these two novels, "*So Long a Letter*" by Mariama Bâ (1979) and "*Pratham Pratishruti*" (The First Promise) by Ashapurna Devi (1964), the condition of women in the domestic world is a key element to analyze the contemporary societies of these two nations. In the previous chapters, we have already discussed how patriarchy exploits women within the four walls. Now in this chapter, we will try to find a possible solution to this domination. Women's education is the key that can open this cage and free women from the darkness of these four walls.

The protagonists of these two novels have led us on this path to freedom. They taught the next generations to break the chains of patriarchy that held women back.

In the novel "*So Long a Letter*," Mariama Bâ (1979) emphasizes the importance of education in our lives through various examples she provides. Through her characters, Ramatoulaye and Aissatou, she effectively describes how education can help a person (a woman in this context) to shape a broader perspective of life and approach a situation rationally with an open mind (Bâ,1979).

The novel begins as Ramatoulaye uses her pen to paint the miserable condition of Senegalese women. Thus, she makes her pen her greatest weapon to break this prolonged chain of silence. And only education can offer this greatest weapon in the world. After completing their primary education at the Koranic school, Ramatoulaye and Aïssatou enlightened themselves with Western education. The equality that results from friendship and education are important elements in "*So Long a Letter*." The two women were educated in French convent schools in Senegal, very rare for women of that time. Ramatoulaye respects Aissatou's choice to divorce her husband because she is educated and understands how the choices made are to assert her own sense of being. Even though Ramatoulaye took a different path, there is respect present due to her education. "*So Long a Letter*" features these educated women, aware of their social situation, and therefore, they tend to react to their situation. Ramatoulaye, a teacher, and her friend Aïssatou, an interpreter, are very educated women. They are victims of traditional customs: yet, these two women react differently to victimization. While Ramatoulaye seems adapted to live within her community, Aïssatou chooses to pursue her studies and move to another country. Another rebellious woman in this novel is Daba. Bâ represents Daba, Ramatoulaye's daughter, in the class of courageous and determined women in their idea of life; for example, she marries a man who respects her and treats her as an equal. Both use their power of education to liberate their minds. The impact of education on an individual's opinions is quite evident as Ramatoulaye asserts that "For your opinions to carry weight, they must be based on knowledge supported by diplomas. A diploma is not a myth." (Bâ, 1979, p.73). Daba is an educated woman who reasons about everything, and this representation reflects a generation of thinking women who base their determination on reason. She is an independent woman, representing the new African wave that does not reject culture but approaches it with sense and reason. Through education, it is women who conquer and stand out from their diverse experiences. It is education that helps Aissatou to leave her husband, rise, and be a powerful source of support for her friend Ramatoulaye. It is this education that gives Ramatoulaye the alternative to ignore marrying Tamsir, her husband's elder brother. It is what gives her the insouciance to refuse the beautiful offer from Daouda Dieng. It is education that gives young Nabou the chance to acquire better reasoning skills and become a determined woman. Thus, comprehensive education inspires women and is an asset to every family and society. An educated woman can manage on her own and reach higher heights.

It is the open-mindedness brought by education that made Ramatoulaye accept with great understanding her daughter's premarital pregnancy. It is education that gives women financial independence that allows them to courageously reject patriarchal oppression. Education serves as a leitmotif in this story and can explain every action and decision that Ramatoulaye and Aïssatou have taken throughout their lives. The fact that the story revolves around a woman in Senegal writing a letter to her friend concretizes the theme of education as a fundamental element of the narrative.

This is why the faithful of patriarchy are always against women's education, and religious tradition does not allow women to study, as expressed by Bâ in the text through the character of Mawdo's Aunt's mother who teaches her niece that "a woman doesn't need too much education" (Bâ, 1979, p.30). It is evident that women's education is a threat to patriarchy.

Similarly to Bâ's "*So Long a Letter*," in the novel "*Pratham Pratishruti*" (The First Promise) by Devi, we can also see that Hindu religion of that period denies women's right to education and approves girls' marriages. But like the protagonists of "*So Long a Letter*," Ramatoulaye and Aïssatou, Satyabati, the protagonist of Devi, manages to enlighten herself with the light of education.

"*Pratham Pratishruti*" (The First Promise) is set in a period stretching from the late 18th century to the late 19th century. Ashapurna Debi, self-taught and confined to the antahpur for much of her life, professed to write the story of the domestic space, apparently muted, of the daily life of women within the antahpur, a domain she rightly pointed out, had been systematically ignored by historians. Ashapurna can thus be considered the pioneer of everyday history.

Satyabati, the protagonist of "*Pratham Pratishruti,*" was married off by her parents at the age of eight, the ideal age for girls' marriage at that time, but she returned to her parents' home shortly after the ceremony. In her parental home, Satyabati enjoyed considerable freedom, if not unrestrained, and although she was deprived of education, she secretly learned the alphabets herself. When her father, Ramkali Chattopadhyay, a traditional physician with a roaring practice, and the great patriarch of the Chattopadhyay family, discovered her taste for education, he decided to give her lessons himself. Satyabati's inquisitive and rebellious spirit, and her anguished awareness of the gender order of things, manifested from an early age. She protested the mistreatment of women. She strongly protested discriminatory norms that considered it a sin for a woman to simply touch the palm leaves on which men wrote. She categorically argued with her cousin Neru that she would like to know what happens if a woman touches this leaf. So many women read and write in Calcutta (Devi, 1964).

She turned a deaf ear to the fustigated criticisms from older women in her own extended family, and often retorted, giving them a piece of her mind. In Calcutta, freed from the control of her in-laws, Satyabati finally gained the freedom to organize her family life as she pleased. Her sons received the best education and eventually settled into high-level jobs. Satyabati enrolled her daughter, Subarnalata, in school. She (Satyabati) started teaching in an adult women's school, initially without her husband's knowledge. She moved freely around the city in a carriage and even visited the house of the Brahmo leader, Keshabchandra Sen, to attend a congregation. Satyabati is not as formally educated as the protagonists, Ramatoulaye, Aïssatou in the novel "*So Long a Letter,*" (Bâ, 1979) but we can find Satyabati more rebellious than any of these protagonists. Satyabati also enlightens herself with Western education when she learns English. This denotes that in these two novels, Western education has had an impact on the protagonists.

But unfortunately, the modern education she dreamed of for her children in Calcutta does not protect her sons from the legacy of patriarchal mentality, nor does it prevent Subarnalata's marriage at a young age.

It is education that nurtures Satyabati's rebellious spirit. As she fails to obtain formal education, she tries to fulfill it by admitting her daughter to a reputable school in Calcutta. Although Subarnalata is also forced to marry very young by her grandmother and without her mother's knowledge, she continues to educate herself within the four walls like her mother. Satyabati thus becomes a leading figure for all the rebellious women against such conformism that Ashapurna gradually began to see around her. As more and more ordinary middle-class women left home to study and work, they also became the mothers of a generation of more enlightened men. Satyabati reminds me of Maya Angelou's poem, "Still I Rise" (Angelou, 1978):

"With your bitter, twisted lies, You may trod me in the very dirt / But still, like dust, I'll rise. / Out of the huts of history's / shame I rise / Up from a past that's rooted in pain / I rise..."

And it is Satyabati's education that makes her always rise (Angelou, 1978).

Education as a means of liberation is the common thread in these two novels. In these two novels, through protagonists like Satyabati, Aïssatou, or Ramatoulaye, or her daughter Daba, the writers have proven that through education, women triumph and stand out in their different experiences. In several situations, it is demonstrated that the concept of education gives women options to get out of any difficult situation they find themselves in. Education acts as a key or a vehicle given to women; the door it opens or the direction it heads becomes her choice. The more educated a woman is, the more options she has: the more keys she has, the more options she has of places to drive. The decisions she then makes depend entirely on her.

Mariama Bâ's "*So long a Letter*" and Ashapurna Devi's "*Pratham Pratishruti*" unearth the hidden tales of women in rural India and Africa, casting a radiant light on their often-overlooked lives. Within the intimate spaces of home, they unravel the intricate tapestry of female existence, showcasing the dual roles women play in both upholding and dismantling patriarchal traditions. These narratives are a testament to the indomitable spirit and inner strength of women, revealing the transformative power of education and sisterhood in breaking the chains of oppression. Bâ and Devi's stories resonate with a profound truth: amidst the quiet corners of domestic life, women's resilience and solidarity ignite a revolution of change.

CONCLUSION

In conclusion, it emerges from this study that although these two pioneers of modern feminism, Mariama Bâ and Ashapurna Devi, belong to different continents, their feminist consciousnesses merge without any barriers of time, culture, and language. Both professed to write the history of the domestic space, seemingly in silence, detailing the daily lives of women within the four walls, a domain they rightly pointed out had been systematically ignored by historians. Bâ and Devi can thus be considered pioneers of the history of everyday life. Bâ wrote an extensive epistolary letter to make us hear the prolonged cry of African women, while Devi wrote a trilogy, a masterpiece illuminating the condition of women in the undivided Bengal society.

These two novels (Bâ, 1979; Devi, 1964) have shed much light on the darkness of the domestic world governed by the adherents of patriarchy and have shown us how religion, in the name of tradition and culture, dominated within the four walls. Furthermore, these two novels have also proven that not only men were the cultivators of patriarchy, but there were also older women who tried to establish the patriarchal tradition more fervently than men. Yet, there is always a glimmer of hope, perhaps in the form of a small firefly at the end of each tunnel. Here, it is education that gives women the courage to follow unexplored paths. Through these novels, Bâ and Devi describe that to succeed in life, women must identify with, have confidence in their own abilities, and overcome the multiple challenges of life with the blessing of education. It is education that has helped these protagonists to break the prolonged silence of the domestic world because, through education, they understod that silence hides violence.

To conclude, it can be said that these modern feminists are like Prometheus, who brought fire to enlighten the next generations. And their protagonists are like the phoenixes that rose from the ashes of sadness, raised their voices, and through their rebellious voices, revolutionized the spring amid winter.

REFERENCES

Angelou, M. (1978.). Still I Rise. Poets.org. Retrieved April 1, 2023, from https://poets.org/poem/still-i-rise

Bâ, M. (1979). *So long a letter*. Les Nouvelles Éditions Africaines du Sénégal.

Bandopadhyay, S. (2013, August). Kabuliwalar Bangali Bau. Amar Boi. Retrieved April 1,2023,from https://www.amarboi.com/2013/08/kabuliwalar-bangali-bau-susmita-bandopadhay.html

Devi, A. (1964). *First promise*. Mitra & Ghosh Publishers.

Devi, A. (2013.). Satyabati trilogy.

Devi, A. Feminist Writer. (2019, March 11). Feminism in India. Retrieved April 1,2023,from https://feminisminindia.com/2019/03/11/ashapurna-devi-feminist-writer-bengali/

Devi, A. (n.d.). Sociology Group. Retrieved April 3, 2023, from https://www.sociologygroup.com/ashapurna-devi/

Devi, A. The Voice of Women. Get Bengal. Retrieved April 1, 2023, from https://www.getbengal.com/details/ashapurna%20devi%20the%20voice%20of%2women

Escape from Taliban. (2003). Wikipedia. Retrieved April 4, 2023, from https://en.wikipedia.org/wiki/Escape_from_Taliban

Kampa, D. (2020, June 27). So long a letter: So long a history—Mariama Ba, Muslimness, and women's rights. Reading Under the Olive Tree. Retrieved April 4, 2023, from https://readingundertheolivetree.com/2020/06/27/so-long-a-letter-so-long-a-history-mariama-ba-muslimness-and-womens-rights/

Larrier, R. (2000). *Francophone women writers of Africa and the Caribbean*. University Press of Florida.

Mutunda, S. Women subjugating women: Re-reading Mariama Ba's So long a letter and Scarlet song. Retrieved April 5, 2023, from https://escholarship.org/content/qt4j47346d/qt4j47346d_noSplash_f8ea2e0a8f21692d94777ca2fe506bd4.pdf

Protishruti, P. (n.d.). Wikipedia. Retrieved April 2, 2023, from https://en.wikipedia.org/wiki/Prothom_Protishruti

Une si longue lettre. (1979). Wikipedia. Retrieved March 28, 2023, from https://fr.wikipedia.org/wiki/Une_si_longue_lettre

Chapter 10
Empowerment Through Legislation: A Legal Analysis of Women's Sexual and Reproductive Rights in India

Subin Thomas
https://orcid.org/0000-0003-4228-8594
Alliance University, Bangalore, India

Smita Satapathy
https://orcid.org/0000-0001-5777-0290
Alliance University, Bangalore, India

ABSTRACT

Women can be empowered by laws that safeguard and promote their sexual and reproductive rights. The concept of reproductive rights specifies the rights of women to make decisions about their own bodies and health, highlighting the right of choice in reproductive matters. The social and cultural context in India presents a complex and multifaceted view of the concept of reproductive rights of women. Childbearing is viewed as a societal responsibility of women in Indian society. This restricts women's right to make decisions about their own lives by putting pressure on them to have children even if they are not ready. In India, abortion is also frequently seen as a women's issue, and women who choose to have an abortion are stigmatized and discriminated against. Authors propose actionable policy recommendations for an equitable and just society where women's rights are both legally protected and culturally esteemed.

DOI: 10.4018/979-8-3693-6069-9.ch010

INTRODUCTION

The concept of women's empowerment is not limited to the process of mainstreaming women or providing them with the necessary tools to participate in the economic and political process; it also encompasses the process of providing women with the tools to have control over every aspect of their lives (Pillai & Anatoliy, 2021). Sexual and reproductive rights are fundamental human rights that have a profound impact on women's empowerment and overall well-being. The sexual and reproductive rights of women are essential human rights that are crucial for their autonomy, health, and overall well-being. In India, the laws regulating these rights have developed over time, mirroring shifts in societal values, medical progress, and global human rights norms.

The Medical Termination of Pregnancy (MTP) Act of 1971 is a significant piece of law in this field that, when it was passed, provided groundbreaking legal justification for abortion under certain circumstances. However, the Act has demonstrated notable shortcomings in light of growing ethical concerns and developments in medical technology. Due to a variety of cultural, political, and social backgrounds, the legal systems that oversee these rights differ greatly around the world. In addition to examining how Indian laws either support or obstruct women's empowerment and access to reproductive healthcare, the authors hope to provide a thorough legal analysis of women's sexual and reproductive rights in India. The primary goals of this comprehensive legal analysis are to provide light on the development of women's sexual and reproductive rights in India and, more significantly, to further the continuing discourse on gender justice and empowerment. In India, recognizing women's sexual and reproductive rights leads to a more just and equal society. Reproductive rights legislation and policies, simple access to healthcare, a robust health infrastructure, and school-based sexual education all contribute to women's empowerment. To paint a complete picture of the current legal environment, the chapter draws on a variety of sources, including as laws, court rulings, and policy papers. Additionally, they provide a sophisticated legal analysis that highlights the critical need to enhance and progress the legislative framework in order to better support and empower women in India.

Even though the Medical Termination of Pregnancy (MTP) Act of 1971 was revolutionary at the time, it now has to be thoroughly reevaluated to ensure that it still meets the needs of the modern world. The Act's absence of clear guidelines for monitoring and authorizing innovative abortion techniques causes people affected to suffer for a longer period of time. The Medical Termination of Pregnancy (Amendment) Act of 2021 was put into effect by the Indian government to address these shortcomings. While this amendment aims to fix some of the shortcomings of the original Act, it leaves out several crucial points. These issues include the practical aspects of putting the law into practice, like making sure safe abortion services are affordable and easily accessible, ensuring that healthcare personnel receive enough training, educating the public, and enforcing the law.

The objective of this research is to analyze and compare the legal frameworks that govern women's sexual and reproductive rights in India in order to assess the impact on reproductive health outcomes, gender equality, and women's empowerment. The authors tried to identify the most effective approaches and insights gained from other nations to offer policy recommendations for India. Thus, the study proposes legislative changes emphasizing empowerment and improved healthcare access to raise women's sexual and reproductive rights in India.

MISSION AND CONCERNS

There have been notable global advances in the area of women's sexual and reproductive rights, but there are still differences in the legal systems and how well they are applied in various nations. This chapter seeks to examine the many chances and challenges that women face in trying to exercise their reproductive and sexual rights within the sociocultural contexts of India. The chapter not only identifies the primary barriers and disparities that hinder women from exercising their rights, but also aims to analyze the impact of legislation in empowering women in regard to their sexual and reproductive health. This research attempts to contribute to the complete discourse on gender equality and women's rights by shedding light on the efficacy of legal frameworks in empowering women and influencing their reproductive choices through the performance of a legal analysis. Legislation has a significant impact on the rights and status of women in society. Its objective is to understand the manner in which specific laws and regulations in India affect the empowerment of women, their capacity to make decisions, and their access to sexual and reproductive healthcare broadly. This chapter aims to analyze and comprehend the scope and effectiveness of Indian laws that influence the autonomy of women with respect to their bodies and reproductive choices.

IDENTIFICATION OF RESEARCH GAPS

Medical Termination of Pregnancy (MTP) Act, 1971, No. 34, Acts of the Parliament, 1971(India).

The Medical Termination of Pregnancy (MTP) Act, 1971 needs to be thoroughly examined and possibly reformed in order to make sure that it is in line with contemporary medical advancements and ethical considerations. This is especially important in cases where the existing legal framework is unclear and could cause the affected parties to suffer and be distressed for a long time. The Act lacks clarity on how new abortion methods should be approved and regulated.

Medical Termination of Pregnancy (Amendment) Act, 2021, No. 8, Acts of the Parliament, 2021(India).

The Medical Termination of Pregnancy (Amendment) Act, 2021 does not address concerns related to the practical application of the law, such as healthcare provider education, public awareness campaigns, accessibility and affordability of safe abortion services, and enforcement of the law.

OBJECTIVES

1. To evaluate the legislative frameworks that regulate the sexual and reproductive rights of women in India.
2. To evaluate the impact of the legislative measures on gender equality, reproductive health care, and women's empowerment.
3. To identify best practices and insights learned from other countries that can be used to inform policy recommendations in India.
4. To propose policy recommendations for enhancing women's sexual and reproductive rights in India, with a focus on empowerment and improved healthcare access.

METHODOLOGY

Authors employ the doctrinal research approach and try to provide a thorough and logical analysis of the legal framework and case laws pertaining to the reproductive rights of women in India. A detailed examination of the legal frameworks in India will be conducted to identify similarities, differences, and trends in women's sexual and reproductive rights legislation. This analysis will encompass laws related to contraception, abortion, maternal healthcare, and gender-based violence.

THE CONCEPT OF REPRODUCTIVE RIGHTS

Human rights are the basic rights inherent to every human being, with certain human rights protected by law as fundamental rights. Among these, reproductive rights for women are considered a basic human right of women that should receive legal sanction and protection. Reproductive rights are intrinsically connected to the right to life. Although there is no formal legal definition of sexual and reproductive rights, these rights empower individuals to procreate and have offspring. Reproductive rights are regarded as the latest category of personal non-property rights, known as the fourth-generation rights. These rights are closely tied to fundamental human rights, including the inalienable right to life, the respect for human dignity, and personal integrity (Aneesh & Anatoliy, 2021).

Reproductive rights include women's physical, emotional, and social well-being in all facets of sexual and reproductive health. This wide definition affirms women's rights to make autonomous body and reproductive decisions beyond disease absence. Safe abortion, sexual freedom, and childbearing choices are among these rights. To guarantee the safety of both the mother and the infant during pregnancy and childbirth, reproductive rights necessitate informed access to family planning methods and sufficient healthcare services (Sharma, 2015). Reproductive rights are acknowledged as a component of human rights in numerous national jurisdictions by the International Conference on Population and Development, 1994. These rights allow each couple to determine the quantity and timing of their children and to manage their pregnancy. Furthermore, these reproductive rights encompass the right to access sexual and reproductive health care services and reproductive services (International Conference on Population and Development, 1994). Reproductive rights are the ethical and legal principles that guarantee that individuals have the autonomy to make decisions regarding their reproductive health. In addition to contraception, these rights include access to secure and legal abortion, fertility treatment, and comprehensive sex education. The fundamental aspect of reproductive rights is the autonomy to determine whether, when, and how to have children, free from coercion, discrimination, and violence. It is imperative to guarantee reproductive rights to empower individuals to make informed decisions about their bodies and futures, promote gender equality, and improve health outcomes. Advocating for reproductive rights frequently involves confronting obstacles such as restrictive laws, inadequate healthcare services, and societal stigmas (Center for Reproductive Rights, 2021). As stated by Pascale Allotey, "Reproductive and sexual health should never be marginalized in healthcare, but rather acknowledged as essential to the general well-being of individuals and the communities in which they reside" (WHO, 2024). Reproductive rights safeguard the autonomy of women in making reproductive decisions and eliminate gender-based discrimination, enabling them to engage in society without fear of coercion, violence, or discrimination. In order to achieve global gender parity and empower women, women's reproductive rights must be upheld and implemented (Sharma, 2015).

Reproductive health refers to the overall condition of well-being, encompassing physical, mental, and social aspects, in relation to the reproductive system and its functions. It goes beyond the absence of illness or weakness. It encompasses the ability to engage in a fulfilling and secure sexual experience, the capacity to procreate, and the autonomy to choose whether, when, and how frequently to engage in sexual activity. This encompasses the entitlements of both males and females to be educated and can utilize secure, efficient, reasonably priced, and socially acceptable means of family planning, along with other lawful approaches for managing reproductive capacity. Furthermore, it includes the entitlement to obtain suitable healthcare services that guarantee secure pregnancy and childbirth, affording couples the optimal opportunity to have a robust newborn. Reproductive healthcare encompasses a variety of procedures, strategies, and services designed to enhance reproductive health and address associated concerns. Additionally, sexual health encompasses more than only reproduction and sexually transmitted diseases. It emphasizes the improvement of personal relationships and overall quality of life (United Nations, 1994). Ensuring protection from sexual violence is a critical component of women's reproductive rights, highlighting the essential requirement to preserve their sexual autonomy. The recognition of the inherent connection between sexual health rights and the preservation of sexual autonomy emphasizes the crucial need of these safeguards. To guarantee women's reproductive rights, it is crucial to protect them from any type of sexual coercion, as this is essential for preserving their sexual well-being and independence. The comprehensive approach is crucial in creating an atmosphere where women can freely and independently make choices regarding their reproductive health without any form of pressure or violation (Vijay & Sharma, 2002).

CONTEXT OF WOMEN'S SEXUAL AND REPRODUCTIVE RIGHTS IN INDIA

About 60 years ago, India was at the forefront of numerous groundbreaking efforts related to sexual and reproductive health rights (SRHR). India became a major actor in the worldwide movement for family planning and reproductive health when the International Planned Parenthood Federation (IPPF) was established during a meeting in Bombay in 1952 (Mathew, 2021). Avabai Wadia and Dhanvanthi Rama Rau established the Family Planning Association of India (FPAI), which played a significant role in supporting family planning and collaborating with the Indian government to shape laws. In 1951, as a result of the FPAI's work, India's first Five Year Plan included family planning, making it the first country to do so. The Third International Conference on Planned Parenthood, which took place in Bombay and witnessed the foundation of the IPPF (FPAI), was also organized in large part by this group. FPAI's programs have grown over the years to include efforts to increase awareness of SRHR, strengthen women's groups called Mahila Mandals, and provide youth with accurate and up-to-date information on sexuality and reproductive health. Although there have been some positive developments, there are still many social, economic, and political obstacles that women in India face when fighting for reproductive rights. Women often lack access to basic facilities, healthcare, and education, with societal norms dictating significant aspects of their lives (Mathew, 2021). The Indian government's approach to reproductive rights has been largely driven by population control rather than recognizing women as the primary decision-makers regarding their bodies. The 1971 reconsideration of abortion laws, for instance, was motivated more by population control than by women's autonomy (Saheli). Women's groups continue to advocate for reproductive rights, emphasizing that birth control is an individual right rather than a state-imposed mandate. The battle for reproductive rights in India is ongoing, needing continuous efforts to ensure that women can exercise control over their bodies as a fundamental right, not a privilege (Mathew, 2021).

The reproductive rights of women and girls in India remain at risk, despite the implementation of groundbreaking abortion and contraceptive legislation. India's women are plagued by inadequate health services and diminished decision-making authority. Population control has been prioritized over women's autonomy in Indian reproductive policies, which typically necessitate spouse approval. Despite regulations prohibiting child marriages and providing maternity healthcare, India has the highest rates of maternal fatalities and child marriages in the world. Several international human rights organizations have encouraged India to enhance reproductive health services and establish greater healthcare equity. Indian courts play a crucial role in enforcing reproductive rights based on international and constitutional law (Center for Reproductive Rights, 2021).

CHALLENGES IN REPRODUCTIVE RIGHTS IN INDIA

There are many obstacles that prevent women in India from using their reproductive and sexual rights. Many obstacles, rooted in cultural, legal, and social concerns, threaten reproductive rights in India. The gender imbalance, which is a result of long-standing prejudices, is a major obstacle. This disparity is made worse by behaviors that violate women's reproductive rights by giving preference to male kids, such as sex-selective abortions and female infanticide (Bhat & Bhatt, 2017). Furthermore, India's population plans have come under fire for being coercive, frequently targeting marginalized women with hazardous contraceptive methods and mass sterilizations, further restricting their right to choose.

Reproductive justice in India necessitates a broader definition that considers women's social and economic circumstances in addition to their legal rights. Critics point to the Medical Termination of Pregnancy Act of 1971 and other legal frameworks that fail to provide women with complete health services and use discriminatory terminology. Reconciling reproductive rights with reproductive justice, according to activists, necessitates coalitions across different political and non-governmental groups to tackle problems including maternal mortality and domestic abuse (Madhok et al., 2014).

Equally important in protecting reproductive rights has been the role of litigation. The Shanti Devi verdict and other landmark cases show how important it is to use the legal system to make the government respect people's right to health and life. Article 21 of the Indian Constitution recognizes the right to procreate, but there is still a lack of consistency in its practical application, which is typically caused by socioeconomic obstacles and insufficient healthcare facilities (Kaur, 2012). These obstacles highlight the importance of all-encompassing laws that protect women's right to choose and ensure they have access to high-quality reproductive healthcare. India faces a significant reproductive health issue with a maternal mortality rate of 45,000 per year, equivalent to one death every twelve minutes. Most of these fatalities are caused by abortions that are not safe. According to studies conducted in India, around one-third of all pregnancies end in abortion, and half of those pregnancies are unwanted. However, just a quarter of these abortions take place at accredited medical centers. Legal and social hurdles make it difficult for many women, particularly those who are young and single, to get safe abortion services. Doctors often require parental or spousal consent, even though it is not legally required, forcing women to seek unsafe abortions (Wilson, 2018).

The Medical Termination of Pregnancy (MTP) Act, 1971, allows termination only up to 20 weeks, requiring court approval for later terminations, which is a difficult process. The law does not consider non-medical reasons, resulting in unsafe practices. The stigma around abortion contributes to many deaths and conceals broader issues like adolescent access to reproductive health services.

FACTORS INFLUENCING THE DEVELOPMENT OF REPRODUCTIVE RIGHTS

Reproductive rights in India are influenced by social, cultural, and economic factors that influence the comprehension and recognition of women. The high rates of female neonatal mortality and unequal sex ratios demonstrate that patriarchal norms and gender roles restrict women's reproductive preferences. Due to their inferior status within their families, societal attitudes, inadequate education and resources, and cultural norms, women are subjected to discrimination. Health care programmes generally prioritize basic care, maternity and child health, and family welfare, even though they emphasize health education and institutional deliveries. In contrast, women's requirements are often overlooked (Sharma, 2015). Unsafe parenthood continues to be a significant concern, particularly in rural areas where women frequently develop gynecological disorders without receiving appropriate medical treatment due to a lack of provision of prenatal and postnatal care. Most medical institutions are inadequately outfitted to provide comprehensive reproductive health management, with an emphasis on immunization and supplement delivery rather than pregnancy and postpartum care. The economic and domestic responsibilities of women also restrict their access to healthcare. Population growth, which is the result of unwanted fecundity and the desire for additional children, as well as behavioral factors such as entrenched gender inequality and restricted autonomy, continue to affect the reproductive health of women (Sharma, 2015).

PERSPECTIVES ON ABORTION IN INDIAN SOCIETY

Abortion is one of the most sensitive and emotional issues because it deals with vulnerable human life. Abortion is a multi-dimensional global issue which gets various opinions from different perspectives. Since India is a land of 'unity in diversity,' the legal, social, religious, moral, and cultural dimensions strongly influence abortion or the right to live. Proponents of the right to life claim that life begins at conception and that, thus, abortion is tantamount to murder. They believe that the fetus is a person with inherent rights and that abortion is an act of violence against an innocent human being. Advocates for women's reproductive rights argue that women should have autonomy over their own bodies and health care decisions, including the choice to terminate a pregnancy. According to them, it is a violation of women's fundamental rights to autonomy and privacy to prevent them from accessing safe and legal abortion services. Many people and organizations may have different opinions and views about abortion and life, which can make the debate difficult and emotionally intense. There is a great deal of overlap between the rights of pregnant women and their unborn children and other issues like healthcare, social justice, and human rights, and it is critical to acknowledge this. Tackling the obstacles to safe and legal abortion services in India requires a comprehensive strategy that considers social, legal, and health-related factors (Datta & Misra, 2000).

It was a crime to have an abortion in India under the regulations that were put in place by the British in the nineteenth century unless the woman's life was in imminent danger. Reforms in India's abortion regulations were prompted by the liberalization of similar policies in the United States and Europe during the 1960s and 1970s. The MTP Act, 1971 was passed in response to the recommendations made by the Shantilal Shah Committee, which sought to legalize abortion to reduce maternal mortality and morbidity. The committee stressed that legalizing abortion for demographic reasons could be detrimental to family planning techniques, even though some states saw the legislation as a means of population control (Shah, 1967)

LEGAL STATUS OF ABORTION IN INDIA

The legality of abortion in India is governed by the Medical Termination of Pregnancy (MTP) Act of 1971, as amended in 2021. Under some circumstances, such as when the woman's life is in danger or when the pregnancy is the result of rape, abortion is legal up to twenty weeks into the pregnancy, as per the MTP Act of 1971. In rare and exceptional circumstances, a medical board may authorize an abortion beyond the twentieth week of gestation. The Act specifies the circumstances that allow for the termination of a pregnancy in Section 3 (2). It states that abortions can only be carried out by licensed medical professionals in specially licensed facilities. According to the Medical Termination of Pregnancy (MTP) Act of 1971, the abortion cannot take place without the woman's agreement; in the case of a minor, the consent of her parents or legal guardian is also required. The most current version of the Medical Termination of Pregnancy Act, 2020 (Medical Termination of Pregnancy (Amendment) Act, 2021), which raises the legal limit for abortion from 20 weeks to 24 weeks under specific conditions. A progressive law that recognizes the reproductive rights of women and grants them more independence is the MTP (Amendment) Act, 2020. When it comes to women's reproductive health, abortion is considered crucial.

According to the World Health Organization's assessment, the MPT Amendment Act of India expands access to comprehensive abortion care while protecting women's rights to privacy, autonomy, and dignity (WHO, 2021). Keep in mind that the term "marital rape" was not included in the amendment. Having sexual relations with a spouse without their consent is considered marital rape. This is still a highly debated topic in Indian law. The modification makes it so that a licensed medical practitioner's discretion is the only factor that counts when deciding whether an abortion can be performed between 20 and 24 weeks of pregnancy. Also, this is a major critique.

Rules and regulations controlling abortion services in India have been set by the country's Ministry of Health and Family Welfare. The goal of these regulations is to make sure that women can get the abortions they need in a safe and effective manner. Standardization of abortion facilities, training of medical staff in the correct use of medical and surgical abortion procedures, reporting of abortion numbers and demographics to the government, and counseling for women seeking abortions are among the most important recommendations and regulations.

The Bharatiya Nyaya Sanhita (BNS), 2023 makes it clear that abortion is a crime under Sections 88 to 92. Anyone who induces a miscarriage or abortion without the woman's agreement or in cases when the pregnancy is not in her best interest is punishable by law according to these sections, which declare abortion a criminal act. If a qualified medical professional performs an operation on a pregnant woman with her consent and in good faith in order to preserve the mother's life, then the act is not criminal under Section 88 of the Bharatiya Nyaya Sanhita (BNS), 2023. Similarly, it is not considered a violation if the conduct is done in a good faith effort to prevent the woman from suffering severe physical or emotional anguish (Bharatiya Nyaya Sanhita (BNS), 2023). Legal abortion is permitted under the Medical Termination of Pregnancy Act of 1971 and the MTP (Amendment) Act of 2021, with certain exceptions and conditions.

LEGAL AND POLICY BARRIERS TO SAFE ABORTION CARE IN INDIA

Abortions in India are governed by the Medical Termination of Pregnancy Act of 1971. The Act, which was urgently needed to address the problem of unsafe abortions, was first viewed as progressive. But its provisions have grown insufficient due to the changing needs of society. The purpose of introducing the Medical Termination of Pregnancy (Amendment) Act, 2021 was to revitalize the legislative framework. The Amendment is ineffective for many stakeholders because, despite this effort at modernization, it does not represent the country's social, economic, and legal progressions adequately. (Singh, 2021)

Until the 20th week of gestation, abortions in India are legal according to the Medical Termination of Pregnancy (MTP) Act, 1971. This Act was amended in 2021 to increase the allowable duration to 24 weeks in specific circumstances, including for incest and rape survivors. There was a lot of applause for this modification because it strengthened confidentiality protections and helped women feel safer and more valued. Under the MTP Act, a doctor's note is needed for any abortion, but it's especially important when the pregnancy poses a risk to the mother's mental or physical health or could cause major birth defects in the unborn child. Another valid reason for abortion is when other methods of birth control fail, leading to an unplanned pregnancy and the accompanying psychological suffering (Singh, 2021).

Despite more than 30 years of progressive abortion legislation in India, significant barriers to safe abortion care persist (Hirve, 2004) The history of abortion law and policy in India reveals ongoing challenges, as highlighted in a comprehensive review published in Reproductive Health Matters (Ministry of Health and Family Planning, 1966). Initially, abortion was governed by the Indian Penal Code of 1862, which criminalized the procedure except to save the woman's life (Berer, 2000). However, influenced by global trends towards liberalization in the 1960s, the Shah Committee was established and recommended the legalization of abortion in 1966 to reduce maternal mortality due to unsafe procedures (Phadke, 1998). This led to the enactment of the Medical Termination of Pregnancy (MTP) Act in 1971, which allowed abortions under specific conditions up to 20 weeks of pregnancy.

Amendments to the MTP Act in 2002 and 2003 aimed to improve access by decentralizing regulatory authority to district committees and introducing punitive measures against unsafe providers (Government of India, 2003). These amendments also rationalized facility requirements for early abortions and approved medical abortion methods (Barge, 1997). However, persistent barriers such as poor regulation, restrictive physician-only policies, low registration of rural clinics, and inadequate awareness of the law continue to limit access to safe abortion services (Ganatra & Hirve, 2002). Additional challenges include unnecessary spousal consent requirements, informal fees, and the continued use of outdated techniques like dilatation and curettage (D&C) (Government of India, 1996).

The relationship between abortion laws and sex determination further complicates the legal landscape. The goals of the MTP Act and the Prenatal Diagnostic Techniques (Regulation and Prevention of Misuse) Act of 1994 are frequently confused, which can lead to misuse and ethical problems (FOGSI, 2002). The purpose of the former is to prevent sex-selective abortions, while the latter seeks to regulate and prevent their abuse. The review proposes that in order to improve access to safe abortion services, more providers should be trained, registration procedures should be made simpler, policies should be in line with current technology and best clinical practices, and standards should be applied consistently in the public and private sectors (Government of India, 2000). These suggestions stress the need of having political will and sufficient funding to put these plans into action (Hirve, 2004).

It is crucial to update abortion legislation in India to reflect current medical practices and human rights standards, as these findings are relevant to women's sexual and reproductive rights in the country. By bringing these two perspectives into harmony, we can chart a course for legislative reforms that will empower women and guarantee that all women in India have access to safe, legal abortions (Hirve, 2004).

Especially in borderless societies and those with social norms that are vulnerable to constitutional examination, the most dangerous combinations in a "normal" social order are the predominance of power-laden values that limit freedom of choice. The inability of the individual carrying the life within her to exercise autonomy in making decisions on medical termination of pregnancy has been one of the most difficult challenges for feminists in the last many years. A particularly unfortunate aspect of medical termination of pregnancy is the constraints disguised as medical liberalizations under the Medical Termination of Pregnancy Act, 1971. Thus, it is essential to evaluate whether the Medical Termination of Pregnancy (Amendment) Act, 2021 truly grants the right to abortion or if it remains a privilege. We must analyze whether the Right to Reproduction recognized by the Indian Constitution is being upheld in its truest form and spirit. Are the challenges faced by rape survivors and those with unwanted pregnancies being overlooked to uphold the patriarchal construct of society? Are women solely responsible for taking care of new life? (Petchesky, 1991).

Throughout history, women have employed various methods of birth control and abortion. Since abortion is not merely a techno-medical issue but a fulcrum of broader ideological conflicts involving family, state, motherhood, and young women's sexuality, these practices have sparked strong moral, ethical, political, and legal debates. Women have resorted to abortion overtly or covertly, but their access to services has been countered by social and legal restrictions. The legal rules governing abortion have been constantly reshaped to fit the historical and social contexts in which they are implemented, often ignoring women's right to choose their sexuality, fertility, and reproduction (Karkal, 1991). The Medical Termination of Pregnancy (Amendment) Act, 2021, is evaluated against the backdrop of essential human rights established in Part III of the Indian Constitution. Historical documents, government reports, and progressive judgments have been referenced to support the argument that despite seemingly permissive rules, abortion in India exists more as a privilege than a right (Anna, 1989).

Legal Lacune in the MTP Act

The Medical Termination of Pregnancy (MTP) Acts 1971 and 2021 contains several legal lacunae that impede its effectiveness in ensuring comprehensive reproductive rights. These gaps not only restrict access to necessary services but also fail to address evolving societal norms and medical advancements in reproductive health care.

Cut-Off for Termination of Pregnancy Prior to the MTP (Amendment) Act, 2021

Before the 2021 amendment, the MTP Act allowed termination of pregnancy up to twelve weeks with one doctor's approval and up to twenty weeks with the approval of two doctors. However, pregnancies beyond twenty weeks due to fetal abnormalities or rape often required judicial intervention. The judiciary, in several cases, permitted termination beyond the statutory limit, recognizing the evolving needs of society since the 1970s (Shah, 1967).

MTP (Amendment) Act 2021: Restrictions Under the Veil of Medical Liberalization

The MTP (Amendment) Act, 2021, introduced significant changes but still places the decision-making power primarily in the hands of medical practitioners rather than women. It allows termination up to twenty weeks with one doctor's approval and between twenty to twenty-four weeks with two doctors' approval. For pregnancies beyond twenty-four weeks, a Medical Board must diagnose substantial fetal abnormalities. Despite these changes, the Act does not fully grant women the right to choose but rather maintains it as a medical intervention subject to approval (The Medical Termination of Pregnancy Act, 1971)

Medical Termination of Pregnancy: Still Not a Right but a Privilege

The MTP Act, 1971, and its amendments do not provide a basic right to induced abortion but restrict the conditions under which women may obtain abortion services. The decision to terminate a pregnancy remains largely dependent on medical practitioners' opinions, highlighting that abortion is treated more as a privilege than a right (The Medical Termination of Pregnancy (Amendment) Act, 2021)

A Woman's Bodily Autonomy Still Primarily Rests with the State

The MTP Act allows abortion on medical and humanitarian grounds but still requires women to justify their decision within the Act's criteria. This situation demonstrates that abortion remains tied to state-sanctioned conditions rather than women's rights. Despite progressive judgments supporting women's reproductive rights, the power to decide remains with medical practitioners rather than the women themselves (Priyanka Shukla v. UOI & Ors., 2017).

The Intersection of the MTP Act with Other Statutes

The Pre-Conception and Pre-Natal Diagnostic Techniques Act, 1994 (PCPNDT Act), the Protection of Children against Sexual Offences Act, 2012 (POCSO), and the Indian Penal Code 1860 (IPC) are among the statutes that are entwined with the Medical Termination of Pregnancy Act. These statutes work together to regulate abortion while addressing issues such as sex-selective abortions and sexual offences. Yet, medical professionals and women seeking abortions frequently face uncertainty and obstacles as a result of these conflicting rule (Meera Santosh Pal v. UOI 2017).

WHY AMENDMENTS TO THE MTP ACT DO NOT GO FAR ENOUGH

The most recent revisions to the Medical Termination of Pregnancy (Amendment) Act, 2021, were a positive step, but they do not address the intricate issues of gender inequality and abortion stigma. It is evident that a more significant overhaul is required to resolve these issues, with an emphasis on patient-centric treatments and the removal of antiquated regulatory challenges.

Lack of Time-Bound Decision-Making by the Medical Board

Abortions performed after twenty-four weeks of gestation must be approved by a Medical Board in the event of severe fetal abnormalities, according to the Medical Termination of Pregnancy (Amendment) Act, 2021. The Act, however, does not lay out a timeline within which the Medical Board must make a determination. Significant delays can occur as a result of this omission, which increases the risk to the pregnant woman's health. Complications can cause serious bodily and psychological injury, so it's crucial to get medical help quickly. The goal of the amendment, which is to ensure that women experiencing serious health problems have access to safe and timely abortion services, is undermined by the lack of a specified deadline.

Ambiguity in Inclusion of Transgender Persons

The Medical Termination of Pregnancy (Amendment) Act, 2021, leaves open the question of whether transgender and non-binary people who may also require abortion services are considered "pregnant women" in its main text. Transgender males and non-binary people can now become pregnant and have access to abortions thanks to medical advancements. These persons face additional obstacles to receiving healthcare because the Act does not specifically acknowledge them. Everyone, regardless of their gender identity, has the right to reproductive healthcare and this exclusion ignores that fact.

Shortage of Qualified Medical Practitioners

The accessibility of competent medical professionals to carry out abortions is a critical component in the MTP Act's efficacy. Community health facilities in India are severely understaffed when it comes to obstetricians and gynecologists, according to the All-India Rural Health Statistics (2018-19). Healthcare services are already scarce in rural regions, making this shortage much more severe. Furthermore, according to the National Health and Family Survey (2015-16), a large number of abortions are conducted by unqualified individuals, auxiliary nurse-midwives, nurses, or registered medical professionals, accounting for only 53% of all abortions. The need for safe and legal abortion services is hindered by a severe lack of skilled experts, which in turn raises the risk of maternal death and morbidity caused by unsafe operations.

Inconsistencies in Judicial Decisions

The MTP Act has been applied and interpreted inconsistently by the judiciary, leading to a range of outcomes for women seeking abortions. Lack of consistency in rulings is a result of courts' divergent readings of the Act's terms. Some courts have upheld requests to terminate pregnancies beyond the legal limit due to health concerns or fetal abnormalities, while others have rejected such cases. Unfair outcomes can arise from this mismatch if various jurisdictions grant varying degrees of access to abortion services to women in similar situations. The inconsistency and lack of consistency in court decisions might make it difficult for women to make informed decisions about their pregnancies, which can lead to undesired pregnancies.

Lack of Accountability Provisions

When a woman dies because her abortion request was denied, the MTP Act, 2021 fails to hold medical professionals or organizations accountable. There is presently no system in place to hold those responsible for the denial of an abortion accountable in cases where the legal requirements are fulfilled, and damage or death occurs as a consequence. There are no consequences for not providing essential medical care, which can lead to negligence and malpractice due to the lack of responsibility. Protecting the rights of women seeking abortions and ending preventable maternal deaths requires holding those responsible to account. To guarantee that all people in need have access to safe, timely, and affordable abortion services, it is crucial to address these concerns in the MTP Act, 2021.

RIGHT TO CHOOSE

Everyone has the inherent right to make their own decisions, and women are no exception. Upholding individual rights, such as the right to life, liberty, and the pursuit of happiness, is consistent with recognizing a woman's freedom to choose whether to have an abortion. Bodily rights, which refer to a woman's autonomy over her own body, are inviolable. A woman's reproductive and sexual health greatly affects her overall happiness, which in turn affects her choices about having children and her path through life. The right to privacy and the right to control one's own body are among the human rights violated when someone is coerced into having an abortion when one is not wanted or when one gets pregnant against one's will. The freedom to choose whether or not to have children and the means to ensure one's own reproductive health are all parts of the reproductive rights package. The right to obtain reproductive health services, use contraception, participate in public sex education, terminate a pregnancy, and plan a family are all examples of what may be included in this (Rani 2022). The US Supreme Court's decision in *Roe v. Wade* upheld a woman's right to choose whether to have an abortion. A woman's right to privacy, according to the Court, encompasses the option to have or not have an abortion. At least throughout the first trimester of pregnancy, the Supreme Court laid out a framework that states' laws that prohibit or limit access to abortion are unconstitutional (Roe v. Wade, 1973).

The United Nations International Conference on Population and Development in 1994 acknowledged reproductive rights as an essential component of individual autonomy and human rights. When it comes to matters of the body and its reproduction, these rights cover a lot of ground. Access to and use of methods of contraception for the purpose of planning and spacing out pregnancies is a reproductive right. The right to an abortion, when a person or couple determines it is essential, includes the right to a safe and legal abortion. The reproductive rights include the guarantee of equal access to sexual and reproductive health care for LGBTQ+ individuals and couples (Ghosh and Khaitan, 2017).

RIGHT TO LIFE VERSUS RIGHT TO ABORTION UNDER ARTICLE 21 OF THE INDIAN CONSTITUTION

Every citizen of India has an inherent right to life and liberty guaranteed by Article 21 of the country's constitution. The right to abortion is one of many new rights that have been added to this article's scope over the years. The right to an honorable life and autonomy over one's own body are components of the right to exist. Subject to reasonable limitations in the sake of public health and welfare, the right to abortion is an expansion of the right to life and personal liberty. The right to choose one's own reproductive path is granted to women in article 21. In addition, it protects autonomy, privacy, and dignity when it comes to having children (Nehra & Rajput, 2019). The right to choose whether to have a child is an inherent part of a woman's personal liberty, which the Indian Supreme Court has affirmed includes the option to end a pregnancy. According to the court's ruling in *Devika Biswas v. Union of India*, Article 21 guarantees a person's right to health care, which includes the freedom to procreate (Devika Biswas v. Union of India, 2016). In the case of *Suchita Srivastava v. Chandigarh Administration*, the same principle was confirmed. The highest court in India reaffirmed Article 21's status as a constitutional guarantee of a woman's autonomy over her reproductive health. According to *Suchita Srivastava v. Chandigarh Administration*, one of these rights is the autonomy to choose one's own reproductive choices, including the timing, location, and number of children (Suchita Srivastava v. Chandigarh Administration, 2009).

A woman's autonomy ensures that she cannot be coerced into having children. There is no room for the exploitation of women as mere reproductive organs. A person's right to health, privacy, autonomy, and dignity is guaranteed by the Constitution. A woman's right to decide for herself whether to carry a pregnancy to term is a part of this right to reproductive autonomy. Article 21 of the Indian Constitution states that the right to abortion is not an absolute right, and that the government has the authority to impose reasonable restrictions on this right for the sake of public health and the protection of other people's rights.

THE REPRODUCTIVE RIGHTS OF WOMEN AND JUDICIAL RESCUE IN LANDMARK CASE LAWS IN INDIA

Several landmark decisions in India have shaped the country's abortion laws. The Indian Supreme Court upheld the freedom of women to make decisions about their own bodies and health in the case of Suchita Srivastava v. Chandigarh Administration, finding that article 21 of the Indian Constitution guarantees women the right to reproductive choice. In the case of *Suchita Srivastava v. Chandigarh*

Administration (2009), the court ruled that a woman's right to privacy would be violated if her autonomy over her own body were to be limited (Suchita Srivastava v. Chandigarh Administration, 2009).

The Supreme Court upheld a woman's right to an abortion at 24 weeks after conception in Meera Santosh Pal v. Union of India because the unborn child had a serious heart defect. The court ruled that the right of the mother to choose her own body supersedes the right of the unborn child to live. Key considerations, according to the Court in *Meera Santosh Pal v. Union of India*, were the woman's right to bodily integrity and consent, as well as her right to take whatever measures necessary to safeguard her life from unnecessary injury (Meera Santosh Pal v. Union of India, 2017).

In the case of *Dr. Nikhil Datar v. Union of India*, the Bombay High Court upheld a woman's right to terminate her pregnancy beyond the 20-week mark, if she could prove that continuing the pregnancy would put her life in jeopardy. The emotional and physical toll that a denial of the choice to continue or end a pregnancy following a diagnosis of many fetal abnormalities takes on a woman was the focus of this case. *Datar v. Union of India* further shows the moral distress experienced by doctors when they are bound by law not to do what is best for their patients (Datar v. Union of India, 2020). The MTP Act of 1971 was amended because of this case. Rather than a "crime" or a "killing," Dr. Nikhil Datar argues that abortion is a matter of individual autonomy and privacy. Every citizen of this nation possesses the power to choose" (Datar, 2020).

A fundamental right under the Indian Constitution is the right to privacy, according to the Supreme Court of India's ruling in *K.S. Puttaswamy v. Union of India*. This right includes the freedom to make decisions about one's own body and health, including the option to have an abortion (K.S. Puttaswamy v. Union of India, 2017). The right to privacy is a defense of autonomy that seeks to maintain an individual's bodily integrity and autonomous decision-making ability (Narain, 2020). The Puttaswamy case confirms this right and provides a comprehensive endorsement of individual liberty.

The Medical Termination of Pregnancy Act, 1971 (MTP Act) is being challenged in the case of *Swati Agarwal & Ors. v. Union of India* by a public interest litigation (PIL) on the grounds that it violates Articles 14 and 21 of the Indian Constitution. The sections in question are 3 (2) (b), 3 (4) (a), and 5. Reducing maternal mortality was the stated goal of the petitioners, who sought the opening of safe abortion clinics. In their 2019 case against the Union of India, Swati Agarwal and others contended that women's reproductive health and autonomy depend on access to safe and legal abortions, and they wanted to legalize these services in order to reduce maternal mortality (Swati Agarwal & Ors. v. Union of India, 2019).

The significance of a woman's autonomy in relation to her reproductive choices, body autonomy, and control over her own body was highlighted in the 2016 case of High Court on its own motion v. The State of Maharashtra (Suo Motu PIL). In the case of High Court on its own motion v. The State of Maharashtra (2021), the court emphasized the importance of physical autonomy and individual rights, stating that no one else should prescribe these decisions to a woman (High Court on its own motion v. The State of Maharashtra, 2021). The importance of the judiciary's role in upholding and protecting fundamental rights, such as equality and personal liberty, are highlighted in these instances.

The petitioner in Mamta Verma v. Union of India sought a medical abortion since the unborn child's anencephaly was a danger to her life. The Supreme Court of India granted her request. A severe and incurable disorder known as anencephaly occurs when the fetal skull bones do not develop (Mamta Verma v. Union of India, 2017). In the instances of *Tapasya Umesha Pisal v. Union of India* and *Meera Santosh Pal v. Union of India*, the Supreme Court reached similar conclusions (Tapasya Umesha Pisal v. Union of India, 2018 & Meera Santosh Pal v. Union of India, 2017). These cases featured pregnancies that were longer than 20 weeks, and the fetuses in those cases showed a variety of illnesses and deformities, posing serious hazards to both the mother and the fetus (Mittal, 2023).

In *X v. Principal Secretary Health and Family Welfare Department*, the Supreme Court of India has ruled that all pregnant people in India have the right to make their own reproductive decisions, including transgender and gender variant people. Only the person who is actually pregnant has the legal right to end the pregnancy. The right of women to make decisions about their reproductive options independently, free from interference from the government (X v. Principal Secretary Health and Family Welfare Department, 2022).

The Supreme Court of India has taken progressive steps to support women's reproductive rights. Notable rulings, such as the decision decriminalizing homosexuality and adultery in *Navtej Singh Johar v. Union of India* case (Navtej Singh Johar v. Union of India, 2016), and upholding women's right to sexual autonomy and reproductive choices as guaranteed by Article 21 of the Constitution in Puttaswamy case (K.S. Puttaswamy v. Union of India, 2017). Girls' reproductive rights were upheld in the Supreme Court decision in the case of *Independent Thought v. Union of India*, which highlighted the girls' human rights regardless of their marital status (Independent Thought v. Union of India, 2017). The importance of safeguarding women's access to safe abortions as a fundamental human right is underscored by these verdicts.

Discussions about public health and democracy should center on the importance of safe and legal abortion access for sexual and reproductive equality. Those working in development and civil society must speak out on these matters. Issues like as trafficking, maternal health, and abortion-related fatalities persist as roadblocks to development, even as gender equality has made strides. “The condition of women must be improved before one can contemplate the welfare of the world,” remarked Swami Vivekananda.

RECOMMENDATIONS FOR ENHANCING REPRODUCTIVE RIGHTS

The purpose of these suggestions is to help people in India better understand and exercise their reproductive and sexual rights. Ensuring women have the power to make decisions about their own health and reproductive rights is of utmost importance, as it is a fundamental right that every woman should have to live a dignified life. Now more than ever, it is critical to combat and ultimately eradicate the societal stigma that surrounds abortion and women's reproductive choices. A society where all women can fully exercise their rights can be achieved by governmental legislation, with the help of society and families.

Inclusion of Sexual Education in the School Curriculum

For young people to have a better grasp of and regard for their reproductive and sexual rights, it is crucial that schools include comprehensive sex education in their curricula. Early childhood education on sexual health, physiological changes, and gender-specific issues equips children with the knowledge they need to address sexual difficulties and concerns effectively. This education helps people comprehend the unique experiences and challenges that each gender faces, which promotes mutual understanding between the sexes. Furthermore, sex education plays a crucial role in educating the public about women's reproductive rights, which in turn encourages acceptance and activism. Additionally, it helps advance gender parity in general by making discussions about sexual health and rights more commonplace, which ensures that people of all genders are knowledgeable and considerate about these important issues. So, sex education in schools isn't only a tool for teaching; it's a crucial step in building a more intelligent and fair society.

Access to Safe Abortions at Primary and Community Health Centers

Abortion services must be included into primary and community health facilities. Women, especially those living in rural or otherwise underprivileged regions, would greatly benefit from these services becoming a part of the Reproductive and Child Health (RCH) Program, which would greatly increase their access to abortion care.

Increase Investments in Public Facilities

To guarantee strong and competent public abortion services, substantial investments in public healthcare facilities are required. All women, regardless of their financial situation, should be able to afford a safe abortion, and additional money for healthcare would make that possible.

Promote Safer Abortion Methods

We must prioritize the promotion of safer abortion methods, including medical abortion and vacuum aspiration techniques. Healthcare providers should be educated and trained to use these techniques instead of the riskier curettage procedure.

Broaden the Base of Abortion Providers

The scarcity of qualified abortion providers can be alleviated by teaching paramedics to carry out abortions during the first trimester, as is done in nations such as Bangladesh and South Africa. This can help make sure that more women, especially in places like rural areas where there might not be as many doctors, can get safe abortions.

Engage Medical Associations in Training

Training abortion providers, particularly those in the private sector, should be an active role for medical societies. Providers can be assured that their patients are receiving safe abortion services if they participate in standardized training programs that are developed and implemented through these associations.

Promote Apprenticeship

As a training method, apprenticeship has the potential to be very effective. To guarantee competence and assurance in offering safe abortion services, it is important to encourage apprenticeship programs where novice abortion physicians can obtain practical experience while learning from seasoned experts.

Re-skill Traditional Providers

It is imperative that conventional medical professionals undergo retraining to better facilitate women's access to safe abortion services. Improving access to safe abortion services can be greatly enhanced by equipping them with new skills and giving them alternate duties, such as counseling and providing information.

Strengthen Regulations for Abortion Facilities

It is of the utmost significance to establish minimal accreditation requirements and provide exceptional care at all abortion facilities. We can ensure that all abortion clinics are dependable and secure for patients by adhering to stricter regulations, which will maintain a high standard of healthcare.

Change Mindsets of Clinic Directors

The best way to get clinic directors to embrace universal clinic certification is to reach out to them through professional associations. If we want certification requirements to be widely accepted and followed, which will improve the quality of treatment at abortion facilities, we must first change the thinking of these leaders.

Display Certification Status

Transparency and trust can be enhanced by mandating that all abortion facilities openly display their accreditation status. Women feel more secure about the quality and safety of the services they receive when they realize that the facility has certification.

Promote Safe Contraception

The key to lowering the rate of unintended pregnancies is increasing awareness of the need of using safe methods of contraception to space out births. The need for abortions can be reduced if more people are able to use effective contraception to reduce the number of unwanted pregnancies.

Educate Providers on Ethics of Sex Determination

Healthcare professionals must be made aware of the morality of sex determination and the significance of abiding by the Prenatal Diagnostic Techniques (PNDT) Act's regulations. Providers must be aware of the moral and legal ramifications to stop prenatal diagnostic methods from being abused for sex-selective abortions.

Increase Awareness Among Women

Raising women's consciousness on reproductive rights and access to safe abortion services should be a priority. Giving women the tools they need to make healthy choices regarding their reproductive health is the goal of many community outreach and education initiatives.

Maintain Confidentiality

Improving young people's access to sexual and reproductive health care and resolving the identified barriers requires comprehensive policies that should be incorporated into the National Health Policy. The unique needs of young people in terms of counseling and services related to sexual and reproductive health should be acknowledged and addressed by these regulations. In order to foster an atmosphere of trust where young people feel safe asking for help and opening up about delicate health issues, it is crucial to ensure confidentiality.

In order to ensure that young people feel heard and that their issues will be adequately addressed, it is necessary that health service workers receive training in providing sympathetic, non-judgmental, and confidential support. The reproductive rights of the youth can be respected and promoted by a supportive healthcare framework that is established by prioritizing these changes (Santhya, & Jejeebhoy, 2007).

Ensure Youth Friendly Care

Promoting youth-centered care that educates and empowers young people to exercise their reproductive and sexual health rights must be a top priority. Sexual and reproductive health care should be a right, not a privilege, for all people, including young people, and this strategy promotes that idea. In order to guarantee that teenagers may get the medical treatment they need, providers should undergo training in polite and transparent communication. It is the responsibility of both communities and schools to educate youth about their rights and available resources through various educational programs. A welcoming, rights-conscious space can empower young people to make educated choices regarding their reproductive and sexual health without fear of judgment or stigma (Santhya & Jejeebhoy, 2007).

Acknowledge Abortion as a Fundamental Right

All women should have the freedom to choose how they want to handle their reproductive health, and that includes the right to have an abortion. There would be stronger legal protections and fewer obstacles to safe abortion services if abortion was recognized as a basic right. These suggestions are an attempt to build a more effective system that helps women's reproductive health and remove the major obstacles to safe abortion treatment in India.

CONCLUSION

The chapter addressed the complex relationship between legal frameworks and the empowerment of women through the safeguarding of their sexual and reproductive rights. It has significant implications for both policy and practice. This allowed for a deeper understanding of the impact of laws on women's sexual and reproductive rights, gender equality, and empowerment. The study determined which are the best techniques in India, which may be used as a model by other areas facing similar difficulties. This work is expected to provide a considerable viewpoint on the impact of laws on women's empowerment, including the benefits and drawbacks of legal arrangements. The chapter offers a deeper understanding of the way laws impact women's sexual and reproductive rights as well as their empowerment.

The findings lead the authors to offer specific policy recommendations for the nation, and they may serve as a model for other nations hoping to increase women's empowerment through legislative measures. The chapter aims to advocate for change and identify areas where legal reforms are required to contribute to women's empowerment. This chapter recognized some progress in Indian legislation, but it also highlights the ongoing inequality and barriers that women face while trying to exercise their reproductive and sexual rights to the fullest. It is essential to implement a comprehensive strategy that includes the reinforcement of legal frameworks, the assurance of their effective implementation, and the education of the public in order to create an environment that is more conducive to the empowerment of women. The establishment of a society in which women's rights are not only legally protected, but also actively advocated for and valued, will enable them to lead healthier, more independent lives.

The legislative frameworks that regulate women's sexual and reproductive rights in India, particularly the Medical Termination of Pregnancy (MTP) Act of 1971 and its 2021 amendment, have significantly impacted the discourse on women's autonomy and health. However, these regulations must undergo ongoing modifications to keep up with the dynamic nature of medical technology and cultural norms to guarantee that these laws remain relevant and effective. These legislations need to be updated to fully consider new medical advances and ethical issues, even though they were progressive when they were first passed. According to the study, even with the most recent changes, the law's practical application still exhibits notable gaps. These comprise issues with public knowledge, the suitability of healthcare providers' training, the availability and expense of safe abortion services, and the law's application. These gaps have to be closed if we are to assist women in their sexual and reproductive rights as well as in their empowerment and welfare. The authors analyze international best practices and lessons learned in order to propose policy changes to the current legislative framework. These suggestions assure affordability, improve public awareness and education, and increase access to healthcare by stressing strongly these three aspects. India would be able to link its legal system with ethical values and modern medical innovations by implementing these recommendations, therefore fostering gender equality and enhancing the results of women's reproductive health.

REFERENCES

Anna, G. J. (1989). The Supreme Court, Privacy and Abortion. *The New England Journal of Medicine*, 321(17), 1200–1203. DOI: 10.1056/NEJM198910263211720 PMID: 2677728

Banerjee, D., & Rao, T. S. (2021). Sexual and Reproductive Rights: Time to Ring the Bell? *Journal of Psychosexual Health*, 3(4), 301–304. DOI: 10.1177/26318318211053900

Barge, S. (1997), Situation analysis of medical termination of pregnancy in Gujarat, Maharashtra, Tamil Nadu and Uttar Pradesh (Paper presentation), MTP workshop Ford Foundation, Mumbai.

Berer, M. (2000). Making abortions safe: A matter of good public health policy and practice. *Bulletin of the World Health Organization*, 78(5), 580–592. PMID: 10859852

Bhat, A., & Bhatt, M. (2017). Skewed Sex Ratio and Women's Reproductive Rights in India: Issues and Challenges [IJSR]. *International Journal of Scientific Research*, 6, 860–863.

Center for Reproductive Rights. (2021), https://reproductiverights.org/our-issues/adolescent-sexual-and-reproductive-health-and-rights/

Center for Reproductive Rights. Reproductive Rights in Indian Courts, www.reproductiverights.org

Datar, N. (2020), India Moves to Liberalization of Abortion Rules, *Businessline*, https://www.thehindubusinessline.com/specials/pulse/india-moves-to-liberalisation-of-abortion-rules/article31189432.ece

Datar v. Union of India: Why This case is Important, (2020) HRLN, www.reproductiverights.org/sites/crr.civicactions.net/files/documents/Datar_v_India.pdf

Datta, B., & Misra, G. (2000). Advocacy for Sexual and Reproductive Health: The Challenge in India. *Reproductive Health Matters*, 8(16), 24–34. DOI: 10.1016/S0968-8080(00)90184-8 PMID: 11424247

Devika Biswas v. Union of India, (2016) 10 SCC 726.

Federation of Obstetrics and Gynecological Societies of India (FOGSI). (2002), Safe abortions save lives: understanding the MTP Act (Paper Presentation), MTP Committee, Medico-Legal Committee FOGSI, New Delhi.

Ganatra, B., & Hirve, S. (2002). Induced abortions in a rural community in Western Maharashtra: Prevalence and patterns. *Reproductive Health Matters*, 10(19), 76–85. DOI: 10.1016/S0968-8080(02)00016-2 PMID: 12369334

Ghosh, A., & Khaitan, N. (2017). A Womb of One's Own: Privacy and Reproductive Rights. *Engage!*

Government of India. (1966), Report of the Shah Committee to study the question of legalization of abortion, Ministry of Health and Family Planning, 1966.

Government of India. (1996), Family Welfare Programme in India: Yearbook 1994–95, Ministry of Health and Family Welfare, 1996.

Government of India (1996), The Prenatal Diagnostic Techniques (Regulation and Prevention of Misuse) Act, 1994, Gazette of India, 1996.

Government of India. (2000). *National Population Policy, 2000*. Department of Family Welfare, Ministry of Health and Family Welfare.

Government of India (2003), Medical Termination of Pregnancy Rules and Regulations, Vide GSR 485(E) and 486(E), Gazette of India, 2003.

. High Court on its own motion v. The State of Maharashtra, (2021) 901- SMP 1. 6518 Bom.

Hirve, S. (2004). Abortion Policy in India: Lacunae and Future Challenges. *Reproductive Health Matters*, 12(24, Supplement), 114–121. DOI: 10.1016/S0968-8080(04)24017-4 PMID: 15938164

Hirve, S. S. (2004). Abortion Law, Policy and Services in India: A Critical Review. *Reproductive Health Matters*, 12(24), 114–121. DOI: 10.1016/S0968-8080(04)24017-4 PMID: 15938164

Karkal, M. (1991). Abortion Laws and the Abortion Situation in India. *Issues in Reproductive and Genetic Engineering: Journal of International Feminist*, 4(3), 223–230. PMID: 11651217

Kaur, J. (2012). The role of litigation in ensuring women's reproductive rights: An analysis of the Shanti Devi judgement in India. *Reproductive Health Matters*, 20(39), 21–30. DOI: 10.1016/S0968-8080(12)39604-3 PMID: 22789079

K.S. Puttaswamy v. Union of India, (2017) 10 SCC 1.

Madhok, S., Unnithan, M., & Heitmeyer, C. (2014). On reproductive justice: 'Domestic violence', rights and the law in India. *Culture, Health & Sexuality*, 16(8), 1025–1039. DOI: 10.1080/13691058.2014.918281 PMID: 24927257

Mamta Verma v. Union of India, W.P.(C) 627/2017

Mathew, S. (2021). History vs. Reality: The State of SRHR in India. *The YP Foundation*. Retrieved from https://theypfoundation.org/resources/history-vs-reality-of-srhr/

Medical Termination of Pregnancy (Amendment) Act. 2002, No.64, Acts of the Parliament, 2002 (India).

Medical Termination of Pregnancy (Amendment) Act. 2021, No. 8, Acts of the Parliament, 2021(India).

Medical Termination of Pregnancy (MTP) Act. 1971 No. 34, Acts of the Parliament, 1971(India).

Medical Termination of Pregnancy (MTP) Act. 1971, No. 34, Acts of the Parliament, 1971(India).

Meera Santosh Pal v. Union of India, (2017) 3 SCC 462.

Meera Santosh Pal v. Union of India, 2017 AIR SC 787.

Mittal, R. (2023), India's progressive approach towards abortion laws, *Legal-bites,*https://www.legalbites.in/topics/articles/indias-progressive-approach-towards-abortion-laws-345469?infinitescroll=1

. Narain, S., (2020), Aftermath of the Puttaswamy Judgement: Reproductive Rights in India, *Vivekananda Journal of Research,* 9 (1).

Nehra, S. S., & Rajput, A. S. (2019). The Legal Personality of an Unborn Child: A Comparative Analysis of USA & India. *Amity International Journal of Juridical Sciences*, 5, 104.

Nowicka, W. (2011). Sexual and reproductive rights and the human rights agenda: Controversial and contested. *Reproductive Health Matters*, 19(38), 119–128. DOI: 10.1016/S0968-8080(11)38574-6 PMID: 22118146

Petchesky, R. P. (1991). *Abortion and Women's Choice: The State, Sexuality and Reproductive Freedom, The Northeastern Series in Feminist Theory*. Northeastern University Press.

Phadke, S. (1998), Pro-Choice or Population Control: A Study of the Medical Termination of Pregnancy Act, Government of India, 1971, http://doccenter.org.in/docsweb/writers/shilpa-phadke1.htm

Pillai, A. V., & Kostruba, A. (2021), Women's Reproductive Rights And Their Scope Under International Legal Frameworks, *Entrepreneurship, Economy and Law*, 8, 18-28, https://hal.science/hal-03397078.

Priyanka Shukla v. UOI & Ors., (W.P.(ST) No. 36727/2017).

Rani, S. (2022), Reproductive Rights of Women in India: An Overview, *IJCRT*, 10.

Roe v. Wade, (1973) 410 U.S. 113.

Saheli Women's Resource Centre. Reproductive Rights In The Indian Context, https://www.saheliwomen.org/health/reproductive-rights-in-the-indian-context

Santhya, K. G., & Jejeebhoy, J. S. (2007). *India Young people's sexual and reproductive health in India: Policies, programmes and realities*. Population Council.

Sharma, S. (2015). Reproductive Rights and Women's Health: A New Prospective. *International Journal of Humanities and Social Science Invention*, 4(4), 51–56.

Shri Shantilal, H. S. (1967). *Report of The Committee to Study the Question of Legalization of Abortion*. Ministry of Health and Family Planning.

Singh, A. (2021). Abortion Law in India in the light of Medical Termination of Pregnancy (Amendment) Act, 2021. *Nyaayshastra Law Review*, 1(3), 1–13.

Suchita Srivastava v. Chandigarh Administration, (2009) 14 SCR 989, (2009) 9 SCC 1.

Swati Agarwal & Ors. V. Union of India, W.P.(C) 825/2019

Tapasya Umesha Pisal v. Union of India, 2018 (12) SCC 57.

The Bharatiya Nyaya Sanhita (BNS). 2023, No. 45 Acts of the Parliament, 2023 (India).

The Medical Termination of Pregnancy (Amendment) Act, 2021 (Act 8 of 2021).

United Nations. (1994), Population and Development: Programme of Action adopted at the International Conference on Population and Development Cairo, 5–13 September 1994, Department of Economic and Social Affairs.

Vijay, P., & Sharma, U. S. (2002). Conceptual Contours of Human Rights and Reproductive Freedom of Women in India. *Journal of Coastal Life Medicine*, 1(11), 594–600.

WHO. (2021), India's Amended Law Makes Abortion Safer And More Accessible, https://www.who.int/india/news/detail/13-04-2021-india-s-amended-law-makes-abortion-safer-and-more-accessible

WHO. (2024), https://www.who.int/news/item/16-05-2024-sexual-and-reproductive-health-for-all-20-years-of-the-global-strategy

Wilson, K. (2018). For reproductive justice in an era of Gates and Modi: The violence of India's population policies. *Feminist Review*, 120(1), 23–39. DOI: 10.1057/fr.1982.13

X v. Principal Secretary Health and Family Welfare Department, (2022) 2022 SCC OnLine SC 1321.

Chapter 11
Women at the Crossroads of Climate Change:
Advancing Gender Equality Through Climate Action

Tanmeet Kaur Sahiwal
SVKM's Narsee Monjee Institute of Management Studies, Bengaluru, India

Jay Kumar Gupta
SVKM's Narsee Monjee Institute of Management Studies, Bengaluru, India

ABSTRACT

"We are still living with the results of millennia of patriarchy that excludes women and prevents their voices from being heard," - António Guterres. In underprivileged countries including Ghana, Egypt, Costa Rica, and Malawi, women face such challenges and risks regularly as per the IMF Report. Unique challenges have often been faced by women and girls when it comes to climate disasters because of deep-rooted inequalities, especially towards girls and women which make them much more vulnerable to its effects. Issues in health care, food security, earning for living, and being forced to migrate often show up because of disasters. Hence, it would be correct to say tackling climate change needs to be equal from the ground up, or everyone suffers. Including women and girls in the climate policies will make solutions more practical and relevant. It will consider the needs of everyone, not just a select few. Empower girls and women to be not just victims, but as an active participant in shaping a sustainable future and will make the UN Climate Change Conference (COP 26 & 33) successful.

DOI: 10.4018/979-8-3693-6069-9.ch011

1. INTRODUCTION

It will not be incorrect to say that the global climate crisis isn't just an environmental issue but it's a stark example of injustice. The recognition of this disparity is crucial as the burden is further amplified for marginalized groups within these communities, particularly women (United Nations, 2009) and girls despite they are those who contributed least to greenhouse gas emissions (U.N. Women,2023) but bearing the highest of its impact (U.N. Women, 2024).

Prioritizing and addressing the gendered impacts of climate change becomes an essential pillar as the UK takes on the responsibility of leading COP26. It should be done alongside emission reduction, financial support, and collaborative action. The fundamental to achieving a successful COP26 and building a climate-resilient future is to empower women and girls through education, economic opportunities, leadership roles, and decision-making. Women face unique challenges and heightened risks due to climate change in countries including Ghana (U.N.D.P., 2019) Egypt, Costa Rica, and Malawi.

Studies like "Adapt Now" from the Global Commission on Adaptation further stress how ignoring gender inequalities in adaptation efforts intensifies the existing vulnerabilities and creates new forms of exclusion because for the clear reason that the path forward demands a paradigm shift.

It requires addressing the broader context of inequality, discrimination, and sociocultural barriers that disproportionately burden marginalized groups to tackle climate change effectively.

A recent UN policy report opens the window to a new way to face this crisis. i.e., putting women and diverse voices at the heart of the solution. The report highlights how countries are finally starting to achieve gender-neutral climate policies. Countries are working on climate adaptation plans that listen to women's voices and consider their specific needs and strengths. The report further highlights that even LGBTQ+ communities and other diverse communities often face their own set of challenges but they hold their reservoirs of resilience (U.N. Women Watch, 2024).

2. PROBLEM STATEMENT

The intensifying climate change has its adverse impacts which disproportionately burden women and girls and promote existing gender inequalities. Systemic barriers, including resource limitations, decision-making exclusion, and unequal access to financing, hinder their full participation in climate action strategies despite their unique knowledge and leadership are crucial for building climate resilience. Suppose, A heavy storm hitting a village causes heavy rain flooding the fields and washing away the crops. In such a situation, a man will struggle to secure livestock and boats while women will worry about the protection of their children and feeding their families from the rising water. This can be understood as one of the examples of how climate change can impact people differently based on their gender.

It presents a critical opportunity to address this injustice by integrating gender-responsive solutions into all aspects of climate change mitigation, adaptation, and financing as decided in COP26 and beyond. Failure of such will further marginalize women and girls further jeopardizes the effectiveness of climate action, and impede progress towards a just and sustainable future for all.

This problem statement highlights the urgency of addressing gender inequality in the context of climate change by emphasizing the need for concrete action at COP26 and beyond and also identifies key barriers women face and emphasizes the potential benefits of gender-responsive approaches.

3. GENDER EQUALITY AND CLIMATE CHANGE IN DECISIONS AND CONCLUSIONS OF THE UNITED NATIONS FRAMEWORK CONVENTION ON CLIMATE CHANGE

The decision made in COP7 in 2001 marked the first official recognition of the critical need for gender equality in climate governance. It specifically advocates for increased female participation and representation in bodies established under the UNFCCC and Kyoto Protocol (The Marrakesh Ministerial Declaration, 2001, p. 3).

The decision made in COP16 in 2010 highlighted the crucial role of gender equality and women's active participation in effectively addressing all aspects of climate change which marks a vital recognition that inclusive solutions require diverse perspectives and experiences to tackle this complex global challenge. (Cancun Agreements, 2010)

The decision made in COP18 in 2012 highlighted the need to boost women's involvement in UNFCCC negotiations and achieve their equal representation in bodies established under the Convention and Kyoto Protocol.(Doha Climate Gateway,2012) This step was aimed at creating climate change policies that effectively address the needs of both genders and recognize that diverse perspectives and experiences are essential for crafting truly impactful solutions by laying down more inclusive and equitable climate policies that benefit everyone.

In 2014, COP20 under "Lima Work Programme on Gender." took a significant leap forward to double down on efforts to increase women's participation in the UN Climate Change process and integrate gender considerations into the development and implementation of climate policies (Lima Call for Climate Action,2014). Moreover, Countries will collaboratively develop a two-year work program focusing on gender equality and climate action and establish a clear definition of "gender-responsive climate policy" to optimize its practical application.

4. GENDER EQUALITY AND CLIMATE CHANGE IN THE GLOBAL CONTEXT

The New Climate Institute (2015) claimed that the countries have been making promises for over 20 years, aiming to keep global warming in check and achieve sustainable development but yet CO2 levels skyrocketing and temperatures still rising dangerously (Le Quere et al, 2012). Current plans are not enough to tackle climate change as it already hurting our planet and jeopardizing our goals for a better future.

As per the Report of the UNDESA (2015), Developed countries need to step up and lead the way, just like they agreed in the 2030 Agenda and the Addis Ababa Action Agenda. It's about our future which is more than just about numbers and targets. The unplanned actions can't afford to keep going down the same path as it's time for real action, real commitment, and a real chance to build a sustainable world for everyone. The next COP meeting must be a turning point, not another missed opportunity.

Women are often the silent superheroes on the front lines when it comes to tackling climate change as they possess valuable knowledge about managing resources sustainably, be it at home, in their communities, or even on a national and global scale. They're the ones leading the charge for sustainable practices and spearheading responses to climate disasters like droughts hold them back and lack of access to resources, education, and decision-making power. The OECD report (2015) claims that it is the inequality that limits their ability to cope with climate change and contributes to solutions. There's already a solid foundation to build on as over 50 international agreements recognize the importance of gender equality in climate action.

Our current economic system, with its business practices, consumption habits, and governance structures, is widening the gap between the rich and poor, and hurting the environment along the way, hence the upcoming COP comes at a critical time when overlapping problems are hitting harder than ever before. It expects unpaid care work from women, while valuing paid work and money-making much higher in the the way our economy works depends on and fuels gender inequality. It's like multiplying existing problems, making our already unequal world even more unfair (United Nation Climate Change, n.d.).

Women in developing countries, who often live in risky situations and take care of families during crises are the ones who get hurt by it more. Climate change throws fuel on this fire and people already struggling are hit the hardest, and women often face even greater risks due to discrimination and other factors (TechnoServe, n.d.)

One positive out of it could be that climate change doesn't have to worsen existing inequalities, especially those affecting women. The upcoming COP26 and 33 climate agreement could be a springboard for positive change, a chance to fix our broken systems and create a fairer future for everyone despite it won't be an easy task (UN Women, 2023).

There need to make gender equality and human rights the guiding principles of this new agreement and every aspect, from reducing emissions to adapting to change, from financing projects to building capacity, needs to be deeply interwoven with these principles (Sébastien Duyck e.t. all (n.d.))

It is evident that every climate change policy impacts women differently, hence there is a need for solutions that both cut emissions and tackle unfairness. One could take different paths including we could stick with the "business as usual" approach, hoping markets and good intentions will fix everything, or trying a "greener than usual" approach, encouraging businesses to go "green." But we need a radical shift if we truly want to achieve sustainable development, end poverty, and protect our planet.

As per Utlities One (2023), Fixing climate change and creating a just society aren't separate problems - they're two sides of the same coin. We're at a crossroads. A complete rethink is required of how we run the world, moving towards fairness, human rights, and sustainability. It might be achieved by reducing our reliance on fossil fuels and other harmful energy sources and building new economic and government systems that prioritize equality, justice, and the environment.

5. KEY ISSUES FOR GENDER RESPONSIVE CLIMATE ACTION FOR SUSTAINABLE DEVELOPMENT

Women often face the brunt of the climate change impact due to limited access to resources, decision-making power, and social networks. Understanding these inequalities and how they play out in areas like resource access, governance, and technology is important to tackle this issue hence gender mainstreaming is required.

It's about integrating gender considerations into every aspect of climate action, ensuring policies and solutions are not only effective but also fair and equitable for everyone. Our approaches become more efficient, impactful, and beneficial for all when we consider women's needs and perspectives. But it's not just about numbers and efficiency as climate action needs a human face and also one can't lose sight of the real people affected by this crisis. So, it advocates for a human rights-based approach, prioritizing the well-being of all, regardless of gender (Bridget Burns and Amy Lee, 2015).

Women are key players in fighting climate change but turning good intentions into real action needs a boost. Governments at all levels need to clearly state their commitment to gender equality in climate action.

As per the Report of UN Women Watch (1997), While political will at all levels is crucial, a strong intergovernmental mandate specifically within the UNFCCC framework would send a powerful message and accelerate progress as effective gender mainstreaming in climate action requires more than just good intentions. The other essential includes clear institutional policies with teeth, meaning measurable accountability and consequences for inaction. The misconception that simply including "women" equates to effective mainstreaming persists, and current metrics primarily focus on process and outcomes rather than fulfilling rights obligations or considering intersectional factors like age and class. Moreover, Vague nods towards gender equality are evident, but prioritizing it as both a goal and a means to achieve other objectives often falls short.

Fragmented national policy frameworks further hinder systematic integration across sectors and levels. It requires a shift away from isolated women-focused projects towards policy-driven approaches deeply embedded within both national climate change and gender equality strategies.

Research shows that ignoring gender in climate action leaves women and girls behind and this gap between promises and reality stems from a lack of systematic gender mainstreaming, even as support for it grows within institutions. Government representatives and gender champions inside the UN climate process are pushing for more focus on gender issues and sharing practical examples. Even, the UN's action plan for gender equality advocates the need for measurable progress (Eleonore Soubeyran, Kamya Choudhary 2023).

However, Hurdles include limited expertise, human resources, and funding, especially within the UN climate secretariat which highlights the critical need for capacity-building in gender mainstreaming to support climate action that includes everyone. Every successful gender mainstreaming strategy from other development areas can be adapted to climate action by sharing knowledge on how to use this approach in policy analysis, research, technical assistance, and monitoring can provide valuable guidance.

As per the Report of United Nations New York(2002), New platforms or expanded existing ones could focus on climate-specific issues while leveraging local platforms to address gender in the context of climate action. This can be understood with an example of how low-emissions capacity-building processes, which allow countries to analyze their specific situations, could also be used for more effective gender integration.

Women's voices are crucial for effective climate action, but their technical expertise on both climate change and gender equality often falls short which limits their advocacy and ability to influence policies. Collaboration is required to fill this gap.

Firstly, it is better to bring together climate and gender experts which would boost efforts to mainstream gender into climate action plans by ensuring that these partnerships would ensure that everyone's needs are considered. Secondly, national gender equality bodies need support to build their capacity and provide crucial leadership in this field. It is essential to integrate gender and climate change strategies for true progress. Finally, a formal expert group, including UN and civil society representatives, could act as a powerful force for good by sharing knowledge, tackling common challenges, and developing practical solutions. Moreover, it could also support countries on the ground and adapt to emerging issues. The UN climate secretariat has a critical role to play, but it shouldn't be the sole driver of change and hence an expert group, empowered and connected, ensuring everyone benefits from a healthy planet, can be the cornerstone for transformative action (Eric Zusman e.t. all, 2024).

A new approach that considers gender issues as "co-benefits" of climate solutions, meaning positive side effects alongside emissions reductions has raised concerns about the growing consensus to address climate change through a lens of human rights and gender equality (UNICEF e.t. all,2022). The UNFCCC (2023) says on one side, some see this as a way to promote gender equality through climate action, but on the other side women's groups have expressed fear of marginalization. It has been argued by them that framing gender equality as a bonus, rather than a central element, undermines its importance and women's agency.

It is to ensure that the fight for gender equality remains central to climate action, not just a convenient byproduct hence it's crucial to analyse and refine this approach carefully. These concerns should be addressed to build a climate-resilient future that benefits everyone equally.

6. THE UNEQUAL EFFECTS OF CLIMATE CHANGE

a. Human Endowments

The Unequal Burden of Climate Change on Health and Education casts a particularly long and harsh shadow on women and girls. It poses a threat to their health, education, and overall well-being (Cecilia Sorensen e.t. all, 2018).

As per World Health Organization (2023), There is an increase in climate-related deaths, with an additional 250,000 fatalities per year by 2050 from factors like malnutrition, malaria, diarrhea, and heat stress moreover, even in this grim scenario, marginalized groups, especially women, bear the brunt of the suffering.

When climate-induced resource scarcity hits, then food shortages and malnutrition disproportionately impact women and children. In such scenarios, girls face increased pressure to drop out of school or bear the burden of fetching water and fuel.

It not only compromises their education but also intensifies their workload and limits their opportunities as women often face greater constraints in responding to crises, due to ingrained social disadvantages.

The impact on education is more explicit, with girls more likely to be pulled out of school or experience reduced schooling during droughts or disasters as this can be driven by both a need for additional household labor and discriminatory societal norms (Christina Kwauk e.t. all, 2019).

Girls tend to be disproportionately burdened with additional household chores, limiting their time for learning and further entrenching gender inequalities even when boys and girls are equally affected by weather events in terms of test scores and cognitive skills.

b. Economic Opportunities

Women are often overrepresented in jobs linked to nature like agriculture, fisheries, and forestry. Moreover, these sectors tend to be low-paying, insecure, and highly vulnerable to climate fluctuations. Hence, in the cases of droughts, rising sea levels, and extreme weather events wreak havoc on these very resources, stripping away income opportunities and jeopardizing food security for entire communities (Shagun, 2023).

The burden falls especially heavily on women, who already shoulder the responsibility of water and fuel collection in many societies and as these resources dwindle, they're forced to travel further and work longer, leaving less time for education, earning income, and other essential activities.

Women are often left behind to manage precarious situations and provide for their families even when men migrate in search of new opportunities due to climate shocks which further exposes them to risks like exploitation, trafficking, and limited access to healthcare and safety nets (Varsha Torgalkar,2022).

c. Voice and Agency

As communities grapple with the shocks of extreme weather, environmental degradation, and conflicts driven by climate stresses, a growing body of evidence points to a worrying surge in Gender Based Violence (Craig Brierley,2022).

The devastating consequences of climate change can cast a long shadow on the lives of women and girls through increased risk of gender-based violence (GBV) and in desperate situations, families may resort to harmful practices like child marriage to cope with climate disasters, which further jeopardizing the futures of young girls.

As per The Office of the High Commissioner for Human Rights (2022), Such GBV inflicts personal pain and suffering and also weakens communities' ability to adapt to climate challenges and it further creates a vicious cycle where vulnerability begets violence, and violence weakens the ability to cope with further environmental pressures.

7. NATIONAL CASE STUDIES

The case studies of some countries like Mexico, Nepal, Malawi, and Sudan offer valuable insights into how gender mainstreaming affects national climate policies.

a. Mexico

As per the research conducted by UNDP (2024), Mexico has strong legal frameworks for both gender equality and climate change, even including some gender-focused provisions in its climate policies. One of the crucial gaps that can be found is a lack of collaboration between gender and climate experts which means gender wasn't fully considered in climate action.

Mexico fostered collaboration between these experts to ensure gender perspectives were effectively integrated after elaborative consideration. Moreover, the need for a dedicated budget for gender-responsive climate action was recognized. Previously, separate budgets existed for gender equality and climate change which normally used to create a funding gap. Mexico identified this gap and ensured adequate resources for effective gender mainstreaming by establishing a dedicated budget (Sejal Patel et all, 2021).

b. Sudan

A country like Sudan offers practical examples of integrating gender into climate action plans. The country considers gender in its national adaptation plan and REDD+ (forest conservation) programs. This is based on their experience which highlights the need for increased knowledge and capacity on gender issues and applies to everyone involved, from government agencies and stakeholders to local communities (Chingarande et al.,2020).

c. Nepal

According to the research conducted by Asian Development Bank(2021), Nepal focuses on gender in its technology needs assessments, particularly in renewable energy as the combined approach is their key takeaway is the importance. There is support for strong political commitment and policies that temporarily empower women, along with dedicated budgets for implementation and education programs.

Moreover, grassroots inclusion, removing cultural barriers through education and empowerment programs is equally crucial which means actively involving women in needs assessments, decision-making, and project implementation. Eventually, success finds its support in supporting women's agency and challenging traditional mindsets, both among men and women.

d. Malawi

In the further study of the Report by Asian Development Bank (2020), Malawi's give their focus on the forestry sector which demonstrates the importance of tailoring gender mainstreaming to local contexts. The approach involved conducting gender analyses, developing action plans based on specific needs, and seamlessly integrating these plans into project work and emphasizes the need for flexibility and adaptation to ensure interventions are relevant and impactful.

Such case studies showcase the multifaceted nature of successful gender mainstreaming in climate action.

8. ROLE OF CIVIL SOCIETY ORGANIZATIONS AND PRIVATE SECTOR IN GENDER INCLUSIVITY

Civil society organizations (CSOs) and the private sector have major roles in ensuring the success, accountability, and gender inclusivity of climate initiatives. Moreover, Climate change demands more than just government action as developmental organizations at various levels, from national to community-based, can be powerful supports for mainstreaming gender into climate action as they can push for equitable climate financing that prioritizes women-led initiatives and projects addressing women's specific needs.

- Building capacity with a gender lens is crucial, equipping communities, particularly women, with the knowledge and skills to adapt to climate impacts and lead resilient practices.
- Monitoring and reporting violations of gender rights within climate action ensures accountability (Halima Ansari, Smarinita Shetty, 2023).

Women are not simply beneficiaries, but key players in driving climate solutions hence, by supporting and empowering women climate entrepreneurs, we unlock their potential to develop innovative solutions and lead positive change within their communities.

9. CASE STUDY: BHUNGROO: EMPOWERING WOMEN, GREENING FARMS IN GUJARAT

As per the study conducted by United Nation Climate Technology Centre & Network(2019), In the parched lands of Gujarat, India, an innovative tool called Bhungroo is transforming lives and landscapes. It is designed by NGO TANs. It is not considered just a piece of technology but an empowerment tool for rural women.

Bhungroo is an irrigation system that harvests and stores rainwater, enabling year-round cultivation even during dry seasons. TANs train and empower local women's self-help groups (SHGs) to manage and own these systems apart from not just offering the technology and it not only provides much-needed water security but also grants women crucial agency and economic independence.

"*Even though women lack land rights, they now have rights to the tools that make the land useful.*" - Trupti Jain.

When they own and operate the Bhungroo system, it strengthens their bargaining power within their families and communities. Moreover, it allows them to participate in decision-making about agricultural practices.(United Nation Climate Change, 2023)

9.1 Case Study: Women Shgs Trained as Climate Entrepreneurs

In the heart of Rajasthan, eighty women from four self-help groups (SHGs) are stepping into the powerful new role of "solar engineers."

Under the Rajasthan Gramin Aajeevika Vikas Parishad (Aajeevika) initiative (2011), these women have received training to install small-scale solar power systems in homes and build solar lamps. Such a step will bring light and clean energy to their communities because this is more than just a technical skill.

It's an empowerment story which demonstrates how investing in women can drive positive change on multiple levels. Aajeevika is not only promoting renewable energy and combating climate change but also creating economic opportunities for women in rural communities by equipping these women with the knowledge and tools to become "solar engineers."

The PWC Report (2022) further highlights that these newly trained entrepreneurs gain financial independence, leadership skills, and a sense of agency within their communities. Furthermore, wider access to clean energy improves the lives of their families and neighbors, contributing to better health, education, and overall well-being.

10. THE PRIVATE SECTOR: BUILDING A FAIRER AND GREENER FUTURE FOR ALL

The private sector needs to step up and make even bolder commitments toward creating a more inclusive economy, thus truly achieving social justice and upholding human rights. It can be achieved by accelerating initiatives that empower women, promote responsible supply chain management, ensure a climate-just transition, and invest in vital health infrastructure.

It is crucial to support organizations and companies that build resilience across their supply chains because it will ultimately protect vulnerable communities and women across all sectors.

Acknowledging the intricate link between corporate structures and gender equity in climate action is the need of the hour. It will be strategically correct to put women at the forefront of the climate narrative because it will pave the way for more holistic solutions that tackle climate challenges while empowering both women and men from the most vulnerable communities after centering women's voices and experiences. It will create a win-win situation for business, society, and the environment.

To keep global temperatures under 1.5°C, the transition must be there to clean energy by 2050. It is also correct that it will come with a hefty price tag of USD 131 trillion.

Mobilizing private sector investments is equally vital as public funding to ramp up climate action as it presents a unique opportunity, coinciding with a shift in business values towards socially responsible investments. More companies are adopting ESG strategies and pledging net-zero emissions which demonstrates a growing awareness of their environmental impact.

Additionally, there's a strong interest in investing in climate solutions that benefit vulnerable groups, particularly women. However, concerns remain about the private sector's ability to address systemic gender gaps in climate response despite their potential to contribute is undeniable (Yannick Glemarec e.t. all, 2016).

At present many companies may be investing in climate solutions but the questions that should be asked are-

- Are they simultaneously addressing the specific needs and challenges faced by women?
- Can their investments truly empower women and promote gender equality in a meaningful way?

Women in leadership positions have a unique opportunity to champion gender-responsive climate action as businesses increasingly prioritize ESG factors and strive for net-zero emissions. Even various research suggests that women leaders are more likely to prioritize sustainability efforts which make them valuable allies in the fight against climate change.

However, despite their potential, women remain significantly underrepresented in positions of power, particularly within carbon-emitting industries as such underrepresentation presents a double challenge. Here we are missing out on the two possible fronts. Firstly, on the valuable contributions women leaders can make to climate action and secondly on the lack of gender diversity at the decision-making table risks perpetuating the very systems that disproportionately impact women and girls when it comes to climate change.

Therefore, it's the need of the hour to dismantle these barriers and empower women to take their rightful place at the forefront of climate action.

10.1 A Case Study in Gender-Responsive Energy Access

Under United Nations Development Programme Sustainable Energy (2001), the Unted Nations presented case studies about a network spanning 18 countries across Asia and Africa, united by a mission to empower women and illuminate their communities through clean energy. The WEE program, a groundbreaking initiative that focused on developing women's enterprises in the renewable energy sector wasn't just about providing access to electricity but also it was about creating lasting economic opportunities for women.

The IDS Bulletin (2020) states that the WEE Programme was implemented between 2014 and 2018 and the program spanned seven countries: Indonesia, Kenya, Nepal, Nigeria, Senegal, Tanzania, and Uganda. There were Local partners like the Centre for Rural Technology Nepal, Energy 4 Impact, Kopernik Solutions, Practical Action Eastern Africa, and Solar Sister joined forces to make it happen.

The WEE programcarefully considered the unique needs and contexts of each community by providing training, resources, and market access moreover, the program empowered women to become entrepreneurs in the clean energy sector by not just selling solar lamps or installing solar panels, but also building thriving businesses that brought light and economic independence to their communities (Soma Dutta,2019).

10.2 Case Study: Women Lead in Clean Energy Entrepreneurship

In furtherance with the WEE program which wasn't just empowered women but created a ripple effect of positive change by partnering with local organizations. This network provided critical support to over 4,000 women which enables them to launch and grow their clean energy businesses across seven countries.

It will be worth noting that these weren't large corporations, but rather micro-level enterprises, each one a testament to the power of individual initiative the impact they brought was far-reaching as they brought clean energy products and services to over 2.9 million people, primarily in rural and low-income communities.

It led to brighter homes, improved health, and economic opportunities for millions who previously lacked access to reliable energy.

Hence, The WEE program empowered entire communities. It successfully proved even small-scale, women-led initiatives can illuminate the path towards a more sustainable and equitable future.

11. CLIMATE FINANCE IS DRIVING WOMEN'S ENTREPRENEURSHIP – THE MSME WAY

As per Niti Aayog(2022), In India, businesses, with over 20% of all MSMEs being women-owned and employing 24% of the sector's workforce, play a crucial role in empowering women economically and building climate resilience.

MSME's strengths lie in their ability to reach women in rural and urban areas by making them ideal channels for introducing climate-smart technologies and products as their deep understanding of local needs allows them to develop solutions that resonate with women and accelerate their adoption.

It can be better understood with an example, imagine a woman-owned business selling solar lamps directly to rural communities. It will not only provide essential clean energy, but also empower the business owner and create jobs within the community as these businesses can provide vital financial resources, decision-making opportunities, and improved well-being for women, which ultimately strengthens their ability to adapt to climate change.

As per the Annual Report of the Ministry of Micro, Small, and Medium Enterprises (2020-2021), Despite MSME's significant contribution to the Indian economy, most women-owned MSMEs struggle to grow beyond micro-enterprises. Furthermore, the Report of the International Finance Corporation World Bank Group (2020) claims that women are also vastly underrepresented in the early-stage entrepreneurial landscape and this disparity stems from multiple challenges.

One major hurdle is the lack of reliable data on climate-friendly products and services offered by women-led MSMEs which makes it difficult to identify and support successful models secondly, securing funding remains a significant obstacle as women face a staggering finance gap of INR 6.37 trillion, often resorting to informal and unreliable sources due to limited access to traditional financing mechanisms (Buteau S, 2021).

Factors like limited awareness of women-owned businesses, perceived higher risk profiles, and difficulties providing collateral contribute to the same as these roadblocks not only stifle women's entrepreneurial dreams but also hinder their potential to contribute to climate solutions.

We advocate for a multi-pronged approach that ensures improving data collection, facilitating access to formal funding avenues, and addressing biases against women entrepreneurs. These crucial steps unlock a powerful force for both economic growth and sustainable development and also steps toward building a more inclusive and resilient future.

The critics say that securing climate finance can feel like scaling a mountain for women entrepreneurs in India because accessing resources specifically designated for climate action presents even greater hurdles than traditional funding options (Parnika Jhunjhunwala, 2023).

Most climate finance in India flows through formal channels like national and international funds, private investments, and government budgets. Moreover, these mechanisms often come with strict rules and limited flexibility which make it difficult for individual MSMEs, especially those led by women, to qualify for direct funding also the challenges are there for complex procedures and requirements attached to many such schemes (Kezia Yonzon and Prakhar Bisht,2023).

The recent report of World Bank highlights that entrepreneurs might need to navigate extensive paperwork, upfront costs, and energy audits before and after receiving funding as these hurdles are particularly daunting for women-owned MSMEs, who often lack the technical expertise and financial resources to comply. Even eligibility criteria for accessing climate finance vary inconsistently across different schemes, creating further confusion and frustration.

The current system of climate finance suffers from a fragmented policy framework, lacking a clear roadmap for channeling funds towards MSMEs as this, coupled with the limited awareness among women entrepreneurs, especially those in remote areas, about existing climate-focused schemes and programs, creates a significant barrier.

Even when information is available, low financial literacy makes it difficult to leverage these opportunities effectively. It is essential to bridge this gap and empower women to contribute to climate action.

For that clear and accessible policy directions with simplified eligibility criteria, designing financial products tailored to the strengths and weaknesses of women entrepreneurs, and enabling microfinance institutions and non-banking financial companies to participate have to be incorporated.

11.1 The Gender-Differentiated Impacts of Climate Action

The PWC report(2022) highlights that the climate action by encompassing both mitigation and adaptation strategies often opens doors for better jobs, green innovations, and the chance to address long-standing inequalities. The shift towards low-carbon economies presents a unique opportunity to tackle climate change while simultaneously closing the gender gap and empowering women.

It is suggested that climate investments can create millions of jobs globally by 2030, offering significant potential for women to gain greater labor market participation. Moreover, a current global gender gap in labor force participation of over 25% highlights the vast room for improvement by further enhancing this potential "just transition" principles that aim to ensure a sustainable and inclusive shift towards eco-friendly economies (Zineb Sqalli et all, 2021).

Such principles support decent work for all, social inclusion, and poverty eradication. It has a specific focus on empowering women and promoting locally-driven economies but gender-blind approaches to climate action can unintentionally widen existing divides as ignoring gender considerations risks solidifying economic and social inequalities. It further disadvantages women across various spheres of life and without dedicated interventions, women are likely to be left behind in the low-carbon transition.

Women are historically underrepresented where the anticipated skills needed for emerging "green jobs," which often align with fields like science, technology, engineering, and mathematics (STEM) – areas. Such underrepresentation could perpetuate patterns of occupational segregation. A recent study shows only 62 women for every 100 men possessing the necessary "green talent" and sectors expected to see significant green job growth, such as energy, manufacturing, construction, and transport, are traditionally male-dominated due to factors like discriminatory policies, societal norms, biased hiring practices, and lack of support systems like childcare (Ada Zakrzewska and Lana Fitzgerald,2023).

A golden opportunity presented by the green revolution to decouple economic growth from environmental harm. However, women entrepreneurs face immense hurdles in accessing the resources needed to contribute. Women-owned businesses are often left out due to persistent gender gaps in financing despite the rapid increase in investments in climate technologies (UNDP, n.d.).

A staggering 32% of the $5 trillion financing gap – a significant chunk of untapped potential is present globally even after women-led businesses constitute 23% of all micro, small, and medium enterprises (MSMEs). Data present more shocking insights, women received only 2% of venture capital funding in 2020 and make up just 30% of the world's scientists which intensifies the stifled ambitions and limited opportunities for women in the green technology space.

It is also evident that there is a lack of gender inclusivity in the development of green technologies and climate solutions. It means they often disregard the specific needs and perspectives of women and communities as this not only hinders women's entrepreneurial potential but also reduces the overall effectiveness of climate solutions by failing to address the vulnerabilities women often face (UN Women Watch, n.d).

12. RECOMMENDATIONS/ SUGGESTIONS

The disproportionate impact of climate change on women calls for immediate corrective measures. Other than the fact that climate change affects women and girls in a magnified deleterious fashion, it also has a 'multiplier' effect. We have seen through this paper how climate change adversely affects the health and economic status of women.

The United Nations Sustainable Development Goals emphasize both – actions to combat climate change, as well as gender equality and the empowerment of women. Cross-integrating both, we can easily conclude that integrating gender perspectives into climate impact mitigation and adaptation techniques is a must. Mitigation of climate-induced impact on women can be categorized into 2 types – macro and micro.

Measures at the macro level concentrate on leadership and policy. It is imperative to not only include women in leadership talks regarding climate and climate change but also to enable women climate leaders to emerge from the grassroots level. Positioning women as leaders in the climate justice movement will mean recognizing their ability to influence a larger section of the population – thereby accelerating the movement toward a more sustainable future.

Additionally, while drafting policies we cannot continue to have a myopic and linear view of the impact of climate change. Our policies cannot be formulated as a 'one-size fits all' solution and shall have to be expanded to include solutions for gender-specific climate impact. The policies can be targeted at climate impact mitigation, as well as alleviation of the consequences of climate injustice. New policies focused on prohibiting anti-environment activities, and are pro-women are the need of the hour.

At the micro level, we can spearhead positive change by recognizing the impact of women as carriers of knowledge. Women traditionally transfer knowledge from one generation to another by building indigenous practices. Since informal economic activities and domestic responsibilities are mostly shouldered by women, they also share knowledge with successive generations by example and by sharing responsibilities. Thus, for any change to be deep-rooted and conditioned in everyday behavior, we must recognize the role of women as knowledge transferors and initiate action to strengthen the same by understanding the challenges faced by women due to unequal climate change impacts.

13. POLICY SUGGESTIONS

Policy Suggestion 1: Gender-Responsive and Inclusive Macroeconomic and Growth Policies

Economic policies need to encourage people and businesses to use fewer resources and produce less waste. At the same time, the fight against climate change and creating a fair future go hand in hand. These policies should create good jobs with decent pay, especially for those who are most in need as everyone deserves a chance to contribute including women and other under-represented groups. They have a lot to offer, and their skills and ideas are crucial for building a sustainable economy that creates plenty of jobs. Efficient economic strategies have the potential to lower carbon emissions (ILO, 2023a). They serve as a guide and improve our readiness for climate-related issues. Furthermore, because these policies promote investment in clean energy, innovative technology, and ecologically friendly manufacturing techniques, they also guarantee a seamless transition to a greener future. Additionally, it can help organizations that provide education and training to provide people with the skills they need to get new employment. Economic initiatives of this kind have the potential to strengthen our economy's resistance to climate change. It promotes gender equality and protects vulnerable people by making the world cleaner and more equitable (ILO,2024).

Adopting A Gender Lens On Macroeconomic And Growth Policies In Just Transition

Making the appropriate investments is essential for a just and sustainable future. It entails allocating funds for public services including developing necessary infrastructure, rehabilitating natural ecosystems, and creating smarter energy networks. People are shielded from climate change by it. To encourage more women to enter the workforce, social initiatives like those that assist in paying for childcare and necessities must also be funded. Everyone needs to consider these climate measures. The impact of carbon emission taxes varies between genders. These levies create money that can be put toward social initiatives or used to incentivize employers to hire more women for environmentally friendly employment. Governments buy a lot of things so they can use their purchasing power to encourage businesses to be more sustainable and promote gender equality. For example, the government can choose to buy from companies that are good for the environment and treat men and women fairly. Buying from such sustainable businesses, and using taxes strategically can help in creating a cleaner economy with more opportunities for everyone, especially women. It will not only help us in fighting climate change but also build a fairer world.

Point of Action Needed at Policy Level: -

Economic policies need to consider the challenges faced by different groups. At present some policies make it harder for women and others for good job opportunities. Such policies often overlook the unpaid work people do at home, like childcare and cleaning. Hence, one needs to take these things into account when designing economic plans by valuing all the work that has to be done. The things women often do for free contribute a lot to the economy, like taking care of family. It must be mandatory to track this work in official statistics in the same way as tracking of paid jobs is being done. One can create new opportunities for women by investing in the right places i.e. by putting money towards infrastructure like building childcare centers, improving public transportation, or investing in renewable energy. It will help people care for their families and protect the environment(ILO,2023a).

Point of Action Needed at Concrete Measures:

Understanding the challenges before taking action must be given utmost importance. It's important to consider how existing inequalities affect different groups, especially women while creating economic policies. It means looking for unfair biases that make it harder for some people to get the resources they need to succeed. Moreover, money must be tracked. Government expenditures on different programs have to be analyzed based on the impact it poses to men, women, and other groups differently which will help us design policies that benefit everyone. Good data to understand the challenges people face have to be collected for better understanding which will include information on how people spend their time, how climate change affects them, and the specific needs of women and other under-represented groups. It's important to listen to the people who will be affected when making economic decisions which includes women and other groups who are often left out of the conversation.

Policy Suggestion 2: Industrial And Sectoral Policies To Enhance Women's Opportunities In Just Transition

Factories and different industries need to be part of the solution to climate change. They can't just focus on making money; they also need to be environmentally friendly. Industries can create new jobs and even grow by investing in green technologies and practices. It is important for places and people who are most affected by climate change. New green policies can help these communities create new businesses and find new jobs in areas like renewable energy or sustainable development which will help protect their livelihoods and make the transition to a greener future smoother. Such policies also help existing businesses become more eco-friendly. Governments and industries can find ways to reduce pollution and protect the environment, while still keeping jobs and growing the economy(ILO, 2022b).

Adopting a gender lens on industrial and sectoral policies in just transition

Balancing a healthy environment, good jobs, and fairness is very challenging. One needs to make sure they create good jobs with decent pay for everyone, not just men which means focusing on areas that can create a lot of jobs without harming the environment. It is recommended to focus on green jobs for women.

New policies can help women get good jobs in areas like clean energy, sustainable transportation, and even caregiving which will help close the gap between the types of jobs men and women typically have and create a more equal workforce (IRENA, 2019). Not everyone starts from the same place and some groups, like women and minorities, often have a harder time getting good jobs. Considering these challenges will make sure everyone has a fair shot while creating green policies. It includes people from all backgrounds in the conversation about how to transition to a greener economy. One can avoid creating new inequalities if we design green industries with fairness in mind from the start. Equality from the beginning is key which will not only be better for everyone, but it can also make these industries more productive(IRENA, 2022).

Working together by involving workers, businesses, and governments in the conversation about green jobs is the best way forward to find solutions that benefit the economy, the environment, and society as a whole. Efforts have to be made to create good green jobs. Many green jobs, especially in areas like recycling, aren't very secure or well-paying. Hence, there is a need for policies to improve these jobs and make them more like traditional jobs with benefits and protections for workers. Some people will lose their jobs as we move away from dirty energy sources like coal. One needs a program to help these workers get new skills and find new jobs in the green economy. It is also necessary to support businesses in growing green industries so they can create even more jobs.

Climate change policies can affect who gets jobs in the new green economy. Hence, to make sure these policies are fair and give everyone a chance to get a good job, not just men, one needs to work together with governments, businesses, and workers to find solutions that benefit everyone.

The source of power supply has changed immensely in recent times. There is a shift from fossil fuels like coal and oil to cleaner energy sources like solar and wind power. Despite it having a positive impact on the environment, some people will lose their jobs in the old energy industries.

It is quite evident that the clean energy sector is creating a lot of new jobs but women and other groups may not get these jobs unless we make a conscious effort to be fair. Hence, it is the need of the hour to design climate policies that create good opportunities for everyone(ILO, 2022h).

Significant hurdles have been faced by women and girls when it comes to energy access and opportunities in the energy sector. In many rural areas, they spend a disproportionate amount of time collecting firewood and other dirty fuels. It is a burden that cuts into their education, work, and safety, especially as travel distances for scarce fuel increase. Household chores like cooking and cleaning become more time-consuming because of a lack of clean energy. Despite clean energy technologies offer a solution, affordability often remains a barrier. It keeps them trapped in a cycle of time-consuming tasks and unhealthy fuels. The energy sector itself lacks female representation. It with even fewer women holding secure, well-paying jobs.

In the primary sector, women are often relegated to more precarious and informal roles where the transition to clean energy displaces workers. Women have limited influence in energy decision-making, with few holding leadership positions or less participation in government bodies that formulate energy policies. Such challenges create a significant barrier for women and girls which benefit from the shift towards clean energy.

The frequent usage of solar panels and other clean technologies can help families reduce pollution and improve their health. It can also free up women's time, as they won't have to spend as much time cooking and cleaning. Such clean energy jobs are a big opportunity for women as the fast-growing clean energy sector has lots of new jobs to offer. The percentage of women working in energy will increase as compared to the old dirty energy industries which are normally male dominant. Skill training for women must be conducted as these new jobs require all sorts of skills, not just physical labor. It is required to have the right skills if one wants more women to work in clean energy. One can also encourage girls to study science, technology, engineering, and math (STEM) in school and also provide training programs for women who want to switch careers.

Women contribute a lot to the clean energy sector by bringing new ideas and perspectives to the table. The transition to clean energy can be sped up by supporting women in the clean energy field. It will create a better future for everyone irrespective of gender. The biggest challenge to overcome is that most clean energy jobs are held by men. Women themselves, might not think these jobs are a good fit for women. Moreover, on most occasions, some girls are discouraged from studying

STEM subjects in school. They may face a tough work environment with lower pay and harassment despite women having the right skills.

Clean energy jobs that welcome women are the need of the hour. The jobs have to be made attractive so the women's percentage in clean energy will increase. The goals should be set for hiring more women, especially in leadership positions. There is a need to change the culture in these workplaces to be more welcoming and supportive of women which includes offering good working conditions, childcare options, and equal pay for equal work.

CASE STUDIES

Making the Renewable Energy Sector More Gender-Responsive in the Marshall Islands.

India's plan for clean energy includes getting more women involved because they have a goal of getting all their electricity from renewable sources by 2050. An important part of this plan is training more women to be electricians. Island Eco is one of the projects that train young women to install and maintain solar power systems on islands. It is great to fight climate change and create good jobs for women at the same time. Island Eco tackles multiple problems and these islands often rely on dirty diesel generators for power. It is expensive and bad for the environment. It replaces these generators with clean solar power. The project also breaks down stereotypes about women in science and technology by showing that they can do these jobs just as well as men. It also helps young women become more financially independent and involved in their communities. It allows islanders to store and sell food that needs refrigeration for the first time. These lessons that have been learned from this project are being used to improve clean energy policies across India(UNFCCC, 2023a).

More Women in The Geothermal Energy Sector in Iceland.

Reykjavík Energy is a green company that values equality. It is a company that is a leader in geothermal energy. It provides clean power and hot water to most of the country. Mostly men used to work there and it wasn't always easy to balance work and family life. The company decided to address these issues by wanting to create a fairer workplace with equal pay for men and women. It will also make it easier for employees to care for their families. They believed this would make everyone happier and more productive. Their plan worked well. The company closed the gender pay gap and created a work environment that's better for everyone. These are a great example of how companies can be successful while also being fair and supportive of their employees. It is even more impressive because Iceland passed a law requiring all companies to pay men and women equally (The Nordic Council, 2023).

Policy Suggestion 3: Occupational Health and Safety in Green Jobs

Climate change is making workplaces more dangerous. Workers face more health risks from heat stress as the world gets hotter which leads to new diseases that people can get at work. Green jobs can be safer but with certain challenges. Jobs can be made safer if one switches to cleaner ways of working. It is using less harmful chemicals. New risks are involved like those associated with making and recycling solar panels and electronic waste. Everyone deserves a safe workplace despite what kind of job someone has. It is needed especially as we move towards a greener economy. One needs strong safety regulations that consider both the dangers of climate change and the risks of new green technologies to tackle it.

Adopting A Gender Responsive Lens on Occupational Health and Safety Concerns

Men and women often face different safety risks at work due to several factors. The kind of jobs they typically have can expose them to different hazards. For instance, many women work in agriculture, which can be dangerous in increasingly hot weather due to climate change. It's not just limited to physical dangers as depending on the job and societal expectations, women may also face more mental stress in the workplace. Safety rules and programs need to take these gender differences into account to keep everyone safe and healthy which means collecting better data on how different jobs affect the health of men and women (ILO, 2014).

One can create targeted solutions once we understand the specific risks which could involve designing safety programs specifically for men and women workers. There must be collaboration between governments, businesses, and workers themselves is essential. Such collaboration can create a work environment that prioritizes everyone's safety and health, especially as we move towards a greener economy with new potential safety challenges. For building a future where everyone can work productively without compromising their well-being, it is required to include everyone in the conversation about workplace safety.

Each and everyone deserve to feel safe at work which will be free from violence, harassment, and discrimination. These problems can be especially bad for certain groups of workers, such as women, minorities, and immigrants. Many factors contribute to workplace violence and harassment which include discrimination, cultural differences, and stressful work environments.

The steps which we can take to prevent these problems are as follows:

a. New International Standards

International organizations must create new rules to protect workers from violence and harassment. These rules must require employers to assess the risks in their workplaces and take steps to make things safer (ILO, 2021a).

b. Healthy and Safe Workplace

Everyone has the right to work in a safe environment, free from violence and harassment because a healthy workplace is a right.

Many countries already have laws in place that require employers to identify and minimize safety risks in their workplaces which includes protecting workers from violence, harassment, and stress. It's important to remember that workers themselves can also contribute by identifying potential hazards. Governments play a crucial role by enacting policies and programs that prioritize both worker safety and gender equality which means specifically including women's needs in safety plans.

Additionally, collecting better data on how different jobs affect the health of men and women is essential for crafting effective safety regulations. At the workplace level, employers need to take concrete steps to prevent violence, harassment, and stress which includes having a response plan in place to address these issues if they arise. Furthermore, employers should consider the potential for increased violence, harassment, and mental health problems when assessing the risks associated with climate change.

Governments, businesses, and workers can create a safer and healthier work environment for everyone, particularly during this critical shift towards a more sustainable future if work together.

14. CONCLUSION

A combination of knowledge transfer, prioritized planning, and women-specific planning, funding, and policy building is sure to go a long way in mainstreaming gender perspectives in climate change initiatives and moving towards a sustainable environmental future at the same time. We can build a just and sustainable future for all by placing inclusion and equal opportunity at the heart of our efforts.

REFERENCES

Aayog, N. (2022). *Decoding government support to women entrepreneurs in India.* Women Entrepreneurship Platform, Microsave.net https://www.niti.gov.in/sites/default/files/2023-03/Decoding-Government-Support-to-Women-Entrepreneurs-in-India.pdf

Ansari, H., & Shetty, S. (2023). *Women can accelerate climate action. idr*https://idronline.org/article/gender/women-can-accelerate-climate-action/

Asian Development Bank. (2020). *Civil Society Brief India.* https://www.adb.org/sites/default/files/publication/879896/civil-society-brief-india.pdf

Asian Development Bank. (2021). *Gender Equality Results Case Study Nepal Gender Equality and Empowerment of Women Project.* ISBN 978-92-9257-479-6 (Print), 978-92-9257-480-2 (e-ISBN) https://www.adb.org/sites/default/files/publication/185563/nepal-gender-equality-and-empowerment-women-project.pdf

Brierley, C. (2022). *Extreme weather and climate events likely to drive increase in gender-based violence, UNDRR*https://www.preventionweb.net/news/extreme-weather-and-climate-events-likely-drive-increase-gender-based-violence

Burns, B. & Lee, A. (2015). *Background Paper – Gender Mainstreaming, EGM/GR-CR/BP.1, UN-Women, UN DESA, UNFCCC Secretariat*https://wedo.org/wp-content/uploads/2016/08/Gender-mainstreaming-practices-in-the-context-of-climate-responses.pdf

Buteau, S. (2021). *Roadmap for digital technology to foster India's MSME ecosystem—opportunities and challenges*, Europe PMC https://europepmc.org/article/pmc/pmc8662980

Chingarande, D. et al. (2020). *Mainstreaming gender into National Adaptation Planning and implementation in SubSaharan Africa, Working Paper No323 CGIAR Research Program on Climate Change, Agriculture and Food Security (CCAFS)* https://cgspace.cgiar.org/server/api/core/bitstreams/16905e9b-6c8a-4a17-8450-fd27d1c475dd/content

Climate Investment Funds. (n.d.). *Gender.* (Accessed on: 17-02-2024) https://www.cif.org/mainstreaming-gender

Decision 1/CP.16, Cancun Agreements: Outcome on the implementation of agreed outcomes related to climate change (FCCC/CP/2010/9 & Add.1)

Decision 1/CP.18, Doha Climate Gateway (FCCC/CP/2012/8 & Add1)

Decision 1/CP.20, Lima Call for Climate Action (FCCC/CP/2014/10 & Add1)

Decision 1/CP.7, The Marrakesh Ministerial Declaration, para3 (FCC/CP/2001/13 & Add.1)

Dutta, S. (2019). *Supporting last-mile women energy entrepreneurs: What works and what does not, Energia*https://www.energia.org/assets/2019/01/Supporting-Last-Mile-Women-Entrepreneurs.pdf

Duyck, S. et al. (n.d.). *Delivering on the Paris Promises: Combating Climate Change while Protecting Rights, Recommendations for the Negotiations of the Paris Rule Book, UNFCCC*https://unfccc.int/sites/default/files/903.pdf

Glemarec, Y. et al. (2016). *Leveraging Co-Benefits Between Gender Equality And Climate Action For Sustainable Development*, UN Women https://unfccc.int/files/gender_and_climate_change/application/pdf/leveraging_cobenefits.pdf

Government of IndiaAnnual Report 2020-21, Ministry of Micro, Small, and Medium Enterprises, https://msme.gov.in/sites/default/files/MSME-ANNUAL-REPORT-ENGLISH%202020-21.pdf

IDS Bulletin. (2020). *Gender and Energy: Opportunities for All,* Volume 51, Number 1, https://www.icrw.org/wp-content/uploads/2020/02/IDS-Bulletin-Gender-and-Energy-Opportunities-for-All.pdf

IEA. (n.d.). *The path to limiting global warming to 1.5 °C has narrowed, but clean energy growth is keeping it open.* https://www.iea.org/news/the-path-to-limiting-global-warming-to-1-5-c-has-narrowed-but-clean-energy-growth-is-keeping-it-open

IFC. (2020), Report: *Financial Inclusion for Women-owned MSMEs in India, International Finance Corporation World Bank Group*, https://www.ifc.org/en/insights-reports/2010/financial-inclusion-for-women-owned-msmes-in-india

ILO. (2014). *The gender dimension: Integrating the gender perspective in OSH policies* (ilo.org)

ILO. (2021a). *Violence and harassment in the world of work: A guide on Convention.* No190 and Recommendation No206

ILO. (2022b). *Care at work: Investing in care leave and services for a more gender equal world of work.*

ILO. (2022h). *Preparing Future Generations of Women for New Job Demands: Skilling, Re-Skilling, Digitalization and Automation,* OP.3_ILO.pdf (unwomen.org)

ILO. (2023). *Achieving a just transition towards environmentally sustainable economies and societies for all.* International Labour Office.

ILO. (2024). *Gender Equality and Inclusion for a Just Transition in Climate Action a Practical Guide.* International Labour Organization. https://www.ilo.org/media/482386/download

IRENA. (2019). *Renewable Energy: A Gender Perspective.*

IRENA. (2022). *World Energy Transitions Outlook 2022: 1.5°C Pathway.*

Jhunjhunwala, P(2023), Bridging the Gaps to Accelerate Women's Entrepreneurship, UNDP https://www.undp.org/india/blog/bridging-gaps-accelerate-womens-entrepreneurship

Kwauk, C. et al. (2019). *Girls' education in climate strategies Opportunities for improved policy and enhanced action in Nationally Determined Contribution, Global Economy & Development Working Paper 133*https://www.brookings.edu/wp-content/uploads/2019/12/Girls-ed-in-climate-strategies-working-paper-FINAL.pdf

Le Quere et al. (2012). *Global Carbon Budget 2012, Global Carbon Project.* https://www.globalcarbonproject.org/carbonbudget/archive/2012/CarbonBudget_2012.pdf

Lioness News Desk. (2016). *5 Top Risks Women Entrepreneurs Face, Lioness*https://lionessmagazine.com/5-top-risks-women-entrepreneurs-face/

Mercy Corps Ventures. (2022). *Understanding MSME Resilience to Climate Change*, Medium, https://medium.com/mercy-corps-social-venture-fund/understanding-msme-resilience-to-climate-change-cd50355bd31b

MSME Finance Gap. (n.d.). *SME Finance Forum.*https://www.smefinanceforum.org/data-sites/msme-finance-gap

New Climate Institute. (2015). *Climate Action Tracker: INDCs lower projected warming to 2.7°C: significant progress but still above 2°C.* https://newclimate.org/news/climate-action-tracker-indcs-lower-projected-warming-to-27degc-significant-progress-but-still

OCHCR. (2022). *Climate change exacerbates violence against women and girls*, United Nations Human Rights Office of the High Commissioner https://www.ohchr.org/en/stories/2022/07/climate-change-exacerbates-violence-against-women-and-girls

OECD DAC Network on Gender Equality (GENDERNET). (2015). *Making climate finance work for women: Overview of the integration of gender equality in aid to climate change*, https://www.oecd.org/dac/gender-development/Making-Climate-Finance-Work-for-Women.pdf

One, U. (2023). *The Role of Renewable Energy in Achieving Sustainable Development Goals*, https://utilitiesone.com/the-role-of-renewable-energy-in-achieving-sustainable-development-goals

Patel, S. et al. (2021). *Tackling gender inequality and climate change through the budget.* International Budget.org https://internationalbudget.org/wp-content/uploads/gender-climate-budgeting-synthesis-april-2021.pdf

PWC. (2022). *Gender equality in climate action: Women at the core*. https://www.pwc.in/assets/pdfs/consulting/esg/gender-equality-in-climate-action.pdf

PWC. (2022). *Gender equality in climate action: Women at the core*. https://www.pwc.in/assets/pdfs/consulting/esg/gender-equality-in-climate-action.pdf

Rajasthan Government. (n.d.). *Rajasthan Gramin Aajeevika Vikas Parishad (Aajeevika)* (Accessed on: 17-02-2024) https://rgavp.rajasthan.gov.in/

Report of World Bank. (2023). *Exploring Opportunities for Women Entrepreneurs Driving Climate Solutions A Discussion Note, International Finance Corporation.* World Bank Group https://www.ifc.org/content/dam/ifc/doc/2023/exploring-opportunities-for-women-entrepreneurs-driving-climate-solutions.pdf

Shagun. (2023). *Climate crisis: Inequality hotspot study shows where women in agriculture are hit hardest, Down to Earth.* https://www.downtoearth.org.in/news/agriculture/climate-crisis-inequality-hotspot-study-shows-where-women-in-agriculture-are-hit-hardest-92849

Sorensen, C. et al. (2018). *Climate Change and Women's Health: Impacts and Opportunities in India,* PMCID: PMC7007102 PMID: 32159002, *National Library of Medicine National Center for Biotechnology Information*https://www.ncbi.nlm.nih.gov/pmc/articles/PMC7007102/

Soubeyran, E., & Choudhary, K. (2023). *Why do we need to accelerate the mainstreaming of gender into climate action?* London School of Economics and Political Science https://www.lse.ac.uk/granthaminstitute/news/why-do-we-need-to-accelerate-the-mainstreaming-of-gender-into-climate-action/

Sqalli, Z. et al. (2021). *Why Climate Action Needs a Gender Focus,* BCG https://www.bcg.com/publications/2021/climate-action-impact-on-gender-equality

Technoserve. (n.d.). *5 Ways Climate Change Disproportionately Impacts Women | Expert Analysis, Techno Serve Business Solutions to Poverty.*https://www.technoserve.org/blog/5-climate-change-impacts-women/

The Nordic Council. (2023). *Bridging the Green Jobs Divide* (norden.org)

Torgalkar, V. (2022). *India's women farm workers suffer job losses from climate shocks.* Thomson Reuters Foundation. https://www.context.news/climate-risks/indias-women-farm-workers-suffer-job-losses-from-climate-shocks

UN CTCN. (2019). *The Women and Gender Constituency, Bhungroo – small women farmers owning and developing an innovative rainwater saving technology, UN Climate Technology Centre and Network UNFCCC Technology Mechanism*https://www.ctc-n.org/resources/bhungroo-small-women-farmers-owning-and-developing-innovative-rainwater-saving-technology

UN Women. (2023). *As climate change pushes millions of women into poverty, UN Women calls for a new feminist climate justice approach,*https://www.unwomen.org/en/news-stories/press-release/2023/12/as-climate-changes-pushes-millions-of-women-into-poverty-un-women-calls-for-a-new-feminist-climate-justice-approach

UN Women. (2023). In focus: *Women's leadership at the COP 28 climate conference,* https://www.unwomen.org/en/news-stories/in-focus/2023/11/in-focus-womens-leadership-at-the-cop-28-climate-conference

UN Women. (2024). *Explainer: How gender inequality and climate change are interconnected.* https://www.unwomen.org/en/news-stories/explainer/2022/02/explainer-how-gender-inequality-and-climate-change-are-interconnected

UN Women Watch. (1997). *Coordination Segment, Chapter IV,* Excerpt from A/52/3 https://www.un.org/womenwatch/osagi/pdf/ECOSOCAC1997.2.PDF

UN Women Watch. (2024). *Women, Gender Equality and Climate Change*https://www.un.org/womenwatch/feature/climate_change/

UN Women Watch. (n.d.). *Gender Equality and Climate Change*https://www.un.org/womenwatch/feature/climate_change/downloads/Women_and_Climate_Change_Factsheet.pdf

UNDESA. (2015). *Report Expert Group Meeting Implementation Of Gender-Responsive Climate Action In The Context Of Sustainable Development*https://unfccc.int/files/gender_and_climate_change/application/pdf/egmreport.pdf

UNDP. (2019). *Ghana Gender*Https://Climatepromise.Undp.Org/Sites/Default/Files/Research_Report_Document/Undp-Ndcsp-Ghana-Gender-Analysis-Final2.Pdf

UNDP. (2001). *Generating Opportunities: Case Studies on Energy and Women.* https://www.undp.org/publications/generating-opportunities-case-studies-energy-and-women

UNDP. (n.d.). *Accelerating Opportunities for Women Climate Entrepreneurs.* https://www.undp.org/india/accelerating-opportunities-women-climate-entrepreneurs

UNFCCC. (2023a). *Implementation of just transition and economic diversification strategies*

UNFCCC. (2023). *Five Reasons Why Climate Action Needs Women,* United Nation Climate Change,https://unfccc.int/news/five-reasons-why-climate-action-needs -women

UNICEF. (2022). *Study On The Role Of Mitigation-Adaptation Co-Benefits For Creating A More Resilient Future For All, Climate Sustainability Working Group (CSWG)* G20 Indonesia 2022 https://www.unicef.org/indonesia/media/17201/file/Study%20on%20The%20Role%20of%20Mitigation-Adaptation%20Co-Benefits %20For%20Creating%20A%20More%20Resilient%20Future%20For%20All.pdf

United Nation Climate Change. (n.d.). *Introduction to Gender and Climate Change-* https://unfccc.int/gender

United Nation Climate Change. (2023) *Bhungroo.* https://unfccc.int/climate-action/momentum-for-change/women-for-results/bhungroo

United Nations. (2009). *Women...In The Shadow of Climate Change From Nos3 & 4, VolXlvi, Special Climate Change Issue: "To Protect Succeeding Generations.* https://www.un.org/en/chronicle/article/womenin-shadow-climate-change

United Nations Development Programme. (2024). *Mexico Case Study, Gender Equality And Women's Empowerment In Public Administration*, https://www.undp .org/sites/g/files/zskgke326/files/publications/es/MexicoFinal%20-%20HiRes.pdf

United Nations New York. (2002). *Gender Mainstreaming: An Overview, Office of the Special Adviser on Gender Issues and Advancement of Women,*https://www.un .org/womenwatch/osagi/pdf/e65237.pdf

USAID From the American People. (2022). *Gender Equality and Climate Finance.* https://www.climatelinks.org/sites/default/files/asset/document/2022-11/Gender%20 Equality%20and%20Climate%20Finance_Technical%20Brief%20Nov%202022.pdf

Women Economic Empowerment. (n.d.). *Global Alliance For Mass Entrepreneurship* https://massentrepreneurship.org/women-economic-empowerment/

World Health Organization. (2023). *Climate change.* https://www.who.int/news -room/fact-sheets/detail/climate-change-and-health#:~:text=Research%20shows %20that%203.6%20billion,diarrhoea%20and%20heat%20stress%20alone

Yonzon, K. & Bisht, P. (2023). *Unlocking Climate Financing for India's Social Enterprises: Challenges and Opportunities*, SAMBODHI https://sambodhi.co.in/unlocking-climate-financing-for-indias-social-enterprises-challenges-and-opportunities/

Zakrzewska, A. & Fitzgerald, L. (2023), *Empowering Women to Power the Green Transition, Cogito* https://oecdcogito.blog/2023/03/08/empowering-women-to-power-the-green-transition/

Zusman, E., Lee, S-Y., Rojas, A., & Adams, L. (2024). *Mainstreaming Gender Into Climate Mitigation Activities, Guidelines for Policy Makers and Proposal Developers*, https://www.adb.org/sites/default/files/publication/217771/gender-climate-mitigation.pdf

Chapter 12
Women's Leadership in Environmental Conservation:
Challenges and Success Stories

Ushaa Eswaran
Mahalakshmi Tech Campus, India

Vivek Eswaran
https://orcid.org/0009-0002-7475-2398
Medallia, India

Keerthna Murali
https://orcid.org/0009-0009-1419-4268
Dell, India

Vishal Eswaran
https://orcid.org/0009-0000-2187-3108
CVS Health Centre, India

ABSTRACT

This chapter explores the essential role of women in environmental conservation, focusing on their challenges and successes. It examines how gender, socioeconomic factors, and cultural contexts impact women's leadership in conservation. Using a comprehensive literature review, case studies, and original research, the chapter highlights the barriers women face—such as discrimination, resource limitations, and cultural norms—and showcases their innovative contributions to conservation. It underscores the need to empower women in leadership roles to achieve sustainable environmental management and advance human rights. The study offers valuable

DOI: 10.4018/979-8-3693-6069-9.ch012

insights for policymakers, conservation organizations, and activists seeking to promote gender equality in environmental stewardship.

1.INTRODUCTION

The interconnected challenges of environmental degradation, climate change, and biodiversity loss have emerged as some of the most pressing issues of our time. As the global community grapples with these complex problems, the role of diverse leadership in crafting effective solutions has become increasingly apparent. Within this context, women's leadership in environmental conservation has gained recognition not only as a matter of gender equality but also as a crucial factor in achieving sustainable and equitable environmental outcomes.

Women, particularly those in rural and indigenous communities, often possess unique knowledge and experiences that are invaluable to conservation efforts. Their close relationship with natural resources, stemming from traditional roles in agriculture, water management, and forest stewardship, provides them with deep insights into local ecosystems and sustainable practices. However, despite this wealth of knowledge and their significant contributions to environmental stewardship, women continue to face numerous barriers in assuming leadership roles within the field of conservation.

These barriers are multifaceted and deeply rooted in societal structures, cultural norms, and economic systems. They range from limited access to education and financial resources to gender-based discrimination in decision-making processes. Moreover, the challenges women face in environmental leadership are often compounded by intersecting factors such as race, class, and geographic location, necessitating an intersectional approach to understanding and addressing these issues.

Despite these obstacles, women around the world have emerged as powerful leaders and change-makers in environmental conservation. From grassroots community initiatives to high-level policy advocacy, women are driving innovative approaches to sustainable resource management, climate change mitigation, and biodiversity protection. Their leadership often emphasizes holistic, community-centered strategies that integrate social, economic, and environmental concerns.

This chapter seeks to explore the complex landscape of women's leadership in environmental conservation, highlighting both the persistent challenges and the inspiring success stories that characterize this field. By examining diverse case studies and drawing on interdisciplinary research, we aim to provide a comprehensive understanding of the current state of women's involvement in conservation leadership and to identify pathways for enhancing their participation and impact.

1.1. Objectives Of the Chapter

1. To analyze the unique contributions and perspectives that women bring to environmental conservation leadership, considering the intersectionality of gender with other social factors.
2. To identify and examine the key challenges and barriers that women face in assuming leadership roles in conservation, across different cultural and socio-economic contexts.
3. To showcase successful models and strategies of women-led conservation initiatives, highlighting their impacts and the lessons that can be drawn from these experiences.
4. To propose recommendations for policymakers, organizations, and communities to support and enhance women's leadership in environmental conservation.

1.2. Organization Of the Chapter

The chapter is structured as follows: First, a comprehensive literature review provides an overview of existing research on women's leadership in environmental conservation, focusing on theoretical frameworks, historical context, and current debates. The methodology section outlines the approach used for data collection and analysis in this study. Subsequently, a series of case studies present real-world examples of women's leadership in various conservation contexts. The research and experimental studies section details original findings that contribute to the understanding of women's roles in conservation. A discussion section synthesizes the key insights from the literature, case studies, and research, exploring their implications for theory and practice. The future trends section examines emerging developments and potential trajectories for women's leadership in conservation. Finally, the conclusion summarizes the main findings and offers recommendations for advancing women's leadership in environmental conservation.

2. LITERATURE REVIEW: WOMEN'S LEADERSHIP IN ENVIRONMENTAL CONSERVATION

The study of women's leadership in environmental conservation encompasses various disciplines, including gender studies, environmental science, anthropology, and development studies. A review of the current literature reveals several key themes and debates that shape our understanding of this topic.

2.1. Ecofeminism and Gender in Environmental Thought

The theoretical foundation for examining women's roles in environmental conservation often begins with ecofeminism, a movement that explores the connections between the exploitation of nature and the oppression of women. This perspective suggests that both forms of domination are interconnected and must be addressed together to achieve true sustainability and gender equality.(Mascarenhas, Oswald AJ, Munish Thakur, and Payal Kumar., 2024)

Ecofeminism emphasizes the unique contributions of women to environmental sustainability through their traditional ecological knowledge. For instance, women's involvement in sustainable agriculture, seed conservation, and biodiversity preservation has been well-documented. This body of work argues that women's close relationship with nature, often due to their roles in food production and resource management, positions them as vital actors in conservation efforts.

Recent scholarship has expanded ecofeminist thought by incorporating intersectional perspectives, which examine how various social categories such as race, class, and sexuality intersect with gender to shape environmental experiences and activism. This intersectional approach has revealed that marginalized women, including women of color, indigenous women, and women from lower socio-economic backgrounds, often bear the brunt of environmental degradation while also leading critical conservation efforts.

2.2.Gender Roles and Natural Resource Management

Gender roles significantly influence natural resource management, with women and men often having different responsibilities, knowledge, and access to resources. The concept of “feminist political ecology” has been instrumental in exploring how gender relations shape environmental knowledge, access to resources, and conservation practices (Pedroza-Gutiérrez, Carmen, et al., 2024).

Studies have shown that women's involvement in community-based natural resource management leads to better conservation outcomes. For example, research on community forestry groups has demonstrated that when women participate in decision-making, there are improvements in forest conditions and more equitable resource distribution. This is partly because women often bring different perspectives and priorities, such as a focus on long-term sustainability and community well-being.

Understanding the local gender dynamics is crucial for designing effective conservation strategies. Women's traditional ecological knowledge, their roles in managing household resources, and their responsibilities in food production position them as key stakeholders in conservation initiatives. Programs that fail to recognize and incorporate gendered knowledge and practices often miss critical opportunities for sustainable resource management.

2.3.Barriers to Women's Leadership in Conservation

Despite their significant contributions, women face numerous barriers to assuming leadership roles in conservation. These barriers include systemic issues such as limited access to education, lack of land rights, and exclusion from decision-making processes.(Sreevas, T. V. T. K., and Prasad Vasantrao Kulkarni., 2024)

Gender biases within conservation organizations also pose significant challenges. Women often encounter discrimination in hiring, promotion, and recognition of their contributions. Studies have revealed that women are underrepresented in leadership positions within conservation organizations and face greater obstacles in advancing their careers compared to their male counterparts.

Addressing these barriers requires structural changes within organizations and broader societal shifts. This includes promoting gender equality in education and training, ensuring women's access to land and resources, and creating inclusive decision-making processes that recognize and value women's contributions.

2.4.Indigenous Women and Traditional Ecological Knowledge

Indigenous women play a crucial role in conservation through their traditional ecological knowledge and practices. This knowledge is often passed down through generations and encompasses sustainable resource management techniques, biodiversity conservation, and climate change adaptation strategies.(Dalyan, M., et al., 2024)

Indigenous women's ecological knowledge is vital for effective conservation efforts. Their practices often emphasize harmony with nature, sustainable use of resources, and community resilience. Recognizing and protecting the rights of indigenous women is essential for preserving this knowledge and ensuring their active participation in conservation initiatives.

The importance of indigenous women's contributions is increasingly being acknowledged in global conservation efforts. Their involvement not only enhances biodiversity conservation but also supports cultural preservation and the well-being of indigenous communities.

2.5. Women's Leadership and Conservation Outcomes

Emerging research suggests that women's leadership can lead to improved conservation outcomes. Studies have shown that women's participation in natural resource management groups is associated with better resource governance and conservation results. Women-led initiatives often prioritize community needs and long-term sustainability, leading to more durable environmental protection.(Kakade, Kiran Shrimant, Anjali M. Kulkarni, and Kasturi R. Naik., 2024)

For instance, in many parts of the world, women-led conservation projects have successfully restored degraded landscapes, increased biodiversity, and improved community livelihoods. These initiatives often adopt holistic approaches that integrate ecological, social, and economic dimensions of sustainability.

Women's leadership in conservation also fosters greater community engagement and participation. Women tend to involve other community members, build consensus, and promote inclusive decision-making processes. This participatory approach enhances the effectiveness and sustainability of conservation efforts.

2.6. Intersectionality in Environmental Leadership

Recent scholarship has emphasized the importance of intersectional approaches to understanding women's environmental leadership. Intersectionality provides a crucial framework for analyzing how multiple forms of oppression interact to shape environmental vulnerabilities and agency.(Dobrich, Emily Olga Rosa, 2024)

Studies have examined how factors such as caste, religion, ethnicity, and socioeconomic status intersect with gender to influence women's roles in environmental conservation. For example, research on water management in South Asia has shown that women's experiences and contributions are shaped by their caste and religious identities, affecting their access to resources and participation in decision-making.

Intersectional approaches highlight the diverse experiences and challenges faced by women in conservation. They underscore the need for inclusive policies and practices that address the specific needs and strengths of different groups of women. By recognizing and addressing these intersecting forms of oppression, conservation efforts can become more equitable and effective.

The literature review reveals a rich and evolving body of research on women's leadership in environmental conservation. Significant progress has been made in recognizing women's contributions and the barriers they face. However, gaps remain in understanding how to effectively support and scale up women's leadership across diverse contexts.

To enhance women's leadership in conservation, it is crucial to address systemic barriers, promote gender-inclusive policies, and recognize the intersectional dimensions of women's experiences. Empowering women as leaders in conservation not only improves environmental outcomes but also advances gender equality and social justice. Continued research and policy efforts are needed to support women's leadership and ensure their vital contributions to environmental sustainability are recognized and valued.

3.METHODOLOGY

This study employs a mixed-methods approach to provide a comprehensive examination of women's leadership in environmental conservation. The methodology integrates four main components: a systematic literature review, case study analysis, qualitative interviews, and a quantitative survey. This multi-faceted approach ensures a robust exploration of the topic by combining theoretical insights, practical examples, and empirical evidence.

Systematic Literature Review

The systematic literature review was conducted using a range of academic databases, including Web of Science, Scopus, and Google Scholar. The review aimed to identify peer-reviewed articles, books, and reports published from 2000 to 2024. Keywords such as "women's leadership," "environmental conservation," "gender and environment," "female environmental leaders," and related terms were employed to ensure a broad and inclusive search.

Process

- **Database Search:** Comprehensive searches were performed in selected databases using Boolean operators to refine the results.
- **Inclusion Criteria:** The review focused on publications that discussed women's roles in environmental conservation, leadership dynamics, gendered approaches to resource management, and case studies of women-led initiatives.
- **Thematic Analysis:** Identified literature was analyzed thematically to identify key trends, debates, and research gaps. This analysis helped in understanding the evolution of women's leadership roles in conservation, highlighting both advancements and persistent challenges.

Latest Findings

- **Emerging Trends:** Recent literature has increasingly focused on the intersectionality of gender with other social categories, such as race, class, and ethnicity, and their combined impact on women's roles in conservation.
- **Innovative Approaches:** New studies highlight the use of digital technologies and social media in amplifying women's voices and facilitating collaborative conservation efforts.

Case Study Analysis

The case study analysis involved selecting six diverse examples of women-led conservation initiatives from various geographical regions and ecological contexts. The selection criteria included the innovation in approach, demonstrable impact, and availability of comprehensive data.

Selection Criteria

- **Innovation:** Initiatives demonstrating novel or pioneering approaches to conservation.
- **Impact:** Projects with measurable and significant conservation outcomes.
- **Data Availability:** Comprehensive and accessible information for analysis.

Data Collection

- **Document Analysis:** Review of project reports, publications, and other relevant documents.
- **Site Visits:** Where feasible, site visits were conducted to gather firsthand insights.
- **Interviews:** Conversations with project leaders and participants provided additional context and depth.

Latest Findings

- **Cross-Continental Insights:** Case studies reveal innovative approaches that integrate traditional ecological knowledge with modern conservation practices, demonstrating the adaptability and effectiveness of women-led initiatives across different contexts.
- **Collaborative Models:** Several case studies highlight successful partnerships between local communities and international organizations, emphasizing the importance of collaborative efforts in achieving conservation goals.

3.1.Mixed-Methods Exploration of Women's Leadership in Environmental Conservation

Qualitative Interviews

Qualitative interviews were conducted with 30 women leaders in environmental conservation, including activists, scientists, policymakers, and community leaders. These semi-structured interviews aimed to explore participants' experiences, challenges, strategies for success, and perspectives on the future of women's leadership in conservation. Participants were selected based on their influence and contributions to environmental conservation. (Cole, Stroma, et al., 2024). The interviews were flexible yet focused on key topics, conducted either in person or via video conferencing, recorded with permission, and transcribed for analysis.

Latest Findings

Interviews revealed personal narratives of resilience and innovation, showcasing how women leaders overcome systemic barriers and drive change in conservation. Strategic insights were shared, including effective leadership and advocacy strategies such as community engagement, capacity building, and policy influence.

Quantitative Survey

A quantitative survey was distributed to 500 individuals working in environmental conservation across 50 countries. The survey aimed to collect data on organizational demographics, gender representation in leadership, perceived barriers to women's advancement, and the impacts of women's leadership on conservation outcomes. It included questions on gender representation, organizational structures, barriers to women's advancement, and the impacts of women's leadership. The survey was administered online, translated into five languages to ensure broad participation, and targeted a diverse sample of respondents from different regions and organizational types.

Data Analysis

Quantitative data were analyzed using statistical software to identify patterns, correlations, and trends. Qualitative data from open-ended responses were coded and analyzed thematically using NVivo software to uncover recurring themes and insights.

Latest Findings

The survey revealed common barriers faced by women in conservation, such as gender bias and lack of support networks, alongside opportunities for advancing women's leadership through mentorship and training programs. Data indicated that organizations with higher female representation in leadership positions experienced more effective and innovative conservation outcomes.

This mixed-methods approach provides a comprehensive exploration of women's leadership in environmental conservation. By integrating a systematic literature review, case study analysis, qualitative interviews, and a quantitative survey, the study captures a multifaceted view of the challenges, successes, and opportunities associated with women's roles in this critical field. The latest findings highlight the evolving nature of women's leadership and underscore the need for continued support and recognition of their contributions to environmental sustainability.

4.CASE STUDIES

4.1.The Green Belt Movement, Kenya

Overview: The Green Belt Movement, founded by Nobel laureate Wangari Maathai in 1977, is a pioneering grassroots initiative in Kenya aimed at combating deforestation and promoting environmental sustainability. The movement engages rural women in reforestation efforts, resulting in the planting of over 51 million trees across the country (Gremaud, Antoine, and Lada Kochtcheeva.,2024).

Impact: The movement has not only achieved significant environmental milestones but has also empowered women by providing education, advocating for land rights, and increasing their participation in democratic processes. Its success has inspired similar initiatives around the world, demonstrating how women's leadership can combine environmental conservation with social and economic empowerment.

Key Themes:

1. **Grassroots Mobilization:**
    - The Green Belt Movement engages local communities, especially women, in its environmental efforts, relying on volunteers for tree planting and educational outreach.
    - This approach ensures sustainable outcomes through active local involvement.
2. **Integration of Environmental and Social Goals:**

- The movement integrates environmental conservation with social development, proving that ecological actions can also advance social justice.
- It aligns ecological objectives with socio-economic benefits, such as enhancing community resilience and living standards.

3. **Empowerment through Education and Advocacy:**
    - The movement offers training on environmental issues, leadership, and advocacy, empowering women to take on leadership roles.
    - It raises awareness about environmental and gender issues, influencing local and national policies.
    - Women are encouraged to become active participants in both environmental and community leadership roles.

Comprehensive Overview

1. **Historical Context and Formation:**
    - The Green Belt Movement was established to address Kenya's deforestation crisis, which was causing soil erosion and biodiversity loss. Wangari Maathai envisioned combining ecological restoration with social progress.
2. **Core Activities and Programs:**
    - The movement organizes tree planting events involving local communities, who participate in planting and maintaining trees.
    - It conducts educational programs focused on environmental conservation, sustainable practices, and leadership development.
    - Advocacy for environmental protection, land rights, and gender equality is a key component of its mission.
3. **Success Stories and Case Studies:**
    - The movement has transformed landscapes in various villages, leading to improved soil health and agricultural productivity.
    - Women involved have assumed leadership roles, driving environmental and social change within their communities.
4. **Global Influence:**
    - The Green Belt Movement's success has served as a model for similar environmental initiatives worldwide, highlighting the global impact of grassroots efforts that merge ecological and social objectives.
5. **Challenges and Future Directions:**
    - Challenges include addressing climate change, sustaining reforestation efforts, and scaling successful models.

- Future goals involve expanding educational programs, increasing community engagement, and strengthening advocacy for both environmental and social justice.

The Green Belt Movement exemplifies how grassroots initiatives can drive significant environmental and social change by integrating ecological restoration with community empowerment.

4.2.Women's Earth Alliance (WEA), Global

Overview: Women's Earth Alliance (WEA) is a global organization dedicated to supporting women leaders who tackle pressing environmental challenges such as climate change, water scarcity, and food security. WEA provides a range of services including training, funding, and networking opportunities to amplify the impact of women's innovations and leadership in these critical areas.(Dube, Musa W, 2024)

Notable Project: In India, WEA has implemented the "Solar Engineers" project, which trains women to install and maintain solar energy systems in remote villages. This project not only provides renewable energy to underserved areas but also challenges traditional gender norms by involving women in technology and energy sectors.

Key Themes:

1. **Network-Based Support:**
    - o WEA creates and nurtures networks of women leaders who are working on environmental issues, offering a platform for sharing knowledge, resources, and strategies.
    - o By connecting women across regions and sectors, WEA enhances collaboration and collective action toward solving global environmental challenges.
2. **Empowerment Through Skill Development:**
    - o The organization focuses on building the skills and capacities of women leaders through targeted training programs.
    - o Skill development initiatives, such as those in the "Solar Engineers" project, enable women to take on technical roles and leadership positions, boosting their confidence and effectiveness.
3. **Addressing Gender Norms in Technology:**
    - o WEA actively works to challenge and change gender norms that restrict women's participation in technology and environmental sectors.

- The "Solar Engineers" project, by involving women in solar technology, disrupts traditional gender roles and demonstrates women's capability and leadership in fields historically dominated by men.

Comprehensive Overview

1. **Historical Context and Formation:**
   - Women's Earth Alliance was established to address the intersection of gender and environmental issues, recognizing that empowering women is crucial for effective environmental stewardship and sustainable development.
2. **Core Activities and Programs:**
   - WEA provides training programs focused on key environmental areas such as climate change adaptation, sustainable agriculture, and renewable energy.
   - The organization offers funding opportunities to support innovative projects led by women, facilitating their work on critical environmental issues.
   - Networking events and workshops organized by WEA help women leaders connect with each other, share best practices, and build collaborative partnerships.
3. **Success Stories and Impact:**
   - The "Solar Engineers" project in India has successfully trained numerous women, who now contribute to their communities by providing clean energy solutions and promoting sustainable practices.
   - Women trained through WEA programs have gained recognition as experts and advocates in their fields, influencing policy and driving local and global environmental initiatives.
4. **Global Influence:**
   - WEA's approach has inspired similar programs and partnerships worldwide, showcasing the importance of integrating gender perspectives into environmental solutions.
   - The organization's work highlights the role of women as vital contributors to addressing global environmental challenges, fostering a more inclusive and equitable approach to sustainability.
5. **Challenges and Future Directions:**
   - Key challenges include scaling programs to reach more women globally and securing sustainable funding for ongoing initiatives.

- o Future goals involve expanding the range of training programs, strengthening global networks, and continuing to challenge and transform gender norms in technology and environmental leadership.

Women's Earth Alliance exemplifies how targeted support and capacity building for women can drive significant progress in environmental sustainability while also addressing gender disparities in technology and leadership.

4.3. Empowering Indigenous Women through Sustainable Conservation: The Impact of AMWAE in the Ecuadorian Amazon

Introduction

The Asociación de Mujeres Waorani de la Amazonía Ecuatoriana (AMWAE) stands as a prominent example of how indigenous women's leadership can drive effective and culturally sensitive conservation strategies. Based in the Ecuadorian Amazon, AMWAE has made significant strides in combining traditional ecological knowledge with contemporary conservation practices to address both environmental degradation and socio-economic challenges. This case study explores the transformative impact of AMWAE's sustainable cacao production initiative, highlighting its benefits for biodiversity conservation and local empowerment.(Commer, Chidimma Ozor., 2024)

Background and Context

The Ecuadorian Amazon, renowned for its unparalleled biodiversity, faces severe threats from industrial activities such as oil extraction. This exploitation has led to deforestation, habitat loss, and disruptions to local ecosystems, impacting the Waorani people's traditional way of life. In response to these challenges, AMWAE was established by Waorani women to protect their land, preserve their cultural heritage, and develop sustainable livelihoods.

Sustainable Cacao Production Initiative

1. **Introduction of Sustainable Practices**

   AMWAE's primary conservation initiative involves transitioning from oil extraction to sustainable cacao production. This shift is a direct response to the environmental and social impacts of oil drilling, which has been detrimental to the Waorani territory. Cacao farming provides an eco-friendly alternative that aligns with traditional agricultural practices while promoting environmental stewardship.

2. **Traditional Knowledge and Modern Techniques**

   The integration of traditional ecological knowledge with modern conservation practices is central to AMWAE's approach. Waorani women leverage their extensive understanding of local ecosystems to implement sustainable farming methods. These methods include:

   - **Agroforestry:** By cultivating cacao alongside other native plants, AMWAE ensures that agricultural activities support rather than replace forest ecosystems. This approach helps maintain soil fertility, prevent erosion, and enhance biodiversity.
   - **Organic Farming:** Avoiding synthetic pesticides and fertilizers, AMWAE uses organic techniques to protect soil health and reduce the ecological footprint of cacao production. This practice aligns with traditional Waorani methods of nurturing the land.

3. **Economic Empowerment and Community Development**

The shift to cacao production has generated multiple benefits for Waorani women and their communities:

- **Economic Opportunities:** Cacao farming provides a sustainable income source, reducing reliance on environmentally damaging industries. Women are trained in cacao cultivation, processing, and marketing, which empowers them economically and enhances their role in local decision-making.
- **Social Empowerment:** AMWAE's initiatives have fostered greater gender equality within the Waorani community. Women's involvement in conservation and agricultural activities has increased their visibility and influence, leading to more equitable representation in community affairs.
- **Capacity Building:** The organization offers training programs and workshops on sustainable farming practices, business skills, and environmental stewardship. This education not only improves agricultural productivity but also strengthens the overall capacity of the community to manage and protect their natural resources.

Impact on Biodiversity Conservation

1. **Protection of Ecosystems**

Sustainable cacao production helps mitigate deforestation and habitat loss by providing an alternative to oil extraction. The integration of cacao farming into agroforestry systems supports the preservation of forest ecosystems, which are vital for maintaining local biodiversity.

2. **Biodiversity Preservation**

AMWAE's approach to cacao farming supports the conservation of native plant and animal species. By avoiding the expansion of monoculture plantations and maintaining forest cover, the initiative protects critical habitats for numerous species, including those that are endangered or endemic to the region.

Challenges and Solutions

1. **Navigating Market Dynamics**
   One of the challenges faced by AMWAE is accessing international markets for their cacao products. To address this, the organization has developed partnerships with fair trade networks and sustainable certification programs, ensuring that their products meet global standards while securing fair prices for their harvest.
2. **Cultural Sensitivity and Adaptation**
   Maintaining cultural relevance while adopting new agricultural practices requires careful adaptation. AMWAE's success is partly due to its emphasis on respecting traditional practices and integrating them with modern techniques, ensuring that conservation efforts align with Waorani cultural values.

Broader Implications and Lessons Learned

1. **Model for Culturally Relevant Conservation**
   AMWAE's success highlights the importance of incorporating indigenous knowledge into conservation strategies. Their approach demonstrates that effective conservation can be achieved by respecting and integrating local cultural practices with modern environmental management techniques.
2. **Empowerment through Leadership**

The organization's impact underscores the transformative power of women's leadership in conservation. By empowering indigenous women and valuing their knowledge and perspectives, AMWAE has created a model for other communities facing similar challenges.

3. **Sustainable Development Goals**

AMWAE's work aligns with several Sustainable Development Goals (SDGs), including No Poverty, Zero Hunger, and Climate Action. The organization's efforts contribute to sustainable development by improving livelihoods, enhancing food security, and protecting the environment.

The Asociación de Mujeres Waorani de la Amazonía Ecuatoriana (AMWAE) exemplifies how indigenous women's leadership can drive impactful and sustainable conservation efforts. Through their innovative approach to cacao production, AMWAE has demonstrated that integrating traditional knowledge with modern practices can address environmental degradation while empowering local communities. Their success offers valuable insights into developing culturally relevant and effective conservation strategies, highlighting the critical role of indigenous women in crafting solutions that address both global environmental challenges and local needs.

4.4.The Coral Triangle Initiative: Women's Leadership in Marine Conservation

The Coral Triangle Initiative (CTI) is a collaborative effort involving six countries—Indonesia, Malaysia, Papua New Guinea, the Philippines, Solomon Islands, and Timor-Leste—aimed at preserving the marine biodiversity of the Coral Triangle region, which is one of the most ecologically rich and diverse marine areas in the world. Women have played a pivotal role in this initiative, leading a range of programs that address critical marine conservation challenges. Their leadership highlights the effective integration of scientific knowledge, community engagement, and sustainable practices.(Pratiwi, Heni, et al, 2024)

Overview of the Coral Triangle Initiative

The Coral Triangle Initiative was launched in 2007 with the goal of safeguarding the Coral Triangle's marine and coastal resources. The region is home to over 3,000 species of fish, 600 species of coral, and supports the livelihoods of millions of people. The CTI focuses on several key areas, including sustainable fisheries management, marine protected areas (MPAs), and climate change adaptation.

Women's Leadership and Contributions

1. Sustainable Fisheries Management

In many Coral Triangle countries, women are actively involved in sustainable fisheries management. Their roles range from community leaders advocating for sustainable practices to scientists and policymakers designing and implementing regulations. Women's involvement often brings a unique perspective to fisheries management, emphasizing community well-being alongside environmental sustainability.

Example: Indonesia

In Indonesia, women have led community-based fisheries management initiatives that integrate traditional knowledge with modern practices. These programs focus on conserving fish stocks and marine habitats while supporting local economies. Women fishers and community leaders work together to establish no-fishing zones and promote sustainable fishing techniques, ensuring long-term benefits for their communities.

2. Establishment of Marine Protected Areas

Marine protected areas (MPAs) are critical for conserving marine biodiversity and protecting ecosystems from overfishing and habitat destruction. Women's leadership has been instrumental in the establishment and management of MPAs across the Coral Triangle.

Example: The Philippines

In the Philippines, women leaders have been at the forefront of creating and managing MPAs. Women's involvement in these initiatives includes advocating for legal protection, monitoring marine environments, and engaging local communities in conservation efforts. Notably, women-led organizations have successfully established MPAs that serve as critical habitats for marine life and support local fisheries.

3. Mangrove Restoration Projects

Mangroves are vital for coastal protection, serving as natural buffers against storm surges and providing important breeding grounds for marine species. Women in the Coral Triangle have led successful mangrove restoration projects that not only protect coastlines but also enhance local livelihoods.

Example: The Philippines

In the Philippines, women leaders have spearheaded mangrove restoration projects that focus on replanting mangrove forests along coastlines. These projects involve local communities in planting and maintaining mangrove trees, which helps prevent coastal erosion, supports biodiversity, and provides resources for local fisheries. The success of these projects demonstrates how women's leadership can drive effective environmental conservation while improving community resilience.

4. Climate Change Adaptation

Climate change poses significant threats to marine ecosystems and coastal communities. Women in the Coral Triangle have played a crucial role in developing and implementing climate change adaptation strategies that address both environmental and socio-economic impacts.

Example: Timor-Leste

In Timor-Leste, women have been actively involved in climate change adaptation initiatives that aim to enhance coastal resilience. These initiatives include training programs for local communities on climate-resilient fishing practices, promoting sustainable land use, and advocating for policies that address climate impacts on marine environments. Women's leadership in these efforts helps bridge the gap between scientific research and community needs, ensuring more effective and equitable adaptation strategies.

5. Community-Based Conservation Strategies

Community-based conservation strategies are essential for the long-term success of marine conservation efforts. Women have led various community-based initiatives that empower local populations and foster collaborative approaches to environmental stewardship.

Example: Solomon Islands

In the Solomon Islands, women have led community-based conservation programs that involve local stakeholders in decision-making processes and conservation activities. These programs focus on protecting marine biodiversity, promoting sustainable tourism, and enhancing community awareness about marine conservation. Women's leadership in these initiatives helps build strong local support for conservation efforts and ensures that conservation strategies are culturally relevant and sustainable.

Impact and Outcomes

The involvement of women in the Coral Triangle Initiative has led to several positive outcomes:

- **Enhanced Conservation Results:** Women's leadership has contributed to the successful implementation of conservation projects, such as the establishment of effective MPAs and the restoration of vital mangrove ecosystems.
- **Improved Community Livelihoods:** Women-led initiatives often include components that address socio-economic needs, such as creating sustainable livelihoods and improving community resilience to environmental changes.
- **Strengthened Local Capacities:** Women's involvement in conservation projects has built local capacities, empowering communities to take active roles in managing and protecting their natural resources.

Challenges and Opportunities

While women's leadership in the Coral Triangle Initiative has achieved significant successes, challenges remain. These include overcoming gender biases in decision-making processes, ensuring equal access to resources and opportunities, and addressing the diverse needs of different communities.

Opportunities for further enhancing women's leadership in marine conservation include:

- **Increasing Support for Women's Roles:** Providing targeted support and resources to women leaders in conservation can amplify their impact and address systemic barriers.
- **Promoting Gender Equity:** Integrating gender considerations into conservation policies and programs can help ensure that women's perspectives are fully represented and valued.
- **Fostering Collaboration:** Strengthening partnerships between women-led organizations, governments, and international agencies can enhance the effectiveness of conservation efforts and ensure sustainable outcomes.

The Coral Triangle Initiative exemplifies how women's leadership can drive impactful marine conservation efforts. Through their involvement in sustainable fisheries management, marine protected areas, mangrove restoration, climate change adaptation, and community-based conservation strategies, women have demonstrated their ability to address complex environmental challenges and contribute to the preservation of one of the world's most biodiverse marine regions. Their leadership not only enhances conservation outcomes but also empowers communities and fosters sustainable development.

4.5.The Barefoot College, India: Empowering Women through Solar Engineering

The Barefoot College, an innovative non-governmental organization based in India, has made a profound impact through its Solar Mama program, which embodies the intersection of women's empowerment and environmental sustainability. Established in 1972 by Bunker Roy, the Barefoot College focuses on promoting grassroots solutions to rural development challenges, with the Solar Mama program standing out as a notable example of its success.(Parvin Shaikh, Neda. "Feminism in Practice, 2024)

The Solar Mama program specifically targets rural women, including grandmothers, providing them with the training needed to become solar engineers. This initiative addresses two critical issues simultaneously: energy poverty and gender inequality. Women from remote and underserved communities are selected and brought to the Barefoot College for training in solar technology. The program's curriculum includes hands-on instruction in installing, maintaining, and repairing solar panels, empowering these women with technical skills that are highly valuable in their communities.

Upon completing their training, these newly certified Solar Mamas return to their villages, where they implement their skills by installing solar panels in homes, schools, and other community buildings. This deployment of renewable energy has several significant benefits. Firstly, it helps to alleviate energy poverty by providing a reliable source of electricity to areas that previously depended on kerosene lamps or had no access to power. This improvement in energy access enhances the quality of life, as families can benefit from better lighting, access to educational resources, and improved health outcomes.

Secondly, the shift to solar energy contributes to environmental sustainability by reducing reliance on fossil fuels. The use of renewable solar power minimizes carbon emissions and lessens the environmental degradation associated with traditional energy sources. This aligns with global efforts to combat climate change and promotes a more sustainable lifestyle in rural areas.

In addition to these environmental and practical benefits, the Solar Mama program also plays a crucial role in challenging and transforming gender stereotypes. By training women as technical experts, the program defies traditional gender roles that often limit women's participation in technical fields. The Solar Mamas become leaders and role models within their communities, demonstrating that women can excel in areas traditionally dominated by men. Their success not only fosters greater gender equality but also inspires other women and girls to pursue careers in science, technology, engineering, and mathematics (STEM).

Furthermore, the Solar Mama program has broader socio-economic impacts. The trained women often become community leaders, using their new skills to contribute to local development and advocate for sustainable practices. Their involvement in solar energy projects often leads to the creation of new job opportunities and the strengthening of local economies.

The Barefoot College's Solar Mama program exemplifies how targeted empowerment initiatives can address multiple challenges at once. By combining women's empowerment with environmental sustainability, the program has created a model that not only transforms lives but also contributes to global goals such as clean energy access and gender equality. The success of this program underscores the potential of grassroots solutions to drive meaningful change and highlights the important role that women play in advancing both environmental and social progress.

4.6. Women's Environment and Development Organization (WEDO): Pioneering Gender-Inclusive Environmental Policy

The Women's Environment and Development Organization (WEDO) stands as a leading force in the integration of gender perspectives into global environmental policy. Established in 1991, WEDO is an international non-governmental organization dedicated to promoting women's rights and environmental sustainability through high-level policy advocacy and strategic influence. Its work highlights how women's leadership can drive transformative changes in environmental governance on the global stage.(Banerjee, Nandita, and Varucha Misra.2024)

WEDO's primary mission is to ensure that gender considerations are woven into the fabric of environmental and development policies. The organization operates at the intersection of gender and environment, advocating for policies that recognize and address the unique challenges and contributions of women in these fields. WEDO's advocacy efforts are aimed at ensuring that women's voices and perspectives are not only heard but also incorporated into policy frameworks that shape global environmental outcomes.

One of WEDO's notable achievements is its instrumental role in the development and implementation of the Paris Climate Agreement. The organization has been actively involved in international climate negotiations, working to ensure that gender considerations are integral to climate policies and agreements. WEDO's advocacy contributed to the inclusion of gender equality as a key component in the Paris Agreement's framework, which emphasizes the need for gender-responsive climate action and recognizes the differential impacts of climate change on women and men.

In addition to its work on the Paris Agreement, WEDO has been pivotal in advancing gender-inclusive policies across various global environmental platforms. The organization has played a significant role in shaping the outcomes of major international conferences and agreements, such as the United Nations Framework Convention on Climate Change (UNFCCC) and the Convention on Biological Diversity (CBD). Through its advocacy, WEDO ensures that these agreements not only address environmental issues but also consider the social dimensions of gender inequality and the role of women in achieving environmental sustainability.

WEDO's approach to policy advocacy involves a combination of grassroots mobilization, strategic partnerships, and high-level lobbying. The organization collaborates with a diverse range of stakeholders, including governments, international organizations, and civil society groups, to promote gender-responsive policies and practices. By building coalitions and leveraging its expertise, WEDO amplifies the impact of its advocacy efforts and drives systemic change in global environmental governance.

Furthermore, WEDO's work extends beyond policy advocacy to include capacity building and knowledge sharing. The organization provides training and resources to women leaders and activists, empowering them to engage effectively in environmental policy discussions and decision-making processes. This support enhances the capacity of women to influence policies at local, national, and international levels, thereby strengthening their role in environmental governance.

WEDO also focuses on monitoring and evaluating the implementation of gender-responsive policies, holding governments and institutions accountable for their commitments. Through its research and analysis, the organization tracks progress and identifies gaps in policy implementation, ensuring that gender considerations are consistently integrated into environmental strategies.

The impact of WEDO's work is evident in the growing recognition of the importance of gender in environmental policy and the increased representation of women in high-level environmental negotiations. By demonstrating how women's leadership can shape global environmental governance, WEDO provides a compelling example of the transformative potential of gender-inclusive policy advocacy.

In summary, the Women's Environment and Development Organization (WEDO) exemplifies how women's leadership at the policy level can drive significant changes in global environmental governance. Through its strategic advocacy, partnerships, and capacity-building efforts, WEDO ensures that gender perspectives are integral to international environmental agreements and policies, contributing to more equitable and effective conservation strategies.

These case studies illustrate the diverse ways in which women are leading environmental conservation efforts across different scales and contexts. They highlight common themes such as the integration of social and economic concerns with environmental goals, the importance of community engagement, and the power of networks and alliances in amplifying women's voices in conservation.

5. RESEARCH AND EXPERIMENTAL STUDIES: KEY FINDINGS ON WOMEN'S LEADERSHIP IN ENVIRONMENTAL CONSERVATION

Our research has produced several important findings regarding the role of women's leadership in environmental conservation:

1. **Leadership Representation:** The quantitative survey revealed a notable gender disparity in leadership within environmental organizations. Although women make up 48% of the overall workforce, they hold only 31% of senior leadership positions. This disparity is more pronounced in larger, international organizations, where women occupy only 24% of leadership roles compared to 39% in smaller, community-based organizations. This suggests that women are underrepresented in top decision-making positions, particularly in larger, more global entities.(Koengkan, Matheus, et al., 2024)
2. **Barriers to Advancement:** Qualitative interviews identified several significant barriers that hinder women's advancement in leadership roles:
    - **Gender Bias and Discrimination:** A substantial 78% of interviewees reported encountering gender bias and discrimination, which significantly affects women's career progression.
    - **Limited Access to Mentorship and Professional Networks:** 65% of participants noted the challenge of limited access to mentorship and professional networks, which are crucial for career development and advancement.
    - **Work-Life Balance Challenges:** 61% of interviewees highlighted difficulties related to work-life balance, particularly for women with care-

giving responsibilities, which can impede their ability to pursue leadership roles.

- **Lack of Visibility and Recognition:** 57% of respondents mentioned that insufficient recognition of their contributions affects their professional advancement.
- **Limited Access to Funding and Resources:** 52% of interviewees pointed to the lack of funding and resources available for women-led initiatives as a significant barrier.

3. **Impact of Women's Leadership:** Organizations with a higher proportion of women in leadership roles (over 40%) reported several positive outcomes:
   - **Community-Based Conservation Approaches:** These organizations were 28% more likely to implement community-based conservation strategies, reflecting a greater focus on local engagement and participatory methods.
   - **Intersectional Programs:** There was a 35% increase in programs that address the intersection of gender and environmental issues, indicating a broader integration of gender considerations into conservation efforts.
   - **Successful Stakeholder Engagement:** Organizations with higher female leadership experienced 22% higher rates of successful stakeholder engagement in conservation projects, demonstrating the effectiveness of women-led initiatives in fostering collaborative relationships.
4. **Leadership Styles and Approaches:** Thematic analysis of interviews revealed that women leaders in environmental conservation often adopt distinctive leadership styles and approaches:
   - **Collaborative Decision-Making:** Women leaders frequently emphasize collaborative and participatory decision-making processes, which foster inclusive and democratic practices.
   - **Holistic Perspectives:** They adopt holistic approaches that integrate social, economic, and environmental concerns, reflecting a comprehensive view of conservation.
   - **Long-Term Perspectives:** Women leaders often focus on long-term, intergenerational perspectives on conservation, prioritizing sustainable and enduring solutions.
   - **Education and Capacity Building:** There is a strong emphasis on education and capacity building within communities, aimed at empowering local populations and enhancing their conservation efforts.
5. **Intersectionality and Leadership:** The research underscored the importance of intersectional factors in shaping women's leadership experiences:

- **Indigenous Women:** Indigenous women leaders reported unique challenges related to cultural preservation and land rights, with 92% of indigenous respondents highlighting these issues.
- **Women from the Global South:** Women from the Global South emphasized the need for decolonial approaches to conservation, with 78% advocating for strategies that address colonial legacies.
- **LGBTQ+ Women:** LGBTQ+ women leaders noted additional barriers related to discrimination and the lack of inclusive policies, with 85% of LGBTQ+ respondents experiencing these challenges.

6. **Strategies for Success:** Successful women leaders identified several strategies for overcoming barriers:
   - **Building Networks and Alliances:** 82% of interviewees highlighted the importance of building strong networks and alliances with other women leaders to enhance support and collaboration.
   - **Mentorship and Training Programs:** Developing mentorship programs and leadership training tailored to women in conservation was mentioned by 75% of respondents as a key strategy for advancing women's careers.
   - **Advocating for Supportive Policies:** 68% of interviewees advocated for organizational policies that support work-life balance and address gender bias, recognizing these as crucial for creating equitable work environments.
   - **Leveraging Technology and Social Media:** 63% of successful women leaders emphasized the use of technology and social media to amplify women's voices and initiatives, enhancing visibility and impact.
7. **Emerging Trends:** The research identified several emerging trends in women's environmental leadership:
   - **Intersection of Climate Justice and Gender Equality:** There is an increasing focus on integrating climate justice with gender equality, highlighting the need for policies that address both environmental and social issues.
   - **Role of Traditional Ecological Knowledge:** The recognition of women's traditional ecological knowledge is growing, emphasizing its importance in effective conservation practices.
   - **Youth-Led, Women-Centered Movements:** The rise of youth-led, women-centered environmental movements reflects a new generation of leaders driving innovative and inclusive approaches to conservation.
   - **Technology and Data-Driven Approaches:** The increasing use of technology and data-driven approaches in women-led conservation

initiatives showcases the potential for leveraging advanced tools to enhance environmental outcomes.

These findings provide valuable insights into the current state of women's leadership in environmental conservation, revealing both the challenges that persist and the significant potential for positive impact. They underscore the need for targeted interventions to support women's advancement and highlight the benefits of increased women's leadership for achieving effective and sustainable conservation outcomes.

6.DISCUSSION

The case studies, literature review, and original research presented in this chapter offer a nuanced picture of women's leadership in environmental conservation. Several key themes emerge from these findings:

1. Transformative Potential of Women's Leadership: The evidence consistently demonstrates that women's leadership in conservation can lead to more inclusive, holistic, and effective outcomes. The success stories from diverse contexts – from the grassroots efforts of the Green Belt Movement to the global policy advocacy of WEDO – illustrate how women leaders often bring unique perspectives that integrate environmental concerns with social and economic issues. This approach aligns with the growing recognition of the interconnectedness of environmental conservation with broader sustainable development goals.(Koengkan, Matheus, et al., 2024)
2. Persistent Barriers and Intersectional Challenges: Despite the clear benefits of women's leadership, significant barriers persist. The underrepresentation of women in senior leadership positions, particularly in larger organizations, points to systemic issues of gender bias and discrimination. Moreover, the intersectional analysis reveals how factors such as race, ethnicity, socioeconomic status, and geographic location compound these challenges. The experiences of indigenous women leaders and those from the Global South highlight the need for approaches that address multiple, intersecting forms of marginalization.
3. Community-Centered and Participatory Approaches: A recurring theme across successful women-led initiatives is the emphasis on community engagement and participatory decision-making. This approach, exemplified by projects like AMWAE in Ecuador, not only leads to more sustainable conservation outcomes but also addresses issues of environmental justice and community empowerment. The higher rates of stakeholder engagement reported by organizations

with strong women's leadership underscore this trend.(Vedam, Saraswathi, et al., 2024)

4. Integration of Traditional and Modern Knowledge: The case studies and research findings highlight the crucial role of women, particularly indigenous women, in bridging traditional ecological knowledge with modern conservation strategies. This integration, seen in initiatives like the Barefoot College's Solar Mama program, offers innovative solutions that are both culturally appropriate and technologically advanced. This approach is particularly valuable in addressing complex environmental challenges that require both local insights and global perspectives.
5. Importance of Networks and Alliances: The success of organizations like the Women's Earth Alliance demonstrates the power of networks in amplifying women's voices and initiatives in conservation. These networks provide crucial support in terms of resources, mentorship, and knowledge sharing. The research findings indicating that building strong networks is a key strategy for overcoming barriers underscore the importance of collective action in advancing women's leadership.
6. Policy and Organizational Change: The research reveals a clear need for policy interventions and organizational changes to support women's leadership in conservation. This includes addressing work-life balance issues, implementing inclusive policies, and actively working to counter gender bias in hiring and promotion. The success of WEDO in influencing global environmental policy highlights the importance of women's leadership at all levels, from local initiatives to international governance.
7. Emerging Role of Technology: The increasing use of technology and data-driven approaches in women-led conservation initiatives represents an important trend. This not only enhances the effectiveness of conservation efforts but also provides new avenues for women to assert leadership and overcome traditional barriers. The use of social media and digital platforms to amplify women's voices in environmental advocacy is particularly noteworthy.
8. Economic Empowerment and Conservation: Many successful women-led initiatives, such as AMWAE's sustainable cacao production, demonstrate the potential for aligning economic empowerment with conservation goals. This approach addresses the often-overlooked economic dimensions of environmental issues and provides sustainable livelihoods for communities, particularly women.
9. Intergenerational and Long-term Perspectives: The research indicates that women leaders often bring long-term, intergenerational perspectives to conservation efforts. This approach, exemplified by the Green Belt Movement's focus on both immediate reforestation and long-term community empowerment, is crucial for addressing complex, long-term environmental challenges like climate change.

10. Challenges in Scaling and Replication: While many successful models of women's leadership in conservation exist, challenges remain in scaling these initiatives and replicating them across different contexts. The research highlights the need for context-specific approaches that consider local cultural, social, and economic factors.

These insights demonstrate that women's leadership is not just a matter of gender equality but a crucial factor in achieving effective and sustainable environmental conservation. The unique perspectives, approaches, and skills that women bring to conservation leadership have the potential to transform how we address global environmental challenges.

7.FUTURE TRENDS

Based on the research findings and emerging patterns in the field, several key trends are likely to shape the future of women's leadership in environmental conservation:

1. Increasing Focus on Climate Justice: As the impacts of climate change become more severe, women's leadership in climate justice movements is likely to grow. We can expect to see more women-led initiatives that address the disproportionate effects of climate change on marginalized communities, particularly in the Global South.(Dobrich, Emily Olga Rosa., 2024)
2. Technology-Driven Conservation: The integration of technologies such as artificial intelligence, remote sensing, and blockchain in conservation efforts is likely to accelerate. Women leaders may play a crucial role in ensuring these technologies are applied in ways that are inclusive and respect local knowledge systems.
3. Rise of Youth-Led, Women-Centered Movements: Following the model of youth climate activists like Greta Thunberg, we can anticipate a surge in youth-led, women-centered environmental movements. These movements are likely to leverage social media and digital platforms to mobilize global support and pressure for environmental action.
4. Expansion of Indigenous Women's Leadership: As the importance of indigenous knowledge in conservation gains recognition, indigenous women's leadership is likely to become more prominent. This may lead to increased integration of traditional ecological knowledge in mainstream conservation strategies.

5. Intersectional Approaches to Conservation: Future conservation initiatives are likely to adopt more explicitly intersectional approaches, recognizing how gender intersects with factors like race, class, and disability in shaping environmental experiences and leadership opportunities.
6. Green Economy and Women's Entrepreneurship: The growth of the green economy is likely to create new opportunities for women's leadership and entrepreneurship in areas such as renewable energy, sustainable agriculture, and eco-tourism.
7. Transboundary and Global Collaboration: As environmental challenges increasingly transcend national borders, we can expect to see more transnational networks of women leaders collaborating on global conservation initiatives.
8. Integration of Mental Health and Well-being: Future conservation efforts may increasingly recognize the links between environmental health and human well-being, with women leaders at the forefront of integrating mental health considerations into conservation strategies.
9. Urban Conservation Leadership: With growing urbanization, women's leadership in urban conservation and sustainable city planning is likely to become more prominent, addressing issues like urban biodiversity, green spaces, and sustainable urban development.
10. Innovative Financing Mechanisms: New financing models, such as gender-lens investing in conservation and women-led green bonds, may emerge to support and scale up women's leadership in environmental initiatives.

These trends suggest a future where women's leadership in environmental conservation becomes increasingly integral to addressing global environmental challenges, driven by technological innovation, intersectional approaches, and a growing recognition of the value of diverse perspectives in conservation efforts.

8.CONCLUSION

This chapter has explored the multifaceted landscape of women's leadership in environmental conservation, revealing both persistent challenges and inspiring successes. Through a comprehensive analysis of literature, case studies, and original research, we have demonstrated the transformative potential of women's leadership in creating more effective, inclusive, and sustainable approaches to environmental conservation.

Key findings include:

1. Women's leadership in conservation often brings unique perspectives that integrate environmental, social, and economic concerns, leading to more holistic and sustainable outcomes.
2. Despite progress, significant barriers to women's leadership persist, including gender bias, limited access to resources, and intersectional challenges related to factors like race, class, and geographic location.
3. Successful women-led conservation initiatives frequently emphasize community engagement, participatory decision-making, and the integration of traditional and modern knowledge systems.
4. Networks and alliances play a crucial role in amplifying women's voices and initiatives in conservation, providing essential support and resources.
5. There is a clear need for policy interventions and organizational changes to support and advance women's leadership in the field of conservation.
6. Emerging trends, including the use of technology, focus on climate justice, and rise of youth-led movements, are shaping the future of women's environmental leadership.

Looking ahead, the field of environmental conservation stands to benefit greatly from increased women's leadership. To realize this potential, concerted efforts are needed at multiple levels:

- Policymakers should work to create supportive legal and policy frameworks that promote gender equality in environmental decision-making and resource allocation.
- Conservation organizations should implement targeted strategies to recruit, retain, and advance women leaders, including mentorship programs, inclusive policies, and efforts to address unconscious bias.
- Funders and donors should prioritize support for women-led conservation initiatives and research on gender dimensions of environmental issues.
- Educational institutions should focus on developing pipeline programs to encourage girls and young women to pursue careers in environmental science and conservation leadership.
- Researchers should continue to investigate the impacts of women's leadership on conservation outcomes, with a particular focus on intersectional approaches and long-term effects.

In conclusion, enhancing women's leadership in environmental conservation is not just a matter of equality; it is essential for developing effective, sustainable solutions to our most pressing environmental challenges. By recognizing and supporting the diverse contributions of women leaders, we can create a more inclusive, innovative, and impactful approach to environmental stewardship. As we face unprecedented global environmental crises, the perspectives, skills, and leadership of women will be more crucial than ever in shaping a sustainable future for our planet.

REFERENCES

Banerjee, N., & Misra, V. (2024). Role of Gender in Sustainable Mitigation for Climate Change in Agriculture. In *The Climate-Health-Sustainability Nexus: Understanding the Interconnected Impact on Populations and the Environment* (pp. 51–69). Springer Nature Switzerland. DOI: 10.1007/978-3-031-56564-9_3

Cole, S., Tallman, P., Salmon-Mulanovich, G., & Rusyidi, B. (2024). Water insecurity is associated with gender-based violence: A mixed-methods study in Indonesia. *Social Science & Medicine*, 344, 116507. DOI: 10.1016/j.socscimed.2023.116507 PMID: 38340386

Commer, C. O. (2024). When Conscious Consumerism Backfires: A Critical Study on Social Exclusion, Partnerships, and White Saviorism. *Sociology*, 14(1), 9–26.

Dalyan, M., Syarifuddin, , Yulandari, , Mastang, , Suma, M., Sosrohadi, S., & Andini, C. (2024). Harmony and Sustainability: Traditional Ecological Knowledge Systems of the Kaluppini Indigenous People. *International Journal of Religion*, 5(6), 82–92. DOI: 10.61707/tdyqck03

Dobrich, E. O. R. (2024). Perspectives on Advancing Gender and Environmental Justice: Implications and Applications of a Conceptual Model for Supporting Women's Environmental Leadership. *Journal of Academic Perspectives Volume*, 3, 130.

Dobrich, E. O. R. (2024). Perspectives on Advancing Gender and Environmental Justice: Implications and Applications of a Conceptual Model for Supporting Women's Environmental Leadership. *Journal of Academic Perspectives Volume*, 3, 130.

Dube, M. W. (2024). African Eco-feminisms: African Women Writing Earth, Gender and the Sacred. In *Ecofeminist Perspectives from African Women Creative Writers: Earth, Gender, and the Sacred* (pp. 3–33). Springer International Publishing. DOI: 10.1007/978-3-031-48509-1_1

Gremaud, A., & Kochtcheeva, L. (2024). The role of non-state actors in promoting environmental justice: A comparative study of Kenya and South Africa. *GeoJournal*, 89(3), 110.

Kakade, K. S., Kulkarni, A. M., & Naik, K. R. (2024). *The Role of Women Leadership in Environmental Management*. Opportunities and Challenges for Women Leaders in Environmental Management.

Koengkan, M., Fuinhas, J. A., Auza, A., Castilho, D., & Kaymaz, V. (2024). Environmental Governance and Gender Inclusivity: Analyzing the Interplay of PM2. 5 and Women's Representation in Political Leadership in the European Union. *Sustainability*, 16(6), 2492.

Mascarenhas, O. A., Thakur, M., & Kumar, P. (2024). Critical Thinking Applied to Ecofeminism. In A Primer on Critical Thinking and Business Ethics (pp. 89-118). Emerald Publishing Limited.

Parvin Shaikh, N. (2024). Feminism in Practice: Learning from the Barefoot "Solar Mamas". *Journal of International Women's Studies*, 26(3), 14.

Pedroza-Gutiérrez, C., Solano, N., Fernández-Rivera-Melo, F., & Hernández-Herrera, I. (2024). What are the factors that enable or hinder gender equality? Lessons and experiences on gender dynamics in Mexican small-scale fisheries. *Marine Policy*, 159, 105944. DOI: 10.1016/j.marpol.2023.105944

Pratiwi, H., Hutabarat, Z. S., Nabila, R., Hevinosa, A. D., Mauli, C., Syaputri, A. H., Nisa, A., Hidayatullah, M. S., Tyaningsih, D. R., Putri, D. I., Fakhira, A. L., Aprilia, M. D., & Al Zahra, F. (2024). Blue Economic Beacon: Red And White Leads The Way Through Niche Diplomacy. *International Journal of Multidisciplinary Sciences and Arts*, 3(1), 244–251. DOI: 10.47709/ijmdsa.v3i1.3947

Sreevas, T. V. T. K., & Kulkarni, P. V. (2024). Women leaders in environmental management. In *Opportunities and Challenges for Women Leaders in Environmental Management* (pp. 173–191). IGI Global.

Vedam, S., Stoll, K., Tarasoff, L., Phillips-Beck, W., Lo, W., MacDonald, K., Metellus, A., Rost, M., Scott, M., Hodge, K., Korchinski, M., van der Pijl, M., Alonso, C., Clark, E., Tatum, A., Olson, R., Xie, K., Decker, M., Wenzel, K., & Hall, W. (2024). The RESPCCT Study: Community-led Development of a Person-Centered Instrument to Measure Health Equity in Perinatal Services. *Journal of Participatory Research Methods*, 5(1), 1. DOI: 10.35844/001c.94399

Chapter 13
Women, Gender Equality, and Environmental Justice

Supratik Guha
https://orcid.org/0000-0001-7447-4469
Shahid Matangini Hazra Government General Degree College for Women, India

ABSTRACT

Environmental justice and gender equality are interconnected issues that highlight the need for an inclusive approach to addressing environmental challenges and promoting social equity. Environmental justice seeks to address the disproportionate impact of environmental hazards on marginalized communities, including women, who often bear the brunt of ecological degradation due to pre-existing social inequalities. Gender equality, on the other hand, aims to ensure that all individuals, regardless of gender, have equal opportunities and rights. The relationship between environmental justice and gender equality is deeply interconnected. This chapter highlights the importance of interdisciplinary research and action in tackling the dual challenges of environmental injustice and gender inequality. By integrating gender perspectives into environmental policies and practices, we can create more resilient and equitable societies that are better equipped to face the environmental challenges of the future.

INTRODUCTION

Environmental justice and gender equality are two critical dimensions of social equity that are deeply interconnected. Both aim to address systemic inequalities and promote fairness, whether in the distribution of environmental benefits and burdens or in the treatment of different genders. Understanding the relationship between these two issues is essential for creating inclusive and effective solutions to the world's

DOI: 10.4018/979-8-3693-6069-9.ch013

most pressing environmental challenges. This chapter explores the intersection of environmental justice and gender equality, examining how these issues affect one another and highlighting the importance of integrated approaches to achieve sustainable development. **Environmental Justice** is the fair treatment and meaningful involvement of all people regardless of race, color, national origin, or income concerning the development, implementation, and enforcement of environmental laws, regulations, and policies. Environmental Justice seeks to ensure that no group of people, especially marginalized communities, bears a disproportionate share of negative environmental consequences. The key Principles of Environmental Justice:

1. **Equity**: Addressing the uneven distribution of environmental benefits and burdens.
2. **Participation**: Ensuring that affected communities have a voice in environmental decision-making.
3. **Access to Information**: Providing transparent and accessible information to all stakeholders.
4. **Policy Integration**: Incorporating EJ considerations into broader policy frameworks.

On the other hand, **Gender Equality** refers to the equal rights, responsibilities, and opportunities of all genders. It emphasizes the need to address systemic inequalities that disadvantage women and non-binary individuals, ensuring equal participation in social, economic, and political spheres. The key Principles of Gender Equality:

1. **Empowerment**: Enhancing the capacity of women and marginalized genders to participate in decision-making.
2. **Equity**: Ensuring equal access to resources and opportunities.
3. **Protection**: Safeguarding against gender-based violence and discrimination.
4. **Inclusion**: Promoting gender-sensitive policies and practices.

Power disparities for women and other marginalized groups in the Global South are perpetuated by historical, market-based, and neoliberal "solutions" to environmental degradation as the climate catastrophe worsens. Marginalized people frequently have fewer resources and are less able to adjust to environmental sabotage and natural catastrophes. Women are disproportionately affected by severe climate change, as they are more likely to be victims of extreme weather events such as droughts, floods, crop loss, and lack of access to clean water. They also have to endure risky post-relief procedures for which they are ill-prepared.

I. INTERSECTION OF ENVIRONMENTAL JUSTICE AND GENDER EQUALITY

Social justice, environmental sustainability, and gender equality are interconnected and mutually reinforcing principles that are essential for building a more equitable and resilient world. Here's how they intersect:

1. Environmental Sustainability and Social Justice

- Environmental sustainability involves ensuring that current and future generations can meet their needs without compromising the ability of future generations to do the same. It requires the responsible use of natural resources, the protection of ecosystems, and the mitigation of environmental degradation and climate change.
- Social justice encompasses the fair and equitable distribution of resources, opportunities, and rights within society. It seeks to address systemic inequalities, discrimination, and marginalization based on factors such as race, ethnicity, socioeconomic status, and gender.
- Environmental degradation disproportionately affects marginalized and vulnerable communities, exacerbating existing inequalities and injustices. Low-income communities, communities of color, indigenous peoples, and marginalized groups often bear the brunt of environmental pollution, climate impacts, and resource depletion, while having limited access to resources and decision-making power.

2. Gender Equality and Environmental Sustainability

- Gender equality refers to the equal rights, opportunities, and treatment of all genders. It recognizes that gender-based discrimination and stereotypes can limit individuals' ability to fully participate in society and access resources.
- Women and girls are often more vulnerable to the impacts of environmental degradation and climate change due to existing gender inequalities. They may have limited access to land, water, and natural resources, as well as reduced decision-making power and economic opportunities.
- Promoting gender equality is essential for achieving environmental sustainability. Research has shown that when women are empowered and involved in decision-making processes related to natural resource management and environmental protection, conservation efforts are more effective and sustainable.

3. Intersections and Solutions

- Addressing the interconnected challenges of social justice, environmental sustainability, and gender equality requires holistic and intersectional approaches. This means recognizing the interlocking systems of oppression and privilege that shape people's experiences and opportunities.
- Policies and initiatives that promote environmental justice, such as equitable access to clean air and water, sustainable energy, and green jobs, can help address both environmental and social injustices. Similarly, efforts to advance gender equality, such as empowering women's leadership and promoting women's rights to land and resources, can contribute to environmental sustainability.
- Intersectional advocacy and activism that center the voices and experiences of marginalized communities are crucial for driving systemic change. By working together to address root causes of injustice and inequality, we can create more inclusive, equitable, and sustainable societies for all.

In summary, achieving environmental sustainability, social justice, and gender equality are interconnected goals that require collaborative and intersectional approaches. By addressing the underlying drivers of inequality and injustice, we can build a more just and sustainable world where everyone has the opportunity to thrive.

Vulnerability to Environmental Hazards

- Women, especially in developing countries, are often more vulnerable to environmental hazards due to their roles in households and communities, such as gathering water, fuel, and food.
- Climate change disproportionately affects women, who may have less access to resources, education, and information necessary to adapt to environmental changes.

Economic Impacts

- Environmental degradation can impact economic activities traditionally performed by women, such as agriculture and informal work.
- Women often have fewer economic resources, making it harder for them to recover from environmental disasters.

Health Implications

- Exposure to environmental pollutants can have serious health impacts, particularly for pregnant women and children.
- Lack of access to clean water and sanitation affects women's health and hygiene disproportionately.

Participation and Leadership

- Women are underrepresented in environmental decision-making processes at local, national, and international levels.
- Empowering women and ensuring their participation can lead to more effective and equitable environmental policies.

II. STRATEGIES FOR PROMOTING BOTH ENVIRONMENTAL JUSTICE AND GENDER EQUALITY

1. **Inclusive Policy-Making**: Incorporating gender perspectives into environmental policies to address the specific needs and contributions of women.
2. **Capacity Building**: Providing education and training to women on environmental issues and leadership.
3. **Research and Data Collection**: Gathering sex-disaggregated data to understand the differential impacts of environmental issues on men and women.
4. **Community Engagement**: Ensuring that women's voices are heard and valued in community-based environmental initiatives.
5. **Integrated Approaches**: Developing programs that address both environmental and gender equity goals simultaneously.

Positive change in the spheres of gender, environment, and human rights is essential for creating a more just, equitable, and sustainable world. Here are some examples of positive changes and initiatives in each of these areas:

1. Gender

Increased representation of women in leadership positions: Efforts to promote gender diversity and inclusion have led to more women occupying leadership roles in politics, businesses, academia, and other sectors. This has helped challenge gender stereotypes, address gender-based discrimination, and promote women's empowerment and participation in decision-making processes.

Legal and policy reforms: Many countries have enacted laws and policies to promote gender equality and protect women's rights, including laws against gender-based violence, measures to promote women's economic empowerment, and policies to support work-life balance and parental leave.

Feminist movements and advocacy: Feminist movements around the world have mobilized to demand gender equality and challenge patriarchal norms and structures. These movements have led to increased awareness of gender issues, changes in public attitudes, and policy reforms to address gender-based discrimination and violence.

2. Environment

Conservation initiatives: Conservation organizations and grassroots initiatives are working to protect and restore ecosystems, conserve biodiversity, and address environmental challenges such as climate change, deforestation, and pollution. These efforts involve community-based conservation, sustainable land management practices, and advocacy for environmental policies and regulations.

Renewable energy transition: The transition to renewable energy sources such as solar, wind, and hydro power is helping reduce reliance on fossil fuels and mitigate climate change. Investments in renewable energy technologies are creating green jobs, reducing greenhouse gas emissions, and promoting energy access and sustainability.

Sustainable consumption and production: There is growing awareness of the need to shift towards more sustainable consumption and production patterns. Initiatives promoting circular economy principles, sustainable agriculture practices, and waste reduction are helping minimize environmental impacts and promote resource efficiency.

3. Human Rights

Legal reforms and accountability mechanisms: Many countries have strengthened legal frameworks and established accountability mechanisms to protect and promote human rights. This includes measures to combat discrimination, ensure access to justice, and hold perpetrators of human rights abuses accountable.

Humanitarian aid and development assistance: International organizations, NGOs, and governments provide humanitarian aid and development assistance to support communities affected by conflict, poverty, and natural disasters. These efforts aim to uphold human rights, provide essential services, and promote sustainable development and resilience.

Advocacy and awareness-raising: Human rights advocates, activists, and organizations work to raise awareness of human rights issues, mobilize public support, and hold governments and institutions accountable for their human rights obligations. Social media and digital platforms have become powerful tools for amplifying voices, sharing information, and organizing campaigns for human rights.

These examples demonstrate that positive change is possible in the spheres of gender, environment, and human rights through concerted efforts, collaboration, and advocacy. By working together to address systemic inequalities and injustices, we can create a more inclusive, sustainable, and rights-respecting world for all.

Women play a crucial role as environmental stewards and champions of sustainable practices across the globe. Their unique positions within families, communities, and various sectors enable them to influence and implement sustainable practices effectively. Here's a deeper exploration of this role:

Leadership in Environmental Movements

Grassroots Activism:

- Chipko Movement (India): Women in the Chipko Movement of the 1970s embraced trees to prevent deforestation, highlighting their role in forest conservation and environmental activism.
- Green Belt Movement (Kenya): Founded by Wangari Maathai, this movement has mobilized women to plant millions of trees, combating deforestation, soil erosion, and environmental degradation while empowering women.

Global Advocacy:

- Climate Change Activism: Women like Greta Thunberg and Vandana Shiva have become prominent figures in the global climate movement, advocating for urgent action against climate change and promoting sustainable development.

Impact on Policy and Decision-Making

Government and International Bodies:

- Policy Influence: Women leaders in government and international organizations have pushed for policies that integrate environmental sustainability and gender equality. Notable figures include Christiana Figueres, who played a key role in the Paris Agreement negotiations.
- Community Leadership: Women often lead local environmental committees and initiatives, ensuring that community-specific issues and sustainable practices are addressed.

Sustainable Practices and Innovations

Renewable Energy:

- Solar Sisters (Africa): This organization empowers women by training them to become solar entrepreneurs, providing clean energy solutions to their communities and reducing reliance on fossil fuels.
- Clean Cookstoves: Women-led initiatives have promoted the use of clean cookstoves, which reduce indoor air pollution and reliance on firewood, improving health and environmental outcomes.

Waste Management:

- Waste-to-Wealth Initiatives: Women's groups in various countries have pioneered waste management and recycling programs, turning waste into valuable resources and creating economic opportunities.

Education and Awareness

Environmental Education:

- Teaching Sustainability: Women educators and community leaders play a crucial role in spreading awareness about sustainable practices and environmental conservation through schools, workshops, and community programs.

- Youth Engagement: Women are often at the forefront of educating and engaging young people in environmental stewardship, ensuring the next generation is equipped with the knowledge and skills to continue sustainable practices.

Challenges and Barriers

Gender Inequality:

- Limited Access to Resources: Women often face barriers in accessing land, financial resources, and education, which can hinder their ability to implement and scale sustainable practices.
- Decision-Making Power: Despite their critical roles, women frequently have limited influence in decision-making processes at higher levels, which affects the integration of their perspectives into broader environmental policies.

Overcoming Challenges

Empowerment Programs:

- Capacity Building: Programs that provide education, training, and resources to women can enhance their capacity to lead and innovate in environmental stewardship.
- Policy Advocacy: Advocating for gender-sensitive policies that recognize and support women's roles in environmental management can help overcome structural barriers.

Policies and Practices Promoting Gender Equality in Environmental Protection

1. **Gender-Responsive Climate Policies:**
   (i) **Paris Agreement**: Recognizes gender equality and women's empowerment as vital to climate action. It encourages parties to consider gender-responsive approaches in national climate plans.

(ii) **Green Climate Fund (GCF)**: Implements a gender policy that mandates the integration of gender considerations throughout its projects and funding proposals, ensuring women benefit equally from climate finance.

2. **Inclusive Land Rights and Tenure:**

   (i) **Forest Rights Act (India)**: Recognizes the rights of forest-dwelling communities, including women, to own, manage, and govern forest resources. This law aims to correct historical injustices and promotes women's leadership in forest conservation.

   (ii) **Ethiopia's Land Certification Program**: Issues land certificates jointly in the names of both husbands and wives, which has improved women's land tenure security and their role in sustainable land management.

3. **Empowerment through Education and Capacity Building:**

   (i) **SEWA (Self Employed Women's Association, India)**: Provides education and training on sustainable agricultural practices and natural resource management, empowering women economically and environmentally.

   (ii) **Global Women's Water Initiative**: Trains women in water management and sanitation, enabling them to lead projects that improve community health and environmental sustainability.

4. **Support for Women's Leadership and Participation:**

   (i) **UN Women's Gender and Climate Change Strategy**: Aims to increase women's leadership in climate decision-making at all levels. It supports initiatives that empower women as key stakeholders in climate resilience.

   (ii) **National Adaptation Plans (NAPs)**: Some countries, like Nepal and Bangladesh, incorporate gender considerations into their NAPs, ensuring women's voices are included in adaptation strategies.

5. **Access to Clean Energy:**

   (i) **Solar Sisters (Sub-Saharan Africa)**: Provides women with the training and resources to become solar entrepreneurs, promoting clean energy access while empowering women economically.

   (ii) **Clean Cookstove Initiatives**: Supported by organizations like the Global Alliance for Clean Cookstoves, these initiatives reduce indoor air pollution and deforestation, improving health and environmental outcomes while often being led by women.

Case Studies and Examples

1. **India's Self-Employed Women's Association (SEWA)**: SEWA supports women in managing natural resources and adapting to climate change, demonstrating the link between women's economic empowerment and environmental sustainability.
2. **Kenya's Green Belt Movement**: Founded by Wangari Maathai, this movement combines tree planting with women's empowerment, showing how environmental and gender goals can be achieved together.
3. **Global Women's Water Initiative**: Focuses on training women in water management, addressing both environmental sustainability and gender equity.

Addressing environmental justice and gender equality together is essential for creating sustainable and equitable solutions to environmental challenges. By recognizing and addressing the specific needs and contributions of women and marginalized groups, we can develop more comprehensive and effective policies that promote both environmental sustainability and social justice. Women are crucial to the pursuit of environmental justice, bringing unique perspectives, knowledge, and leadership to the forefront of environmental protection and sustainability. Addressing the specific challenges faced by women and promoting their active involvement in environmental decision-making are essential for achieving equitable and effective environmental outcomes. By integrating gender-responsive approaches in environmental policies and practices, we can ensure that both environmental justice and gender equality are advanced simultaneously.

Here are some examples of how environmental justice and gender equality can be addressed together:

- Supporting women-led initiatives in environmental protection and conservation.
- Promoting education and training for women on climate change adaptation and disaster risk reduction.
- Ensuring that women have equal access to land, water, and other natural resources.
- Integrating gender equality considerations into environmental policies and programs.

The intersection of gender and environment is a vital area for achieving sustainable development. Women's unique experiences and knowledge make them indispensable to environmental stewardship. However, to fully harness their potential, it is imperative to address the gender-specific challenges they face and promote gender-responsive approaches in environmental policies and practices. By doing so, we can create more inclusive, equitable, and sustainable solutions to the pressing environmental issues of our time. By working together, we can create a world where everyone has a healthy environment to live in, and where everyone has the opportunity to thrive.

III. IMPACT OF CLIMATE CHANGE ON WOMEN

Climate change is not a "gender neutral" issue. The effects of climate change are most felt by women and girls, who also face particular risks to their safety, health, and livelihoods. Climate change also exacerbates already-existing gender disparities. Women are more reliant on natural resources globally, but they also have less access to them. Women are disproportionately responsible for securing fuel, water, and food in many locations. For women in low- and lower-middle-income nations, agriculture is the most important sector of the labor market. As primary procurers and agricultural workers, women put in extra effort to provide supplies and income for their family amid dry spells and unpredictable rainfall. Girls are particularly affected since they frequently have to drop out of school to assist their mothers in carrying the heavier load.

As a "threat multiplier," climate change intensifies social, political, and economic pressures in unstable and conflict-affected environments. Women and girls are more vulnerable to all sorts of gender-based violence, such as child marriage, human trafficking, sexual violence associated to conflicts, and other forms of violence, as a result of climate change driving conflict throughout the world. Due to historical gender disadvantages that have produced differences in information, mobility, decision-making, and access to resources and training, women are more likely to sustain injuries during disasters than to survive them. Women and girls are less likely to receive aid and support in the wake of the disaster, which puts their healing, well-being, and means of subsistence at risk. This creates a vicious cycle of susceptibility to more tragedies.

Climate change and catastrophes pose a threat to women's and girls' health because they increase the risks associated with maternal and child health and restrict access to services and healthcare. Studies reveal that high temperatures raise the risk of stillbirth, and that vector-borne diseases including dengue fever, zika virus, and malaria are spreading more widely due to climate change. These infections are associated with poorer outcomes for both mothers and newborns.

While the effects of climate change are not universal, women and girls are disproportionately affected globally. When considering climate change from the perspective of intersectional feminism—which emphasizes how different kinds of inequality frequently coexist and exacerbate one another—it becomes evident that certain groups are particularly vulnerable to its effects: migrant women, women of color, older women, LGBTIQ+ individuals, women and girls with disabilities, and those who live in remote, rural, conflict- or disaster-prone areas.

In recent years, persistent gender inequality has emerged as a key issue on the global development agenda. It is prominently featured in Sustainable Development Goal (SDG) 5, which calls on national governments to develop policies aimed at attaining 'gender equality and empowerment of all women and girls' by 2030. Although SDG Goal 5 is the only one dedicated to gender equality, it is important to note that no economic, social, ecological, or political issue is gender-neutral (Koehler, 2016). This underscores the notion that gender inequality is 'intersectional' - that it is worsened and aggravated by a variety of causes (Kabeer, 2010). Climate change is expected to disproportionately affect women and children due to current unsustainable development policies that prioritize economic expansion over environmental sustainability (Raworth, 2012). Gender equality must be related to climate justice, as per the underlying premise.

Gender disparity is often attributed to political, economic, and social factors that disadvantage women and girls (Seguino, 2000; Quaranta & Dotti, 2018). These studies highlight women's systematic disadvantage, including deliberate exclusion from assets and persisting views of inferiority to men (Eastin, 2018). Davis et al. (2012) identifies a gap in existing study regarding how gender-specific vulnerabilities created by established power relations and societal structures disproportionately affect women in the face of climate change. Limited data exists on how environmental degradation affects gender norms, especially in emerging and transitional countries (Denton, 2002; Montt, 2018). Climate change has a greater impact on marginalized and vulnerable populations, contrary to popular belief that it affects both genders equally. (Eastin, 2018).

Environmental deterioration has three indirect implications on women's economic activity. Environmental deterioration can negatively impact job-sustaining ecosystems and infrastructure (Montt, 2018). Women in poor nations primarily work in climate-sensitive areas like agriculture, forestry, and fishing (Terry, 2009). However, as greenhouse gas emissions contribute to climate change and global warming, female labour participation may be disproportionately affected by environmental degradation and natural resource depletion. Research indicates that women make up the majority of smallholder farmers on marginal areas, which are vulnerable to climate risks such floods, landslides, and droughts (Koehler, 2016). According to the International Labour Organization (2018), environmental pollution has a direct impact on workers' health and labour hours. Thus, when environmental deterioration reaches critical levels, employees may become ill and call in sick (Montt, 2018). According to Kim et al. (2017), workers' health can lead to a decrease in labour supply over time. According to Aragón et al. (2017), environmental pollution has a third impact on labour supply by affecting the care of dependents with illnesses. Research indicates that children are more vulnerable to the negative health effects of air pollution.

The World Health Organization (WHO) reported that in 2016, around 6 lakhs children died from acute lower respiratory infections caused by contaminated air (WHO, 2018). Women and children are disproportionately affected by household air pollution due to their role in cooking and spending time with their mothers. Air pollution can negatively impact women's labor force participation in underdeveloped nations, particularly in Sub-Saharan Africa, where women are primarily responsible for caring for dependents. This is consistent with existing research, which suggests that the presence of young children reduces women's labour force involvement (Afridi et al., 2012). Environmental deterioration, including air, water, and soil pollution, can hinder women's capacity to work when ill or caring for dependents. However, having to care for dependents may limit the opportunities for other females to enter the labour market.

IV. ROLE OF WOMEN ENTREPRENEURS IN ENVIRONMENTAL SUSTAINABILITY

Women entrepreneurs are at the forefront of promoting environmental sustainability through innovative solutions and sustainable business practices. By supporting and empowering women in green business ventures, we can accelerate the transition to a more sustainable and inclusive economy. Providing the necessary resources, education, and policy support will enable more women to lead in this critical area, driving positive change for the environment and society.

Women entrepreneurs play a vital role in driving environmental sustainability and promoting green business practices. Their leadership and innovation can lead to the development of eco-friendly products, sustainable business models, and practices that benefit both the environment and the economy. Here are some key aspects and examples of how women entrepreneurs are contributing to environmental sustainability:

1. **Innovation in Green Technologies**:

Women entrepreneurs often lead in the development and implementation of green technologies, such as renewable energy, sustainable agriculture, and eco-friendly manufacturing processes. Their innovative solutions can significantly reduce environmental footprints and promote sustainability.

2. **Sustainable Business Practices**:

Many women entrepreneurs prioritize sustainability in their business operations. This includes using sustainable materials, reducing waste, conserving energy, and implementing circular economy principles. These practices not only help the environment but can also improve business efficiency and profitability.

3. **Social and Environmental Advocacy**:

Women entrepreneurs frequently advocate for environmental and social causes, leveraging their businesses as platforms to raise awareness and drive change. They often engage in community-building and educational initiatives to promote sustainable practices.

4. **Empowerment and Inclusion**:

By creating inclusive and equitable work environments, women entrepreneurs contribute to social sustainability. They often provide opportunities for marginalized groups, support local communities, and ensure fair labor practices, which are essential for holistic sustainability.

Examples of Women Entrepreneurs Leading in Green Initiatives

(i) **Yvon Chouinard and Rose Marcario, Patagonia**: While Yvon Chouinard founded Patagonia, Rose Marcario, as the former CEO, significantly expanded its environmental initiatives. Under her leadership, Patagonia became a model for sustainable business practices, including the use of recycled materials, environmental activism, and the promotion of fair trade.

(ii) **Anita Roddick, The Body Shop**: Anita Roddick, the founder of The Body Shop, was a pioneer in the beauty industry for her commitment to sustainability. She promoted cruelty-free products, ethical sourcing, and environmental activism, setting a standard for sustainable and socially responsible business practices.

(iii) **Sara Blakely, Spanx**: Sara Blakely, the founder of Spanx, has integrated sustainability into her business model by focusing on eco-friendly materials and production processes. She also supports various environmental causes through the Sara Blakely Foundation.

(iv) **Lauren Singer, Package Free**: Lauren Singer founded Package Free, a company dedicated to reducing waste by offering zero-waste products and promoting sustainable living. Her business model challenges conventional consumer practices and encourages a shift towards a zero-waste lifestyle.

(v) **Leila Janah, Samasource**: Leila Janah, the founder of Samasource, focused on using technology to provide employment to marginalized communities while promoting sustainable practices. Her work demonstrated the potential of combining social impact with environmental responsibility.

Women's representation in political leadership roles varies widely across G20 countries. While some countries have made strides in electing women to high-level positions, others lag behind. For example, Germany has Chancellor Angela Merkel, while Saudi Arabia has limited female representation in government. In several G20 countries, women hold significant proportions of parliamentary seats. For instance, in countries like Canada, the United Kingdom, and France, women hold around 30% to 40% of parliamentary seats. However, in countries like Japan and South Korea, women's representation in parliament remains low.

Gender quotas can indeed increase both the equality and effectiveness of climate policy interventions. By ensuring that women are adequately represented in decision-making processes, gender quotas help to incorporate diverse perspectives and priorities, which can lead to more comprehensive and inclusive climate policies. Gender quotas are a powerful tool for enhancing both equality and effectiveness in climate policy interventions. By ensuring that women are adequately represented in climate decision-making, quotas help create more inclusive, innovative, and sustainable policies. Implementing and supporting gender quotas through legislative measures, capacity building, and public awareness can significantly contribute to addressing the complex challenges of climate change in a fair and effective manner. Gender quotas ensure that women have a seat at the table in climate policy-making, addressing the historical underrepresentation of women in this field. This leads to more balanced and equitable decision-making processes. Quotas empower women by providing them with opportunities to participate in leadership roles. This inclusion can inspire other women and girls to pursue careers in environmental policy and activism. Women often experience climate change differently from men, particularly in vulnerable communities. Their representation ensures that climate policies address the specific needs and challenges faced by women, leading to more equitable outcomes.

For example, Rwanda implemented gender quotas, resulting in women holding 61% of parliamentary seats as of 2021. This representation has contributed to progressive environmental policies, including initiatives for reforestation and renewable energy. On the other hand, Norway's gender quota law requires that 40% of board members in publicly listed companies be women. This has not only promoted gender equality but also led to better environmental, social, and governance (ESG) practices within these companies. In India, the introduction of gender quotas in local governance (Panchayati Raj institutions) has increased women's participation in environmental management. Women-led councils have shown greater effectiveness in implementing water conservation and sanitation projects.

There has been a rise in women's leadership roles in climate negotiations, including the appointment of women as heads of delegations and facilitators of negotiation tracks. Climate agreements, such as the Paris Agreement, increasingly incorporate gender-responsive language, recognizing the importance of gender equality and women's empowerment in climate action.

Women's participation in UN climate negotiations has increased over the years, but there is still progress to be made in achieving gender balance and ensuring that women's voices are fully heard and integrated into decision-making processes. While there have been positive developments, such as the adoption of gender-responsive language in climate agreements and the establishment of gender focal points within the UNFCCC, challenges remain in addressing gender inequalities and promoting women's leadership in climate action. Tipping the balance requires concerted efforts to enhance representation, amplify voices, mainstream gender, adopt intersectional approaches, provide capacity-building support, and ensure accountability. By empowering women as leaders and decision-makers in climate action, we can foster more inclusive, effective, and equitable responses to the urgent challenges of climate change.

V. GENDER EQUALITY- A FOUNDATION FOR ENVIRONMENTAL JUSTICE

The "triple planetary crisis" of pollution, climate change, and loss of natural resources poses one of the biggest threats to human rights. It directly affects several rights, such as the right to food, health, development, and even life itself. All people are affected by the hazards and threats posed by this crisis, but the extent to which they are affected varies depending on their gender, social status, place of residence, and way of life. The impacts on men and women, households, and communities vary according to their susceptibility and resilience levels. Particularly impacted are women and girls, particularly those who are in crisis-affected situations, reside in rural regions, are members of minority groups, or are indigenous people. On the other hand, women's capacity to effectively exercise their environmental rights and cope with climate and environmental crises and disasters is impeded by gender inequality and their unequal access to land, natural resources, and other assets.

Through the realization of environmental rights and the advancement of the environmental rule of law, the UNDP's approach to environmental justice aims to promote justice and accountability in problems pertaining to the environment and climate change. In this endeavor, gender equality is a cross-cutting priority.

It is appropriate to discuss three important ways that gender equality is essential for climate justice in light of the sixty-sixth session of the Commission on the Status of Women's (CSW66) focus on "achieving gender equality and the empowerment of all women and girls in the context of climate change, environmental, and disaster risk reduction policies and programs":

a) A Supportive and Gender-Responsive Legislative Framework Is Necessary Before Women Can Exercise Their Right To A Safe And Healthy Environment

Research shows that traditional, religious, and customary laws and practices limit women's freedom to claim and protect these assets in 123 countries (roughly two thirds of all the States in the world), despite the crucial role women play in managing land, natural resources, and climate action. A foundation for gender-responsive governance is frequently lacking from national law and policy frameworks on environmental issues, disaster risk reduction, and climate change. Furthermore, laws and programs that do include gender equality frequently portray women as disadvantaged groups rather than as change agents, which limits their capacity to reach their full leadership potential.

Women cannot exercise their right to a safe and healthy workplace until there is a supporting and gender-responsive legislative framework in place. Despite the critical role that women play in managing land, natural resources, and climate action, research reveals that traditional, religious, and customary laws and practices hinder women's rights to claim and safeguard these assets in 123 nations (approximately two thirds of all the States in the world). National legal and policy frameworks on environmental issues, catastrophe risk reduction, and climate change often lack a foundation for gender-responsive governance. In addition, gender equality legislation and initiatives often depict women as members of marginalized groups rather than as agents of change, which hinders women's ability to realize their full leadership potential.

b) To Provide Gendered Responses on Climate Justice, People-Centered Organizations Are Essential

Interministerial cooperation and institutions with adequate capacity to address the gender aspects of climate change are necessary for climate justice. The UNDP is concentrating on increasing the ability of organizations, particularly national decision-makers, to take more gender-responsive climate action. A pilot project in Uganda aims to mitigate the adverse effects of climate change and the heightened likelihood of gender-based violence. The project involves strengthening the ability of district and community-level representatives, such as leaders from the Ministries of Water, Environment, and Agriculture, to address the hazards of gender-based violence, among other things.

Furthermore, it is widely acknowledged that women play a crucial role in the governance of resources. Women's involvement in natural resource management, for example, is linked to improved conservation and resource governance at the local level. In order to support women's leadership in the management of natural resources, UNDP has collaborated with 97 nations. This includes backing the Central Province of Zambia in its efforts to mandate gender parity in local governance groups responsible for overseeing indigenous forests.

c) Having Access to Justice Is Essential to Empowering Women to Defend and Exercise Their Rights

Important connections between livelihoods, violence against women, and climate change were revealed by UNDP's field research. One persistent issue is the rise in violence against female environmental human rights advocates, particularly indigenous women. At least 331 environmental defenders, including 44 women, lost their lives in 2020; 69% of them were fighting for the rights of Indigenous peoples, the environment, and land. Women who are already more susceptible to prejudice in many forms—such as LGBTIQ+ and refugee women—face greater dangers and challenges when trying to use the legal system.

Preventing and responding to violence against women and girls in the context of environmental disasters and crises related to climate change is essential to addressing this crisis. Effective investigations of violations and abuses should result in the accountability of offenders and justice for victims. Women environmental human rights defenders should be provided protection. Victims and survivors should have equitable and unhindered access to high-quality services. The establishment of the first dedicated court for violence against women in the occupied Palestinian territories, the Nablus Court, was made possible in the West Bank thanks to a collaborative UNDP, UN Women, and UNICEF effort.

Gender equality is a prerequisite for a sustainable future for everybody. In order to achieve climate justice on the ground, a more integrated strategy will be built around UNDP's environmental justice strategy, which will prioritize gender equality and the realization of women's environmental rights in advocacy and programmatic initiatives.

VI. CONCLUSION

Women are integral to the pursuit of environmental justice, and their active involvement is crucial for achieving equitable and sustainable outcomes. Environmental justice seeks to address the disproportionate environmental burdens borne by marginalized communities, including women, who are often among the most affected by environmental degradation and climate change. The unique roles, knowledge, and experiences of women make them essential agents of change in the movement toward environmental justice.

To ensure the right to a healthy environment, states should prioritize the needs of those who are disproportionately impacted or at high risk of environmental harm. Gender differences should be considered when evaluating environmental damage experiences. Environmental harms disproportionately affect people who have contributed the least, are vulnerable, and have limited resources to deal with the repercussions. The Paris Agreement, Convention on Biological Diversity, and Convention to Combat Desertification, among other multilateral environmental agreements, prioritize gender equality and women's empowerment. They also recognize the importance of women's involvement in climate change responses, biodiversity conservation, and land preservation. Countries must prioritize addressing the gendered implications of interlocking environmental crises through quick and aggressive measures. To effectively address land degradation, climate emergency, mass extinction, zoonotic diseases, and pollution, a gender-responsive, human rights-based approach is necessary. This approach recognizes the unique contributions of all genders, particularly women and those with diverse gender identities. Addressing the core causes of human rights violations, including poverty, inequality, discrimination, and marginalization, is crucial. To combat gender-based discrimination and violence in the environment, states should alter or repeal laws, policies, and practices that cause or perpetuate gendered environmental consequences. Goals include reducing inequality in land ownership, tenure, and resource access, as well as empowering women in conservation, stewardship, and management.

The governments must include sustainable practices into development policies. The environmental standards must be severely enforced to ensure that economic activities and corporations follow environmental justice and ethical governance.

Policies aimed at reducing environmental degradation should address gender exclusion. To improve environmental quality, policies aimed at improving women's rights should be pursued alongside efforts to reform socio-cultural norms, values, and national laws that put women at risk. Evidence suggests that having more women in decision-making positions can lead to better environmental outcomes. Efforts to improve women's social standing can lessen environmental vulnerability and empower women in vulnerable situations.

Promoting and measuring women's environmental leadership requires a comprehensive approach that includes policy actions to ensure equal opportunities, supportive environments, and effective monitoring mechanisms. Women in government have the potential to be powerful agents of change in environmental protection and anti-corruption efforts. By addressing gender inequalities, empowering women leaders, and promoting integrity and transparency in governance, we can harness the full potential of women's leadership to advance sustainable development and combat corruption in the management of natural resources and environmental policies. Women in managerial and leadership positions within the G20 countries vary significantly in terms of representation and influence. While progress has been made in some countries, gender disparities persist in many sectors, including politics, business, academia, and civil society.

Empowering women and integrating their perspectives into environmental justice efforts is not just a matter of equity; it is a strategic necessity for achieving sustainable development. Women's leadership, knowledge, and advocacy are critical in addressing environmental injustices and fostering a more inclusive and resilient future. By prioritizing gender equality in environmental policies and programs, we can create a world where both people and the planet thrive in harmony.

REFERENCES

Afridi, F., Mukhopadhyay, A., & Sahoo, S. (2012). Female labour force participation and child education in India: the effect of the national rural employment guarantee scheme.

Eastin, J. (2018). Climate change and gender equality in developing states. *World Development*, 107, 289–305. DOI: 10.1016/j.worlddev.2018.02.021

Kabeer, N. 2010. Can the MDGs provide a pathway to social justice? The challenge of intersecting inequalities. Child poverty and inequality new perspectives, 57. https://www.undp.org/content/dam/turkey/docs/Publications/mdgs/MDGs%20and%20Inequalities%20Final%20Report.pdf

Koehler, G. (2016). Tapping the sustainable development goals for progressive gender equity and equality policy? *Gender and Development*, 24(1), 53–68. DOI: 10.1080/13552074.2016.1142217

Montt, G. (2018). Too polluted to work? The gendered correlates of air pollution on hours worked. *IZA Journal of Labor Economics*, 7(1), 7. DOI: 10.1186/s40172-018-0067-6

OECD. (2008), *Household Behaviour and the Environment: Reviewing the Evidence*, https://www.oecd.org/environment/consumption-innovation/42183878.pdf

OECD. (2015). *The Economic Consequences of Climate Change*. OECD Publishing.

OECD. (2017), *Policy Brief on Women's Entrepreneurship*, https://www.oecd.org/cfe/smes/Policy-Brief-on-Women-s-Entrepreneurship.pdf

OECD (2019), "Gender Equality and Sustainable Infrastructure"

OECD (2020), *Policies and Practices to Promote Women in Leadership Roles in the Private*

OECD. (2021), Entrepreneurship Policies through a Gender Lens, OECD Studies on SMEs and Entrepreneurship, OECD Publishing, Paris, DOI: 10.1787/71c8f9c9-en

OECD. (2021), *Gender and the Environment: Building Evidence and Policies to Achieve the* OECD, Gender and Sustainable Development: Maximising the Economic, Social and Environmental Role of Women. Organization for Economic Co- operation and Development (OECD) to the United Nations Commission on Sustainable Development (UNCSD) (2008). https://www.oecd.org/social/40881538.pdf

OECD. 2021. Gender and the environment: Building evidence and policies to achieve the SDGs.https://www.oecd-ilibrary.org/environment/gender-and-the-environment_3d32ca39-en

Quaranta, M., & Dotti Sani, G. M. (2018). Left behind? Gender gaps in political engagement over the life course in twenty-seven European countries. *Social Politics*, 25(2), 254–286. DOI: 10.1093/sp/jxy005

Raworth, K. (2012). A safe and just space for humanity: Can we live within the doughnut. *Oxfam Policy and Practice: Climate Change and Resilience*, 8(1), 1–26.

UNDP. 2022. Climate Promise Progress Report, New York, NY: UNDP. https://climatepromise.undp.org/sites/default/files/research_report_document/Climate%20Promise%20Global%20Progress%20Report%202022.pdf

Chapter 14
Women as Environmental Change-Makers in Civil Society

Shyamashree Roy
https://orcid.org/0000-0001-7137-4534
Sister Nibedita Government General Degree College for Girls, India

ABSTRACT

Women have played vital roles as environmental leaders within civil society across the globe. Women's participation in environmental decision-making promotes gender equality and environmental action. Involving women in political decision-making leads to more aggressive climate objectives and policies. Women in civic society form formidable networks to resist environmental degradation and address climate-related injustices.Women's environmental leadership offers a myriad of benefits, contributing to both environmental sustainability and societal progress. By occupying leadership roles, women inspire others, particularly young women, to pursue careers and activism in environmental fields, thereby fostering greater diversity and inclusivity within these sectors. Women leaders are increasingly influential in shaping environmental policies and agendas at local, national, and international levels. This chapter will aim to look into the various roles of women as vital change makers and leaders in the domain of environmental protection and sustainable development in the civil society.

DOI: 10.4018/979-8-3693-6069-9.ch014

INTRODUCTION

Women play a pivotal role in environmental conservation through their various contributions and activities that promote sustainable development and ecological balance. Empowering women and integrating gender perspectives into environmental conservation efforts are critical for achieving sustainable development goals. Women's unique knowledge, skills, and leadership are indispensable for preserving the environment and ensuring a sustainable future for all. Women have played vital roles as environmental leaders within civil society across the globe. Their contributions span various domains, including advocacy, policymaking, grassroots activism, research, and education. Women often face intersecting forms of discrimination based on gender, race, class, and more. This intersectional perspective enables them to understand environmental issues through diverse lenses and advocate for inclusive solutions that address the needs of marginalized communities. Women bring diverse leadership styles to the table, including collaborative, transformational, and servant leadership, which can be particularly effective in addressing complex environmental challenges that require adaptive and innovative approaches. Women leaders serve as role models and mentors, inspiring future generations of environmentalists and empowering women and girls to pursue careers in STEM fields and environmental advocacy.

Examples abound of women making significant contributions as environmental leaders. For instance, **Wangari Maathai**, founder of the Green Belt Movement in Kenya, empowered rural women to plant trees, combat deforestation, and promote sustainable development. **Vandana Shiva**, an Indian environmental activist, has campaigned against genetically modified organisms (GMOs) and advocated for sustainable agriculture and biodiversity conservation. **Gro Harlem Brundtland**, a former Prime Minister of Norway, Brundtland chaired the World Commission on Environment and Development, which produced the landmark report "Our Common Future" in 1987. This report introduced the concept of sustainable development and called for global action to address environmental and developmental challenges. **Christiania Figueres**, as the former Executive Secretary of the United Nations Framework Convention on Climate Change (UNFCCC), Figueres played a key role in negotiating the Paris Agreement in 2015. Her leadership and diplomacy were instrumental in securing commitments from nearly 200 countries to combat climate change. **Greta Thunberg**, the Swedish environmental activist gained international recognition for her climate strike outside the Swedish parliament in 2018. Since then, she has inspired millions of young people worldwide to join the Fridays for Future movement, demanding urgent action on climate change from political leaders and policymakers.

In India, the Chipko Movement, which originated in the 1970s, was a grassroots environmental movement aimed at protecting trees and forests from deforestation. While it was a decentralized movement with active participation from various communities, several key leaders emerged who played significant roles in its development and spread. Among the leaders, the contribution of women leaders were significant. Gaura Devi, a village woman from the Uttarakhand region of India, is remembered for her courageous act of leading a group of women to hug trees in the Reni village in 1974 to prevent them from being felled by loggers. This act of nonviolent resistance became a defining moment in the Chipko Movement and inspired similar protests across the country. Vimla Bahuguna actively participated in the Chipko Movement alongside her husband, Sundarlal Bahuguna. She played a crucial role in organizing women's groups and encouraging their involvement in environmental activism.

While primarily known for her leadership in the Narmada Bachao Andolan, Medha Patkar has also been associated with environmental and social justice movements, including the Chipko Movement. Her advocacy for the rights of marginalized communities and sustainable development aligns with the principles of the Chipko Movement. Medha Patkar played a pivotal role in the Narmada Bachao Andolan (NBA), which was a social movement aimed at advocating for the rights of people affected by the construction of large dams on the Narmada River in India. Medha Patkar provided strong leadership to the NBA, mobilizing affected communities, activists, and supporters to protest against the construction of dams, particularly the Sardar Sarovar Dam. She traveled extensively to rural areas to understand the grievances of local communities and to build solidarity among them.

Women's environmental leadership offers a myriad of benefits, contributing to both environmental sustainability and societal progress. Women often bring unique perspectives and insights to environmental issues, considering factors such as community well-being, health, and intergenerational equity. Their varied experiences and viewpoints enrich decision-making processes, leading to more holistic and effective solutions. Women's leadership in environmental initiatives serves as a powerful tool for empowerment and representation. By occupying leadership roles, women inspire others, particularly girls and young women, to pursue careers and activism in environmental fields, thereby fostering greater diversity and inclusivity within these sectors. Women are disproportionately affected by environmental degradation and climate change, particularly in developing countries where they often bear the primary responsibility for household water, food, and energy security. Women's leadership can help address these disparities by advocating for policies and projects that prioritize gender equality, women's rights, and social justice. Women leaders are increasingly influential in shaping environmental policies and agendas at local, national, and international levels. Their advocacy efforts contribute to the adoption of gender-responsive policies, climate agreements, and sustainable development frameworks that prioritize environmental protection, gender equality, and social inclusion. Women leaders often exhibit a long-term vision and commitment to intergenerational equity, considering the needs of future generations in environmental decision-making. Their leadership emphasizes sustainable practices that preserve natural resources and ecosystems for future use, promoting a more resilient and equitable world.

Women's leadership in public governance plays a crucial role in advancing environmental conservation and sustainable development. Here's why their presence matters:

Stronger Environmental Policies

Studies show a positive correlation between women in government and stricter environmental policies. Countries with higher female representation in parliament are more likely to:

- Ratify international environmental treaties
- Establish protected areas
- Implement stricter climate change regulations

Women leaders often bring a more long-term perspective to decision-making, prioritizing environmental sustainability alongside economic growth.

Focus on Social Equity

Environmental issues are deeply intertwined with social justice concerns. Women leaders are more likely to champion policies that address:

The unequal impacts of environmental degradation on women and marginalized communities

o The need for inclusive decision-making processes concerning environmental issues

Collaborative and Transparent Governance

- Research suggests that women leaders tend to favor collaborative and transparent governance styles. This can lead to:
    - More open communication and public participation in environmental decision-making
    - Stronger collaboration between different stakeholders involved in environmental issues

Examples of Women Leading in Public Governance

- **Former President of Liberia, Ellen Johnson Sirleaf:** Championed environmental protection alongside peacebuilding efforts in her war-torn country.
- **Former New Zealand Prime Minister, Jacinda Ardern:** Declared a "climate emergency" and implemented ambitious policies to address climate change.
- **Environment Ministers like Aminah Mohammed (Nigeria) and Cristina Narbona (Spain):** Leading national and international efforts towards environmental sustainability.

By amplifying women's voices in public governance, we can create a more sustainable and equitable future for our planet. Women, among many others, have demonstrated leadership, courage, and dedication in addressing environmental challenges and promoting sustainable development. Their contributions have not only raised awareness but also inspired action at local, national, and global levels. Support for women's environmental leadership should start with policy action to address gender discrimination, social norms, and behaviours, from harassment to gender stereotyping, that sustain pervasive leadership gaps. Gender equality and environmental goals are mutually reinforcing, with progress on gender equality supporting environmental action and vice versa.

Women have played significant roles in environmental conservation throughout history, often pioneering efforts to protect and preserve natural resources, advocate for sustainable practices, and address environmental challenges. Their leadership in this field is multifaceted and spans various sectors, including science, activism, policy-making, and grassroots organizing.

1. **Scientific Research**: Women have made groundbreaking contributions to environmental science, conducting research on topics such as biodiversity, climate change, conservation biology, and ecosystem management. Their work has helped deepen our understanding of environmental issues and inform conservation strategies.

2. **Activism and Advocacy**: Women have been at the forefront of environmental activism, leading campaigns to raise awareness about environmental degradation, promote sustainable lifestyles, and push for policy changes. From grassroots movements to international organizations, women have been instrumental in mobilizing communities and driving action on environmental issues.
3. **Policy-making and Governance**: Women have increasingly assumed leadership roles in environmental policy-making and governance at local, national, and international levels. They have been elected to political office, appointed to government agencies, and participated in multilateral negotiations to shape environmental policies and agreements.
4. **Community Engagement and Empowerment**: Women-led initiatives at the grassroots level have been instrumental in promoting sustainable development and conservation practices within local communities. These efforts often prioritize community engagement, capacity-building, and the inclusion of marginalized voices, contributing to more equitable and effective conservation outcomes.
5. **Education and Outreach**: Women have played key roles in environmental education and outreach, inspiring future generations of environmental stewards and fostering a deeper connection to nature. Through teaching, writing, public speaking, and media engagement, they have raised awareness about environmental issues and promoted environmental literacy.
6. **Corporate Leadership and Innovation**: Women have also made strides in corporate leadership within the environmental sector, driving innovation and sustainability practices in industries ranging from renewable energy and green technology to corporate social responsibility and supply chain management.

Overall, women's leadership in environmental conservation is characterized by a holistic and inclusive approach that integrates scientific expertise, grassroots activism, policy advocacy, community engagement, and corporate responsibility. Their contributions are essential for addressing the complex environmental challenges facing our planet and building a more sustainable future for all.

Here are a few notable women leaders who have made significant contributions to environmental conservation and sustainable development:

1. **Wangari Maathai**: The Kenyan environmentalist and political activist was the founder of the Green Belt Movement, which focused on tree planting, conservation, and women's rights. Maathai's work earned her the Nobel Peace Prize in 2004, making her the first African woman to receive this honor.
2. **Jane Goodall**: Renowned for her groundbreaking research on chimpanzees in Tanzania's Gombe Stream National Park, Goodall is not only a primatologist but also a passionate advocate for wildlife conservation and environmental

education. Through her organization, the Jane Goodall Institute, she has promoted conservation efforts and empowered communities to protect their natural resources.

3. **Gro Harlem Brundtland**: Brundtland, a former Prime Minister of Norway, chaired the World Commission on Environment and Development, which introduced the concept of sustainable development in the landmark 1987 Brundtland Report. Her leadership has been instrumental in shaping global environmental policies and promoting sustainable development goals.
4. **Vandana Shiva**: An Indian environmental activist and scholar, Shiva has been a fierce advocate for biodiversity conservation, sustainable agriculture, and farmers' rights. Through her organization Navdanya, she has campaigned against corporate control of agriculture and promoted traditional farming practices that prioritize environmental sustainability and social justice.
5. **Marina Silva**: A former Brazilian Minister of Environment and environmentalist, Silva has been a leading voice for rainforest conservation and indigenous rights in the Amazon region. She has worked tirelessly to combat deforestation, promote sustainable development initiatives, and protect the rights of local communities and indigenous peoples.
6. **Greta Thunberg**: While still young, Thunberg has become a prominent figure in the global environmental movement, inspiring millions of people to take action against climate change through her advocacy and activism. She has sparked youth-led climate strikes worldwide and continues to push for urgent action from world leaders to address the climate crisis.

These women, among many others, have demonstrated visionary leadership, passion, and dedication in their efforts to protect the environment, promote sustainability, and create a more equitable and resilient world for current and future generations. Their contributions serve as inspirations for individuals and organizations around the globe to take action for environmental conservation and sustainable development.

The intersection of gender, environment, and human rights is a complex and interconnected issue that highlights the ways in which gender inequalities intersect with environmental degradation and human rights violations. Here's how these three dimensions are interrelated:

1. **Gender Inequality and Environmental Impact**: Gender inequalities exacerbate environmental degradation and vulnerability to environmental risks. Women and girls often bear a disproportionate burden of the impacts of environmental degradation, including loss of access to clean water and sanitation, food insecurity, displacement due to climate change, and exposure to pollution and environmental toxins. Discriminatory norms and practices limit women's

participation in decision-making processes related to natural resource management and environmental governance, hindering efforts to address environmental challenges effectively.

2. **Environmental Degradation and Human Rights Violations**: Environmental degradation can lead to violations of human rights, including the rights to life, health, food, water, and a healthy environment. Disproportionate exposure to environmental hazards and pollution in marginalized communities, often compounded by socioeconomic inequalities, can result in adverse health outcomes and infringements on the rights of vulnerable populations, including indigenous peoples, minorities, and low-income communities. The loss of biodiversity, deforestation, and land degradation also threaten the livelihoods and cultural rights of indigenous peoples and local communities who depend on natural resources for their survival and well-being.
3. **Gender-Based Violence and Environmental Displacement**: Environmental degradation and climate change exacerbate risks of gender-based violence, particularly in contexts of environmental displacement and migration. Women and girls are disproportionately affected by displacement due to climate-related disasters, conflict over natural resources, and land grabbing, increasing their vulnerability to exploitation, trafficking, and sexual violence. Lack of access to safe and secure housing, sanitation facilities, and livelihood opportunities further compounds the risks faced by displaced women and girls, undermining their human rights and dignity.
4. **Intersectional Perspectives and Social Justice**: Recognizing the intersectionality of gender, environment, and human rights is essential for addressing underlying power dynamics and promoting social justice. Intersectional approaches acknowledge the multiple and intersecting forms of discrimination and marginalization faced by individuals and communities, including those based on gender, race, ethnicity, class, age, disability, and other factors. By centering the experiences and voices of marginalized groups, including women and indigenous peoples, in environmental decision-making and policy processes, we can advance more equitable and inclusive solutions that respect human rights, promote environmental justice, and build resilient communities.

Understanding the interconnectedness of gender, environment, and human rights is essential for addressing the root causes of environmental degradation, promoting gender equality, and advancing human rights for all. By adopting a holistic and rights-based approach, we can work towards sustainable and just solutions that protect both people and the planet.

CHALLENGES FACED BY WOMEN

Women in traditionally male-dominated fields related to the environment face a range of challenges, stemming from societal norms, systemic biases, and structural barriers. Some of these challenges include:

1. **Underrepresentation**: Women are often underrepresented in leadership roles and decision-making positions in environmental organizations, government agencies, and academic institutions. This lack of representation can limit their influence and opportunities for advancement within these fields.
2. **Gender Stereotypes**: Gender stereotypes may perpetuate the perception that certain environmental roles are more suitable for men, leading to implicit bias and discrimination against women. This can manifest in hiring practices, promotion decisions, and workplace dynamics, creating barriers to entry and advancement for women in environmental careers.
3. **Workplace Discrimination**: Women in environmental fields may experience discrimination, harassment, and unequal treatment in the workplace. This can include disparities in pay, access to resources, and opportunities for professional development, as well as instances of sexism and gender-based microaggressions.
4. **Balancing Work and Family**: Balancing career aspirations with family responsibilities can be particularly challenging for women in environmental fields, where fieldwork, travel, and long hours may be common. The lack of supportive policies such as paid parental leave and flexible work arrangements can further exacerbate this challenge.
5. **Lack of Mentorship and Networking Opportunities**: Women in male-dominated fields may face limited access to mentorship and networking opportunities, which are essential for career advancement and professional growth. The absence of role models and support networks can hinder their ability to navigate career pathways and overcome barriers to success.
6. **Stereotype Threat**: Women may experience stereotype threat, which refers to the fear of confirming negative stereotypes about one's gender, leading to self-doubt, diminished confidence, and performance anxiety. This can undermine women's ability to thrive in male-dominated environments and fulfill their potential.
7. **Intersectionality**: Women from marginalized communities, such as women of color, LGBTQ+ women, and women with disabilities, may face compounded challenges due to intersecting forms of discrimination and marginalization. Addressing the unique experiences and needs of diverse women is essential for promoting inclusivity and equity in environmental fields.

8. **Cultural and Societal Norms**: Cultural and societal norms can influence perceptions of gender roles and expectations, shaping opportunities and constraints for women in environmental fields. Challenging these norms and fostering cultural change is essential for creating more inclusive and equitable work environments.

Addressing these challenges requires a multifaceted approach that involves promoting gender equality, challenging biases and stereotypes, implementing policies that support work-life balance, fostering inclusive workplaces, and providing mentorship and professional development opportunities for women in environmental fields. By addressing these barriers, we can harness the full potential of women's leadership and expertise to address pressing environmental issues and create a more sustainable future for all.

ROLE OF FEMALE PARLIAMENTARIANS IN ENVIRONMENTAL POLICIES

It is imperative that women participate in political decision-making in order to accomplish more ambitious and successful climate goals and policies. Their distinct viewpoints, dedication to sustainability, and cooperative leadership philosophies support more all-encompassing and just climate action. In addition to being a social justice issue, encouraging gender diversity in political leadership is also a critical strategic move in the fight against climate change. A more sustainable and resilient future for all can be achieved by elevating the standard and aspiration of climate governance through the empowerment of women and guaranteeing their equal participation in political processes.

Female parliamentarians can indeed make a significant difference in addressing climate change, as they bring unique perspectives, priorities, and leadership styles to the table. Here are several ways in which female parliamentarians contribute to advancing climate action and promoting gender-responsive policies:

1. Representation and Advocacy
   (i) **Voicing Gender Perspectives**: Female parliamentarians often advocate for gender-sensitive climate policies that address the specific needs and vulnerabilities of women and girls. They raise awareness about the gendered impacts of climate change and advocate for policies that promote gender equality and social justice.

(ii) **Inclusive Decision-Making**: Female parliamentarians promote inclusive decision-making processes that engage women and marginalized groups in climate policy development and implementation. They advocate for the meaningful participation of women in climate governance at all levels.

2. Policy Development and Legislation

    (i) **Promoting Gender-Responsive Policies**: Female parliamentarians champion the development and implementation of gender-responsive climate policies and legislation. They advocate for gender mainstreaming in climate adaptation, mitigation, and resilience-building efforts.

    (ii) **Addressing Gender Inequality**: Female parliamentarians work to address underlying gender inequalities that exacerbate climate vulnerability. They advocate for policies that empower women economically, socially, and politically, enabling them to adapt to and mitigate the impacts of climate change.

3. Budgeting and Resource Allocation

    (i) **Allocating Resources**: Female parliamentarians advocate for budget allocations that prioritize gender-responsive climate initiatives, such as investments in renewable energy, sustainable agriculture, and women-led climate adaptation projects.

    (ii) **Ensuring Accountability**: Female parliamentarians oversee the implementation of climate policies and programs to ensure that they effectively address gender disparities and deliver equitable outcomes. They hold governments accountable for their commitments to gender equality and climate action.

4. International Cooperation and Advocacy

    (i) **Global Advocacy**: Female parliamentarians engage in international forums and networks to advocate for gender-responsive climate policies and mobilize support for women's empowerment in climate action. They collaborate with counterparts from other countries to share best practices and lessons learned.

    (ii) **Influencing Global Agreements**: Female parliamentarians contribute to shaping global climate agreements, such as the Paris Agreement, to ensure that they incorporate gender considerations and promote women's rights and participation in climate governance.

Case Studies and Examples

(i) **Costa Rica**: Costa Rica's Parliament adopted a Gender Equity Law in 2015, which requires the inclusion of a gender perspective in all public policies, including those related to climate change. Female parliamentarians played a crucial role in advocating for this law and ensuring its implementation.

(ii) **Rwanda**: Rwanda has one of the highest rates of female representation in parliament globally, with women holding 61% of parliamentary seats as of 2021. Female parliamentarians in Rwanda have been instrumental in promoting gender equality and mainstreaming gender considerations in climate policies and programs.

There is a growing recognition of the connections between environmental sustainability and gender equality, including the ways in which women are disproportionately affected by environmental degradation and the ways in which women's empowerment and increased gender equality may advance environmental action.

Increasing the representation of women in environmental decision-making across all societal domains is a primary goal of the gender-environment nexus.

There are still gender disparities in environmental decision-making in many nations and businesses, with the largest inequalities found in low-income nations and ecologically sensitive sectors. In OECD nations, women make up 39% of environment ministers; but, in nations where women are most at risk from environmental hazards and where climate change is most likely to cause significant harm, their share is far lower. Women are underrepresented in infrastructure ministries as well (OECD, 2019). International environmental negotiations and civil society organizations have the largest percentage of female environmental leaders.

Enhancements in representational justice and heightened awareness of gender-specific environmental repercussions are not the sole benefits of women's environmental leadership. Additionally, it may result in more robust and successful environmental action (OECD, 2021). Research indicates that the involvement of women in decision-making processes can result in improved environmental results. This includes the reinforcement of environmental regulations and the encouragement of sustainable investments in both public and private domains.

Empirical evidence suggests that increased female representation in public sector decision-making entities leads to the implementation of more ambitious climate policies and objectives. Research from the business world demonstrates that women begin to influence climate governance in organizations when they hold at least 30% of board positions (BloombergNEF and The Sasawaka Peace Foundation, 2020). Women are building strong networks all across the world to address climate-related inequality and stop environmental destruction. Particularly in indigenous societies, some of these advocates face violent hostility.

GENDER GAP IN ENVIRONMENTAL MINISTRIES ACROSS THE WORLD

The gender gap in environmental leadership is sustained by the same "sticky floors" associated with social norms, gender stereotyping and discrimination, and "glass ceilings" associated with motherhood penalties that prohibit women from building human capital at the same rate as men (Ciminelli, Schwellnus, and Stadler, 2021). Inherent societal discrimination based on gender, particularly against girls in low- and middle-income nations, impedes the advancement of female environmental leaders. Improving women's representation in environmental decision-making requires addressing the various forms of overt and covert discrimination that women face as well as fostering their leadership potential. It is imperative that women's environmental activism, particularly that of marginalized and indigenous women, be recognized and encouraged in society. Goals related to gender equality and the environment complement one another; advancements in gender equality encourage environmental action and vice versa (OECD, 2021). The Sustainable Development Goals (SDGs) that go along with Agenda 2030 and the United Nations acknowledge the connections between gender and the environment.

The idea that men and women can be affected differently by environmental deterioration is at the heart of the gender-environment debate. Gender differences in physical attributes, economic susceptibility, and social roles and behaviours all influence how differently genders experience and react to environmental influences. For example, indoor air pollution disproportionately affects women globally, and many do not have access to safe water and proper sanitation. The opportunity costs of gathering fuel and water, as well as providing care for those impacted by natural catastrophes, may fall on women (OECD, 2021; OxFAM, 2018). Women themselves also differ in how the environment affects them; rural or indigenous women experience the environment differently than urban women do (OECD, 2021). The fact that men and women behave and think differently while making decisions that have an impact on the environment is another part of the gender-environment relationship. Research indicates that women are more likely to buy eco-friendly products and recycle (Kennedy and Kmec, 2018; Xiao and Hong, 2010). When it comes to driving less, women in OECD countries have a slightly "greener" mindset (OECD, 2008). The usage of alternate forms of transportation is applicable globally (Ng and Acker, 2018). Women tend to be more supportive of local communities, indigenous peoples, and other environmental stewards. They are also more likely to support regulations that increase transparency about product manufacturing, use clean energy to move things, and phase out the use of polluting fuels.

Additionally, perceptions regarding environmental hazards and damages tend to differ between men and women. According to a 2019 Pew Research Center study on global climate change perceptions, women were more likely than males to regard the issue as a significant concern in nine out of the 26 nations that responded (Pew Research Center, 2019).

Research conducted in the United States has also revealed that women perceive environmental issues differently than do men, especially when local environmental challenges also present health risks (Xiao and McCright, 2012). Women view the risks as greater, more urgent, and more likely to directly affect them, even though the proportion of men and women in the US who believe in climate change and understand its anthropogenic sources is similar (Ballew et al., 2018).

According to surveys conducted in developing nations, men and boys are slightly more likely than women and girls to view climate change as an emergency. This difference in opinion may be due to the fact that men in these nations tend to have higher levels of education. Higher educated people often recognize the climate emergency at a much higher degree. This serves as a reminder that addressing gender inequality and raising environmental awareness both depend on equitable access to education.

Numerous advantages that women's environmental leadership offers can improve environmental activity and help the UN reach its Sustainable Development Goals (OECD, 2021). Research demonstrates that women in leadership positions in government agencies, businesses, and community-based organizations contribute to better environmental results (Offermann and Foley, 2020). Evidence suggests that women in positions of political authority are more likely to support gender-sensitive issues and encourage women to participate in politics, which can help elevate women's experiences with the environment (Clayton, Josefsson, and Wang, 2017).

Climate talks are typically headed by ministers and other high-ranking officials; therefore, the gender balance in environmental ministries may have an impact on these talks. When women are able to actively participate in policy negotiations, they fight for more specific and challenging goals and produce agreements that are more beneficial and long-lasting (Krause, Bränfors, and Krause, 2018). It has been demonstrated that the inclusion of women in international negotiations broadens the topics covered and increases the pressure on all sides to find a consensus or return to the negotiating table when negotiations are halted. International environmental treaties are more likely to be ratified by nations with larger proportions of women (Norgaard and York, 2016).

A look at environmental ministries around the world illustrates the persistent gender gap in top decision-making positions. In general, the gap is widest in regions likely to suffer the greatest damage from climate change, and where women are most vulnerable to environmental risk. The narrowest is in OECD countries (where women accounted for 39% of all environment ministers in 2021). There is a persistent gender disparity in top decision-making positions when it comes to environmental ministries across the world (Figure 1). The gender gap is highest in areas where women are affected mostly by climate change and other environmental hazards. The gap is narrowest in OECD countries, where 39 percent of environmental ministries are held by women (OECD, 2022). In Sub-Saharan Africa, a region worst affected by climate change (Swiss Re Institute, 2021), women's representation in ministries is only 28 percent. In South and South-East Asia, women constitute 14 percent of environment ministers. In Latin America and the Caribbean, women comprise on 7 percent of environment ministers. The nations of the Middle East and North Africa, along with Sub-Saharan Africa, will be most negatively impacted by climate change. Women make up just 10 percent of environment ministers, which is also a very low proportion. Women make up just 13 percent of ministers in environment ministries across the remainder of Europe and Asia, which is dominated by men.

Figure 1. Gender share in Environmental Ministries

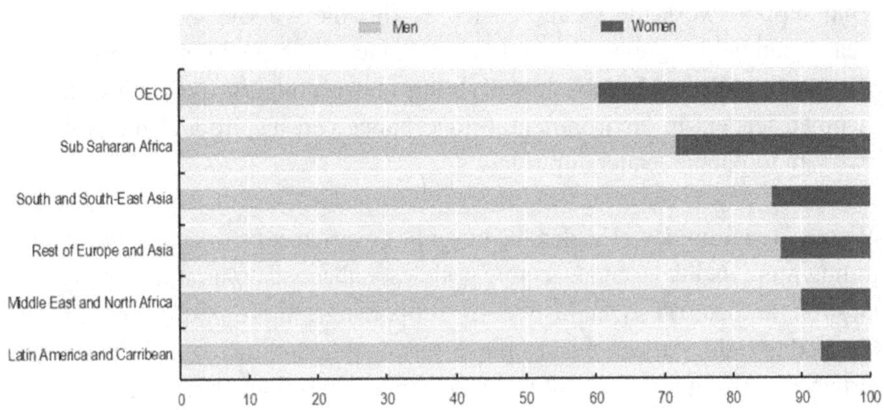

Source: OECD, 2022

Not just in senior leadership roles but in all government ministries, there is a gender gap. For example, women only make up 18 percent of workers in OECD country infrastructure ministries related to the environment (energy, transport, and communications), compared to 38% in socio-cultural ministries (OECD, 2019).

Female parliamentarians play a vital role in advancing gender-responsive climate policies, promoting women's empowerment, and ensuring that climate action is inclusive and equitable. Their leadership and advocacy contribute to more effective and sustainable responses to the climate crisis, benefiting both women and men, as well as future generations. By amplifying the voices and priorities of women in climate governance, female parliamentarians help build a more resilient and just society in the face of climate change.

STRATEGIES TO SUPPORT AND ENHANCE WOMEN'S LEADERSHIP

1. **Gender Quotas and Targets**: Implementing gender quotas and targets in public governance can ensure a minimum level of women's representation in environmental decision-making bodies.
2. **Capacity Building and Training**: Providing training and capacity-building programs for women can enhance their leadership skills and knowledge in environmental governance, enabling them to contribute more effectively.
3. **Mentorship and Networks**: Establishing mentorship programs and networks for women in environmental leadership can provide support, share experiences, and foster collaboration among women leaders.
4. **Supportive Policies and Environments**: Creating policies and environments that support work-life balance, such as flexible working hours and parental leave, can help more women to take on leadership roles in public governance.
5. **Promoting Role Models**: Highlighting and promoting the achievements of women leaders in environmental governance can inspire and motivate other women to pursue leadership roles.

Women's environmental leadership in public governance brings numerous benefits, including more inclusive and effective policies, enhanced community engagement, and a stronger focus on social justice. By supporting and promoting women's leadership in this field, we can ensure more sustainable and equitable environmental governance, leading to better outcomes for communities and the planet.

THE IMPACT OF WOMEN'S POLITICAL PARTICIPATION ON CLIMATE GOALS AND POLICIES

Globally, women are working to stop climate change by making sustainable purchasing decisions and improving control, access, and conservation of resources. Their participation is necessary for policy and implementation initiatives that will benefit both the present and the future. The Women's Environment and Development Organization, or WEDO, was established specifically to influence the 1992 UNCED by promoting the involvement and inclusion of women across the globe. The group is committed to advancing gender parity in decision-making in the interest of social and economic fairness. Due to WEDO's work, the United Nations Framework Convention on Climate Change (UNFCCC 2019) negotiations included the first-ever gender-specific text. These international agreements mandate that all women have a voice in the policies that affect their environment. Women have been leaders, thinkers, experts, strategists, educators, and role models in the environmental movements, and they have made a substantial contribution to their success.

The presence of women in political decision-making is increasingly recognized as a key factor in advancing more ambitious and effective climate goals and policies. Research and case studies from around the world demonstrate that gender diversity in political leadership not only promotes more comprehensive and equitable climate action but also drives innovation and sustainability. This essay explores how women's participation in political decision-making contributes to stronger climate policies and goals, and highlights the mechanisms through which gender diversity enhances climate governance.

THE CONNECTION BETWEEN WOMEN'S LEADERSHIP AND CLIMATE AMBITION

1. **Holistic Perspectives and Inclusivity:**
    - Women often bring different perspectives and experiences to the table, particularly in understanding the social dimensions of climate change. Their unique insights into issues such as health, education, and community well-being enable more holistic approaches to climate policy.
    - Female leaders are more likely to advocate for inclusive policies that address the needs of vulnerable populations, including women and children, who are disproportionately affected by climate change.
2. **Commitment to Sustainability:**
    - Studies have shown that women are generally more supportive of environmental protection and sustainable development. This is reflected in their voting patterns, policy preferences, and leadership styles.
    - Female politicians tend to prioritize long-term environmental sustainability over short-term economic gains, advocating for policies that promote renewable energy, conservation, and climate resilience.
3. **Enhanced Collaboration and Multilateralism:**
    - Women in political leadership are often more adept at fostering collaboration and consensus-building. These skills are crucial for addressing global challenges like climate change, which require coordinated action across multiple sectors and countries.

- Female leaders are instrumental in advancing international climate agreements and initiatives, such as the Paris Agreement, by championing cooperative and inclusive approaches to climate governance.

EVIDENCE OF WOMEN'S INFLUENCE ON CLIMATE POLICY

1. **National Case Studies:**
   (i) **Rwanda:** With one of the highest proportions of women in parliament globally, Rwanda has implemented progressive environmental policies, including ambitious reforestation projects and climate adaptation strategies.
   (ii) **Sweden:** Known for its strong gender equality and environmental policies, Sweden has set some of the most ambitious climate goals, aiming to become carbon neutral by 2045. Female political leaders have played a significant role in shaping these policies.
2. **Local Leadership:**
   (i) **United States:** Women mayors and local officials have been pivotal in implementing innovative climate initiatives at the city level. For example, Mayor Libby Schaaf of Oakland, California, has championed urban sustainability projects, including zero-emission transportation and green infrastructure.
   (ii) **Brazil:** Female leaders in the Amazon region have been at the forefront of efforts to combat deforestation and protect indigenous lands, highlighting the intersection of gender, environmental, and social justice.
3. **International Advocacy:**
   Women have been prominent figures in international climate negotiations and organizations. Christiana Figueres, former Executive Secretary of the UNFCCC, played a key role in the successful negotiation of the Paris Agreement, demonstrating the impact of female leadership on global climate policy.

MECHANISMS ENHANCING CLIMATE GOVERNANCE THROUGH GENDER DIVERSITY

1. **Policy Innovation:**

Women's unique perspectives and priorities can lead to innovative climate solutions that might be overlooked by more homogenous leadership groups. Their focus on social equity, health, and community resilience enriches the policy-making process.

2. **Greater Accountability:**
   Gender-diverse political bodies are often more transparent and accountable, which can lead to stronger enforcement of environmental regulations and policies. Women's leadership is associated with lower levels of corruption, enhancing the credibility and effectiveness of climate action.
3. **Enhanced Public Support:**
   Women's involvement in climate leadership can increase public support for climate policies. Female leaders often engage more effectively with communities and stakeholders, building trust and fostering collective action on climate issues.

Challenges and Barriers

Despite the clear benefits, women still face significant barriers to political participation and leadership in many parts of the world. Socio-cultural norms, gender discrimination, and structural inequalities limit women's access to political power and decision-making roles. Addressing these barriers is essential for leveraging the full potential of women's leadership in advancing ambitious climate goals.

Overcoming Challenges

Despite the benefits, significant gender gaps persist in environmental leadership roles within governments. Here are some challenges:

- **Social and cultural barriers** that limit women's participation in politics
- **Lack of access to funding and support networks** for female candidates
- **Unconscious bias** in selection processes for leadership positions

Moving Forward

- **Promoting policies** that encourage women's participation in government, such as quotas and capacity-building programs.
- **Shifting societal attitudes** to support women's leadership in all spheres, including environmental governance.

- **Highlighting the positive impacts** of women's leadership in environmental decision-making.

CONCLUSION

The presence of women in political decision-making is crucial for achieving more ambitious and effective climate goals and policies. Their unique perspectives, commitment to sustainability, and collaborative leadership styles contribute to more comprehensive and equitable climate action. Promoting gender diversity in political leadership is not only a matter of social justice but also a strategic imperative for addressing the global climate crisis. By empowering women and ensuring their equal participation in political processes, we can enhance the quality and ambition of climate governance, leading to a more sustainable and resilient future for all. Understanding and addressing the interconnectedness of gender dynamics, environmental sustainability, and human rights violations is essential for creating a just and sustainable future. Comprehensive strategies that consider these links can lead to more effective solutions and a more equitable world.

Addressing systemic inequalities and environmental degradation requires a multi-faceted approach that integrates policy changes, community empowerment, sustainable practices, and international cooperation. The government needs to implement inclusive policies. The need is to develop and implement policies that ensure the participation of marginalized groups, including women, indigenous peoples, and minorities, in decision-making processes. This can be achieved through gender quotas, community consultations, and participatory budgeting. The government also need to strengthen the country's legal framework by enforcing laws and regulations that protect the rights of marginalized communities and promote environmental sustainability. This includes land rights for indigenous communities, anti-discrimination laws, and regulations on pollution and resource extraction. It must also be ensured that environmental policies and programs are gender-responsive, addressing the specific needs and contributions of women. This involves gender analysis in environmental impact assessments and the inclusion of women in leadership roles in environmental projects.

Steps must be taken to provide education and training programs focused on sustainable practices, environmental stewardship, and human rights. This can empower communities to manage resources sustainably and advocate for their rights. Economic incentives should also be provided for sustainable practices, such as subsidies for renewable energy, grants for sustainable agriculture, and penalties for environmental violations. Supporting green jobs can also promote both environmental sustainability and economic equity. Moreover, facilitating access to credit and financial services for women and marginalized groups is important to start and expand sustainable businesses. Microfinance institutions and cooperatives can play a key role in this effort.

The governments must also encourage community-led initiatives for managing natural resources, such as community forestry, watershed management, and sustainable fishing practices. Empowering local communities ensures that resource management is aligned with local needs and knowledge. Investing in renewable energy projects, particularly in underserved communities is crucial in the path to sustainable development. Solar, wind, and small-scale hydropower projects can provide clean energy while creating jobs and reducing environmental impact. Conservation and restoration projects that protect biodiversity and rehabilitate degraded ecosystems can be very helpful. This can include reforestation, wetland restoration, and the creation of protected areas.

Along with this, the governments must also ensure that communities, especially vulnerable groups, have access to healthcare services that address environmental health risks. This includes addressing pollution-related illnesses, providing clean water and sanitation, and improving maternal and child health. Implementing social protection programs such as conditional cash transfers, food security programs, and social insurance schemes to support vulnerable populations during environmental crises and economic downturns can be beneficial as well. By implementing these practical solutions and strategies, we can address systemic inequalities and environmental degradation, paving the way for a more equitable and sustainable future.

REFERENCES

Ballew, M., ... (2018), “Gender Differences in Public Understanding of Climate Change”, https://climatecommunication.yale.edu/publications/gender-differences-in-publicunderstanding-of-climate-change/

Ciminelli, G., Schwellnus, C., & Stadler, B. (2021), “Sticky floors or glass ceilings? The role of human capital, working time flexibility and discrimination in the gender wage gap”, *OECD Economics Department Working Papers*, No. 1668, OECD Publishing, Paris. DOI: 10.1787/18151973

Clayton, A., Josefsson, C., & Wang, V. (2017), “Quotas and Women’s Substantive Representation: Evidence from a Content Analysis of Ugandan Plenary Debates”, *Politics & Gender*, Vol. 13/2, pp. 276-304, . https://dx.doi.org/. https://www.oxfamamerica.org/explore/research-publications/researchbackgrounderwash-gender-report-puerto-rico/DOI: 10.1017/S1743923X16000453

Kennedy, E. H., & Kmec, J. (2018). Reinterpreting the gender gap in household pro-environmental behaviour. *Environmental Sociology*, 4(3), 299–310.

Ng, W., & Acker, A. (2018), “Understanding Urban Travel Behaviour by Gender for Efficient and Equitable Transport Policies”, http://www.itf-oecd.org

Norgaard, K., & York, R. (2016). Gender Equality and State Environmentalism. *Gender & Society*, 19(4), 506–522. DOI: 10.1177/0891243204273612

OECD. (2008), *Household Behaviour and the Environment: Reviewing the Evidence*, https://www.oecd.org/environment/consumption-innovation/42183878.pdf

OECD. (2015). *The Economic Consequences of Climate Change*. OECD Publishing.

OECD. (2017), *Policy Brief on Women’s Entrepreneurship*, https://www.oecd.org/cfe/smes/Policy-Brief-on-Women-s-Entrepreneurship.pdf

OECD (2019), “Gender Equality and Sustainable Infrastructure”

OECD (2020), *Policies and Practices to Promote Women in Leadership Roles in the Private*

OECD. (2021), Entrepreneurship Policies through a Gender Lens, OECD Studies on SMEs and Entrepreneurship, OECD Publishing, Paris, DOI: 10.1787/71c8f9c9-en

OECD. (2021), *Gender and the Environment: Building Evidence and Policies to Achieve the* OECD, Gender and Sustainable Development: Maximising the Economic, Social and Environmental Role of Women. Organization for Economic Co- operation and Development (OECD) to the United Nations Commission on Sustainable Development (UNCSD) (2008). https://www.oecd.org/social/40881538.pdf

OECD. 2021. Gender and the environment: Building evidence and policies to achieve the SDGs.https://www.oecd-ilibrary.org/environment/gender-and-the-environment _3d32ca39-en

Offermann, L., & Foley, K. (2020). *Is There a Female Leadership Advantage?* Oxford Research Encyclopedia of Business and Management., DOI: 10.1093/ acrefore/9780190224851.013.61

OXFAM (2018), *The Weight of Water on Women: The Long Wake of Hurricane María in Puerto Rico.*

Pew Research Center. (2019), *How people worldwide view climate change*, https:// www.pewresearch.org/fact-tank/2019/04/18/a-look-at-how-people-around-the -worldview-climate-change/

SDGs, OECD Publishing, Paris, .DOI: 10.1787/3d32ca39-en

Sector, https://www.oecd.org/corporate/OECD-G20-EMPOWER-Women-Leadership.pdf

Swiss Re Institute. (2021), *The economics of climate change: no action not an option*, https://www.swissre.com/dam/jcr:e73ee7c3-7f83-4c17-a2b8-8ef23a8d3312/swiss -re-instituteexpertise-publication-economics-of-climate-change.pdf

The Sasawaka Peace Foundation and BloombergNEF. (2020), "Gender Diversity and Climate UN Environmental Programme. Global Mutilateral Environmental Agreements. https://www.unep.org/explore-topics/oceans-seas/what-we-do/working -regional-seas/partners/global-multilateral

UN General Assembly. (2015). Transforming our world: the 2030 Agenda for Sustainable Development. https://www.un.org/ga/search/view_doc.asp?symbol= A/RES/70/1&Lang=E

UNDP. (2022). Climate Promise Progress Report, New York, NY: UNDP. https:// climatepromise.undp.org/sites/default/files/research_report_document/Climate%20 Promise%20Global%20Progress%20Report%202022.pdf

United Nations Conference on Environment and Development (UNCED), Rio de Janeiro, Brazil, (1992, June). Conferences Environment and sustainable development. https://www.un.org/en/conferences/environment/rio1992

Xiao, C., & Hong, D. (2010, September). Gender differences in environmental behaviors in China. *Population and Environment*, 32(1), 88–104. DOI: 10.1007/s11111-010-0115-z

Xiao, C. and D. Hong (2010), "Gender differences in environmental behaviors in China".

Xiao, C. and A. McCright (2012), "Explaining Gender Differences in Concern about Environmental Problems in the United States",Vol. 25/11, pp. 1067-1084, .DOI: 10.1080/08941920.2011.651191

Chapter 15
Understanding Disasters, Vulnerabilities, and the Resilience:
Impact of River Bank Erosion on Riparian Communities

Sumedha Dey
CSSSC, India

Malay Mandal
https://orcid.org/0009-0002-5866-0514
University of Hyderabad, India

ABSTRACT

Rivers have historically served as lifelines for civilizations, providing essential resources and opportunities for livelihood. However, they also present significant challenges, particularly when turned into disaster propelled by human interventions. The very qualities that make rivers indispensable to the dependent communities also render them vulnerable in many ways. Among the myriad challenges facing riverine communities, river bank erosion stands out as a particularly pressing issue, with far-reaching social, economic, and environmental implications. In recent time, well-being of riparian people in West Bengal, India, is seriously threatened by river bank erosion in the Ganges river basin, especially in parts of Malda and Murshidabad districts have [un]intentionally gotten worse since the Farakka Barrage was built in 1972.The article emphasises how these elements are interrelated and have consequences for the welfare of communities by looking at them through the lenses of geopolitics, feminism, and human rights.

DOI: 10.4018/979-8-3693-6069-9.ch015

INTRODUCTION

Rivers have historically served as lifelines for civilizations, providing essential resources and opportunities for livelihood. However, they also present significant challenges, particularly when turned into disaster propelled by human interventions. The very qualities that make rivers indispensable to the dependent communities also render them vulnerable in many ways. Among the myriad challenges facing riverine communities, river bank erosion stands out as a particularly pressing issue, with far-reaching social, economic, and environmental implications. In recent times, the well-being of riparian people in West Bengal, India, has been seriously threatened by river bank erosion in the Ganges river basin, especially in parts of Malda and Murshidabad districts has [un]intentionally been worsening since the 'Farakka Barrage' was built in 1972. The barrage was meant to control water flow and make navigation easier. However, now it seems to juxtapose the reason it was built as the river changed its natural course and repository capacities. Another additional factor, among many others, contributing to the ecosystem's instability is the growth of sand mining 'mafias' who dug out sand illegally along riverbeds, hence increasing erosion rates and jeopardising nearby ecosystems.

Against this backdrop, it becomes essential to examine the intricate dynamics of disaster and its profound impact on riparian communities. This study is trying to unfold the fundamentals of river bank erosion and its disproportionate effects on nearby communities, making them vulnerable, especially women and children, leading to displacement, child marriage, and human trafficking. Poor governance, inadequate policies, and the influence of syndicate systems further contribute to the perpetuation of these challenges. Amidst this adversity, there are also signs of active resilience, as evidenced by the emerging civil society movements as well as community-led initiatives aimed at providing support and assistance for those who have lost their minimum means of livelihood. By adopting an intersectional approach that considers the interplay of gender, socio-economic status, and political factors, we can gain deeper insights into the differential impacts of river bank erosion and the pathways to resilience.

This work explores the complexity of disaster and resilience in impacted communities by using original data that draws attention to the ways that gender, socioeconomic position, and political processes interact to influence how vulnerable people feel and how resilient they can be. The article emphasises how these elements are interrelated and have consequences for the welfare of communities by looking at them through the lenses of geopolitics, feminism, and human rights. Recognizing the intricacies of disaster and resilience is essential for developing effective strategies to address the broader consequences of environmental disasters. By understanding the intersecting factors that shape individuals' experiences, policymakers and practitioners can design interventions that promote community resilience and mitigate the impacts of river bank erosion. This paper calls for a holistic approach that integrates gender-sensitive, socio-economic, and political considerations into disaster management and adaptation efforts.

EFFECT OF RIVER-BANK EROSION ON HUMAN FABRIC: UNDERSTANDING DISASTER AND VULNERABILITIES

Rivers have been assisting people in numerous ways such as irrigation, transport, generating hydroelectricity, and providing significant source of income. Perhaps that is why rivers do not only seen as water bodies but also hold significant attribution of sacredness across cultures around the world. India is often called as 'land of rivers' because of the extensive river network that is spread all over the country. Composed of seven key rivers together with their more than four hundred tributaries the river system in India plays a pivotal role in the lives of Indian people. However, despite being an important lifeline for human settlement rivers also have some negative impacts as well. Unstable river-bank erosion is one among them. The riverbank word refers to the land besides a river. Geomorphologically it is a result of river gradation. The natural process of riverbank erosion can produce favourable outcomes such as the formation of productive floodplains and alluvial terrain; however, unstable rivers and erosion can become cause for agony for the communities who live near it. As a crucial geo-morphological phenomenon river-bank erosion mostly effects fluvial and coastal environment; though the nature and impact of erosion may vary (Das et al. 2007; Pati et al. 2008). River-bank erosion could be better understood if we look at it as a process in which the bank lines witness gradual shifting. In other words, Rivers and streams are dynamic systems that constantly change. According to experts, there could be various factors that affect river-bank erosion; stream bed lowering or in fill, saturation of banks from off-stream source, redirection and acceleration of flow within the channel, characteristics of river bank and bed materials, pressure imbalance at the bank face, poor vegetation cover, wind or boat waves (Mirza A.T.M. et al. 2014; Tripathy & Mondal, 2019).

Riverbank erosion does not only disrupt ecological balance but also leaves a profound impact on the people who live nearby areas. As an environmental catastrophe it is incomparable with other disasters. Its uniqueness lies in its much powerful destructible nature and long term and permanent sociological, economic and cultural effect on human lives. In other disasters, except earthquake, people do still hold a chance of restoring their previous state of life. However, in the case of riverbank erosion the chances are nearly zero as people lose their land permanently (Islam & Rashid, 2011). It leaves permanent scar on human lives as it dislocates land and human settlements with a consequential effect such as destruction of crop-field, homes, disruption in road linkage, and communication infrastructure (Karim, 2014). It is one of the major causes of destitution in rural areas where hundreds of households could lose their physical assets, become landless and homeless (Rasheed, 2000; Hassan et al., 2017). While other disasters such as floods and cyclones allow people to recover from the damages, river erosion often causes total loss which is often irrecoverable (Howlader & Rahman, 2016). Riverbank erosion not only displaces people but also marginalize them in terms of livelihood patterns and psycho-physical stability. It is one of the most harmful hazards that have the most severe long-standing impacts on human beings (Das et al., 2017).

As a serious matter of concern riverbank erosion is comparatively new in academic discussion. However, despite its new entry, the academic discussion around it has been showing steady growth in terms capturing different perspectives. Scholars are exploring it from the perspective of flood and population displacement (Elahi et al., 2000; Islam & Rashid, 2011; Rabbi et al., 2013; Mollah & Ferdaush, 2015), hazard impact in the floodplain areas and land settlements (Nazem & Elahi, 1990; Rahman, 2010), water diversion and environmental effects (Mirza, 2004), and assessment of channel changes of river system (Dewan et al., 2017). There are numerous scholarly articles and research papers on riverbank erosion. Major contributions of these researches address river morphological changes based on the perspective of geology and geography. There are very few studies that have been devoted to understand the impact of riverbank erosion and its effect on human fabric such as women, children, human lives and livelihoods. Those studies are limited as they are based on quantitative data analysis which is important too as it reveals systematic and structured understanding of the impact of riverbank erosion. However, in order to understand how riverbank erosion is not only a natural phenomenon but a disaster and has major contribution in causing human vulnerabilities we cannot always depend on quantitative data as it might bring limitation to the intended study.

Once we are identifying riverbank erosion as disaster what we are indicating at is the human intervention which is propelling the aggravation of erosion. In the considered case, there are many interventions by human which are turning a natural phenomenon into disaster. For example, the construction of Farakka Barrage that changed the natural course of River Ganga, the activities of sand mafias that is weakening river bed strength, and the callousness of policy makers and government initiatives are the main players behind turning natural phenomenon into a disaster making the riverine lives vulnerable. The word vulnerable comes from, etymologically speaking, the word to signify wounds or vulness to be wounded. The discourse opens up by focusing on bodies and situation in which lives are lived and lives are rendered woundable or injurable. The aesthetics of vulnerability studies vastly looks at different types of vulnerabilities such ageing, sickness, epidemic, industrial disaster, and pandemic vulnerabilities etc. Ecological pricarity or ecological dystopia as a matter of concern is comparatively new addition to the vulnerability studies. This research paper particularly interested in exploring ecological pricarity, which is appearing in the form of Ganga riverbank erosion, rendering lives vulnerable, for the people of Malda. The study is exploring human intervention in riverbank erosion and its impact on human fabric through victimhood, human rights violation, policy making, and gender based vulnerabilities. It also explores the resilience of the erosion-inflicted people in order to understand their ability to recover or adjust despite enduring vulnerabilities.

AREA AND METHOD OF THE STUDY

This study has been conducted in particular regions of Malda district in West Bengal, India. Malda has always been an important district due to its historicity, geographical location, and agricultural production. This district covers an area of 3,733.66 square kilometres over the latitude range 24°40'20" N to 25°32'08" N and longitude range 87°45'50" E to 88°28'10" E. Due to its unique location Malda district is called the gateway of North Bengal as this is the only place that connects north and south part of West Bengal. Malda shares international border with Bangladesh in the east having Jharkhand and Bihar in the west, Murshidabad district in south, and Dinajpur district in north. The river Ganges, which is the main character of our discussion, enters into West Bengal near Rajmahal of Jharkhand state and then flows in south-eastern direction within West Bengal. After flowing through Malda district it splits into two branches near north of Dhulian in Murshidabad district. One branch enters into Bangladesh as the river Padma, while the other one flows through West Bengal in southern direction as Bhagirathi and Hoogly River contributing to a dominant irreparable loss of land each year.

To understand the effect of Ganges river bank erosion on riparian communities this present study has been conducted in selected villages that are spread over two administrative subdivisions: Chanchal and Malda Sadar. Chanchal subdivision has six community development blocks and Malda Sadar subdivision has two municipalities (Old Malda and English Bazar) along with nine community development blocks. The studied villages Rajnagar and Panchanandapur I & II come under Kaliachak II (24°57′49″N 88°05′22″E) community development block, Birnagar-I and Birnagar-II are in Kaliachak III (24.8402°N 87.9776°E) community development block, Gopalpur and Duarni Tafir (Bhutnir Char) are in Manikchak (25.0624410°N 87.9091110°E) community development block, and Mahananda Tola comes under Ratua-I (25.20°N 87.93°E) community development block. These above-mentioned villages are the area that was possible to cover during this research and they are situated near areas of Farakka Barrage. Ganges river bank erosion is visible in other places too but it was not possible to cover every erosion affected village because of time constraints.

Figure 1. Map showing the location of the studied area

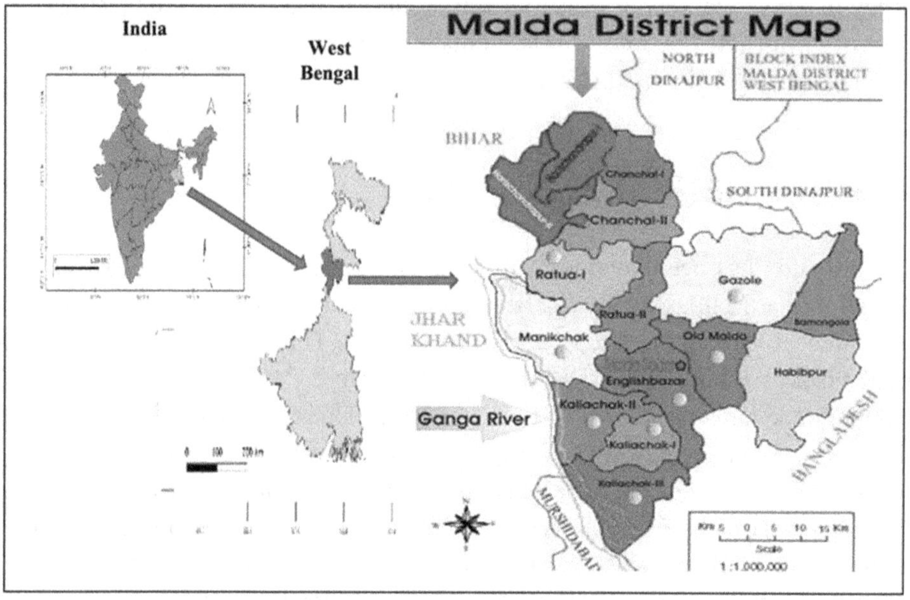

Source: Map generated by using QGIS; District Map obtained from Malda Sahayogita Samiti

This research is ethnographic in its nature and qualitative methodology as applied to make an in-depth study of the experiences of women and children of the affected area. As per 2011 census Malda district had population of 3,988,845 of which male and female were 2,051,541 (51.43%) and 1,937,304 (48.57%) respectively. For this study 2011 census data has been used because the last census of Malda was done in 2011 and next census of 2021 has been postponed or cancelled.

Table 1. Population distribution of studied area

Block	Villages	Population		Total	%	
		Male	Female		M	F
Kaliachak II	Rajnagar	3978	3741	7719	51.54	48.46
	Uttar Panchanandapur I	10205	9600	19805	51.53	48.47
	Uttar Panchanadapur II	7887	7418	15305	51.53	48.47
Kaliachak III	Birnagar I	9618	8997	18615	51.67	48.33
	Birnagar II	11284	10556	21840	51.67	48.33
Manikchak	Gopalpur	7688	7222	14910	51.56	48.44
	Bhutnir Char (consists of 3 panchayets)	46052	42969	89021	51.73	48.27
Ratua I	Mahanandatola	12404	11559	23963	51.76	48.24

Source: Census Data 2011

Data obtained through 2011 census shows women nearly cover half of the total population. So it would be very partial if we only assess riverbank erosion effect from the perspective of men as is has been happening in such way. Whenever a research has been conducted upon riverbank erosion and its impact on human lives we concentrated on loss of lands, homes, and stable settlement. Now, if we pay closer attention to the discussion along the line of above mentioned registers we would be noticing that the analysis coming out are mostly drawn by men as majority of the time patriarchal society makes them the beholder of lands, homes, and earthly belongings. However, what about others who belongs to the marginal position of the patriarchal system? Have we ever heard them closely? We need to, because gender marginality also determines how disaster impact people by shaping vulnerability. Across the board, from life expectancy to education, health, violence, and livelihoods, they are the one impacted by disasters disproportionately.

To know about the inner world of marginal genders such as women and their experiences inductive approach was quite useful for us as it led us to a journey towards an immense amount of information. The researches asked the correspondents open ended questions so they can elaborate their experiences and opinion in their own pace. As a result various themes have emerged during the discussion that helped us to capture the emic or insider's view about how they are experiencing the worsening effect of Ganges engulfment, how they are building their own resilience based on indigenous coping mechanism, what are the are experiences of gender minorities in their everyday lives. To unfold these crucial quarries the researchers have spent significant amount of time in the field collecting data through focused group discussion (FGD), in-depth interviews, and observation. The observation method has been used to observe respondents' language and gestures on problems, present condition of houses, and lands, etc. The researchers also participated in the meetings organised by civil society and took active part in community support. This method helped blurring the boundaries between researchers and the local communities.

The secondary data have emerged through government documents, political pamphlets, and posters that are collected by the researchers. The documents helped the researchers in understanding various schemes and government initiatives towards the betterment of the local communities who are affected by the river bank erosion. It also helped the researcher framing their questions to understand the opinions of local communities towards those government initiatives. Posters and political pamphlets also brought a significant amount of information for the research as those are the primary sources of responses to a particular event. In our case it is Ganges river bank erosion and people's response towards it.

ASSESSING RIVER-BANK EROSION OF THE GANGES: PRECARIOUS EXISTENCE, VULNERABLE LIVES

River bank erosion is a complex phenomenon influenced by a combination of natural processes and human activities. The Ganges bank erosion is much higher in Malda and Murshidabad when compared to other districts due to the general tendency of Ganges River to shift towards left bank above Farakka barrage which is in Malda district, and towards right side of Farakka Barrage which is in Murshidabad and other districts. The construction of the Farakka Barrage has had profound implications for erosion patterns upstream and downstream. By altering the river's flow regime and sediment transport dynamics, the barrage has disrupted the natural equilibrium, leading to increased erosion along vulnerable stretches of the riverbanks. After analysing satellite images from the year 1973 to 2018, it has been observed that the River Ganges continues to shift eastwards and is eroding villages one after another (Das & Samanta, 2023). Till 2008, it was estimated more than 700000 people have been displaced in these two districts collectively due to Ganges river bank erosion (Mukherjee, 2008). Over last three decades of past century, the uncontrollable hunger of Ganges upstream of Farakka Barrage has rendered nearly 4.5 lakhs people homeless in Ratua II, Manikchak, and Kaliachak I, II and III blocks of Malda district (Banerjee, 1999). In last 20 years the amount of displacement has increased as the River Ganges kept engulfing new lands every year. Sadly we cannot find the right estimation as there is a lack in maintaining proper records, except few recent initiatives by non-governmental organizations such as CRY (Child Rights and You) and *Ganga Bhangon Protirodh Action Nagarik Committee* (GBPANC), toward knowing the loss of home or land. Their recent report shows the enduring destruction of River Ganga that has affected more than 40000 families so far along with wiping out trading centre like Panchanandapur, regular weekly markets, sugar mills, 16 primary schools, 14 high schools, several temples, mosques, administrative offices, and lands use for silk cultivation. In last thirty years about 750 square kilometres area of which 269 square kilometres of fertile land particularly in Kaliachak and Manikchak community development block of Malda have been engulfed. According to a respondent, "people have moved from their homes for four to sixteen times in their lives during the last 15 years."

Trying to find the reason behind such a huge land loss and mass displacement led the researcher into age old controversial human achievement i.e. Farakka Barrage. It has been a discourse now that every technological achievement by itself does not translate into real development. The most prominent example of this statement is Farakka Barrage which has affected a whole system such as the ecological process of nature which renews life support system, a river like Ganga with a rich socio-cultural history, the people living on its banks, their lives, livelihoods, and culture. History goes back to time when a decision was made to construct a barrage across Ganga by Indian Government in order to divert 40000 cusecs of water from its prevailing main flow in the River Padma to its moribund branches River Bhagirathi and Hoogly. Thus a site was chosen in Farakka, 160 miles away from north Calcutta, situated on the Bengal-Bihar border near Rajmahal. The major objectives behind the project were: improvement of navigation of Calcutta port, adequate drinkable water supply to the city, and increase inland water transport etc.

Despite having noble causes behind the project there were some juxtaposing facts. For example, need for an Environmental Assessment was ignored and only a geo-technical assessment was done by Geological Survey of India mainly to ascertain the best site for the project. Those who raised their concerns and appealed to the International bodies against the project were branded as either Pakisthanis or traitors of the nation (Banerjee 1999). The construction of the Barrage finally ended in 1971 and after four years it was dedicated to nation on the 21st May in 1975. Since then Farakka Barrage has been standing up there as technological achievement and the pride of national interest. Now it is going to be nearly 50 years of its establishment and majority of the scientific studies have shown that the existence of the barrage has negatively impacted more than its positive roles. In a recent study the amount of erosion and deposition was calculated using ArcGIS 10.6 and it has been seen that a higher amount of erosion between 1965 and 2017 at the east bank or River Ganga where Manikchak, Kaliachak II and Kaliachak III blocks are situated (Sarif M. N. et al. 2021). River Ganga is constantly shifting its flow leftward since 1973 and the changes in the flow character invited severe river bank erosion as the natural process of siltation, and sedimentation, and discharge variation have been changed as the river repeatedly trying to adjust itself. The construction of Farakka Barrage became an obstruction and that is leading the river to make her own way inducing the continuous shift and erosion in the upstream of the barrage (Banerjee & Chakraborty 1983; Banerjee 1999; Mondal & Satpati 2012; Mandal 2017).

The barrage is not the only one reason that is responsible for Ganga river bank erosion and precarity of riparian lives. There are other players too that are equally contributing toward degrading situation. For example, the activities of sand "mafias," further degrading the river ecosystem and heightening the risk of erosion. Unrestricted exploitation of sand resources by illegal sand mining operation is worsening the erosion rates, particularly in areas where the riverbed is mined for construction purposes. The linkage between illegal sand mining and river bank erosion cannot be divorced from the broader political and economic context in which it occurs. In West Bengal, the nexus between local political bodies, sand mining syndicates, and government agencies has perpetuated a system of corruption and impunity that undermines efforts to address erosion effectively. Politically connected individuals and groups often exploit natural resources for personal gain, disregarding the long-term consequences for the environment and local communities. It is not like local people do not know about the nexus and the individuals who are responsible for illegal activities but due to the fear of getting harmed they are pushed towards being mute witness. When researchers were trying to know the matter in details they were strictly instructed not to bring these issues further. However, recent reporting might highlight the issues to a certain extent. A correspondent of renowned online news portal, The Telegraph Online, reports about how an octogenarian MLA (Member of Legislative Assembly) is trying to prevent illegal mining of sand and soil from the riverbed by walking up to the banks of the Fulhar river and spending hours. According to the sitting MLA of Ratua block, "*I have lost faith in a section of local police as they are not doing anything to stop illegal extraction of sand and soil from Fulhar, despite repeated instruction from chief minister and that is why I visit the river bank every day and raise an alert if I spot any such extraction going on*". Earlier he accused few police officers by pointing out the nexus and wrote letters to senior police officials but they declined to comment on the MLA's accusation, though, their illegal sand and soil prevention drive showed a number of vehicles have been seized for carrying sand and soil which taken out from rivers without proper permission. On the other hand local area president of state ruling party accused the MLA, who is also from the same party, of talking in delirium (The Telegraph Online, 2022). Thus we can see there are some activities of sand and soil extraction illegally that are not properly discussed in the open but at the same time its existence cannot be denied.

As the River Ganges erosion is real in so as its effects on the local communities in terms of "forced migration" making their existence precarious. *"There are many people who once had multiple bighas of land, big houses; now they live under a single tarpaulin covered tent after losing everything in the hungry river"*- says one of the victims. River bank erosion and forced migration means a continuous process of displacement, then settlement, then displacement, then again resettlement and so on. This continuous displacement and resettlement has another consequence for the people of studied area in Malda district and that is inter-state border dispute. Some part of the river bank lines are considered as inter-state or inter-country border. River Ganga in Malda district has been border between West Bengal and Jharkhand state of India. Due to continuous changes in river flow the River Ganga is constantly submerging lands on the left bank (West Bengal) and emerging new land (Jharkhand) at the right bank resulting in aggravating precarious existence of local communities in Malda. The new land, which is emerging on the right bank of the river, now belongs to Jharkhand despite the loss at the side of West Bengal. Problem is arising when erosion-afflicted people of lost land move to the newly emerged highland on the opposite side of the river, which is now in Jharkhand. Previously they belonged to West Bengal and had all the documents such as voter identity card, ration card, and Adhar card of this state. Now the new place they moved in does not recognise them as residents of Jharkhand. As a result they are suffering from the deprivation of getting basic civic amenities such as health and education. They even cannot take advantage of different government schemes, like the National Rural Employment Guarantee Scheme, which provides 100 days of work guarantee for the rural unemployed.

First and secondary data from the field have revealed that the marginalized riparian communities of Malda district lack access to adequate resources and representation, exacerbating their vulnerability to the disaster of Ganga River bank erosion. In the absence of robust governance structures and accountability mechanisms, these communities are left to fend for themselves, often bearing the brunt of displacement, loss of livelihoods, and social dislocation. Asking upon the major issues that needed immediate attention the local respondents revealed how health care is seriously failing in the region due to the lack of hospitals and proper medical equipment. The incidents of heart attack and mental health issues are very common among the victims of river bank erosion. The minority cell which is supposed to monitor proper care does not function properly. If someone wants to avail state government-provided schemes such as "*Kanyasree and Rupasree*" they need to spend a hefty amount of money first to bribe officials to get those schemes. Since erosion-afflicted people have no permanent residential address government initiatives like "*Duare Sarkar*," which is supposed to solve any document-related issue in the local area itself, do not function properly. The same is happening with "Asha Workers," as they are reluctant in paying visit in the area of erosion victims. Losing lands making significant amount of people move to other states in search of income. Once the place flourished with adequate income sources and indigenous culture, now they have lost everything to the hungry River Ganga. The hunger which is engulfing everything is not natural; rather, it is human made disaster.

GENDER DIMENSIONS OF VULNERABILITY

The impacts of river bank erosion are not evenly distributed across populations; rather, they intersect with existing inequalities along gender lines. Women and children, in particular, are disproportionately affected by displacement, loss of shelter, and disruptions to social networks. In many cases, the breakdown of traditional support systems intensifies vulnerabilities, leading to increased risks of exploitation and abuse. For women, the loss of land and livelihood opportunities can have profound implications for their economic autonomy and social status. Displacement often forces women into precarious situations, where they may be subject to exploitation, trafficking, or early marriage as a coping mechanism. Similarly, children are at heightened risk of exploitation and trafficking, as the breakdown of social networks and protective mechanisms leaves them vulnerable to predatory actors.

Social equality is a social condition in which every individual or distinct group within a given society has the same status. It means that there should be no boundaries based on socioeconomic class or caste, nor should there be any unjustified discrimination based on a person's gender, age, caste, creed, origin, disability, etc. Article 15 of the Indian Constitution prohibits discrimination on grounds of religion, race, caste, sex, or place of birth and Article 21 ensures protection of life and personal liberty which include right to live a private life. But in words of an informant, the state of affairs in Birnagar village of Kaliachak III block, Malda district in West Bengal, where 1500 houses have collapsed and families who once possessed acres of land and other belongings have quickly had them devastated by Ganga erosion, presents a sad picture of human fragility and social inequity. Just 300 of the 1500 displaced families have obtained *patta* (document or temporary will of a land) certificates; the remaining 1200 families face severe hardships and as the paper gets written, the families live under tarpaulins with no protection or security of life. The community struggles with serious privacy and safety issues in these makeshift shelters. Though men, as conditioned in a patriarchal society like India, still manage it without probably considering their own choices of being exposed to the outer world all the time, however, the crisis hits women most making them exceptionally vulnerable to society. The recent instance of an 11-year-old girl getting pregnant secretly highlights the severe lack of supervision and safety for those who are most susceptible in such a setting (narrative received via local respondents during the field). Women who fall prey to erosion lack basic comforts and security and are compelled to work in low-wage, hazardous professions like tying bidis, labouring in fields, or providing services in homes and hotels. It was brought to the authors' knowledge that the women do not receive the amount fixed from the government as their daily wage. There are middlemen, who take commission out of the women's hard-earned wages. Despite their hard work, they struggle to provide even two proper meals for themselves and their families. This reality pushes these already vulnerable women towards desperate measures, including the flesh trade, where they are further exploited by powerful men, perpetuating a cycle of abuse and poverty.

Their messy and critical circumstances are made worse by a lack of adequate healthcare facilities, which mostly has disastrous results. As per information received from the ground, pregnant women were pushed to death from lack of access to timely transportation to hospitals situated at far-off places. The villages and the areas affected by the Ganga erosion provide the fewest health care services to their children. They are not fully immunised, which leaves them open to a host of serious illnesses in the future.

The social fabric of Birnagar is further torn by the prevalent practice of polygamy, which adds another layer of inequality for women. The reasons as sighted by locals ranged from instances where men migrate to other states for work, often abandoning their families, seeking new relationships and leaving their previous wives and children in even more desperate conditions. This abandonment not only strips women of their partners but also of the economic and emotional support necessary for their survival, thrusting them deeper into helplessness. The toxic cycle of husband/father moving away and forming new relationships creates a dysfunctional and unstable environment that exacerbates the vulnerability of women. These compounded hardships highlight the urgent need for comprehensive interventions, including secure housing, healthcare access, economic opportunities, and social support systems to protect and empower the women of Birnagar. Though there was also mention of such practices amongst women however the ratio is more in the case of men. The respondents added that men are not choosing to commit to one relationship out of their insecurities about their financial position and their reality of not having a permanent home to think further about building a family. These all add up to the women's misery. Addressing these issues is crucial for breaking the cycle of exploitation and building a more just and equitable society.

Additionally, gender-based violence is a pervasive issue that intensifies the vulnerability of women in disaster-hit regions like Birnagar, Kaliachak III block of Malda district, West Bengal. In the chaos following displacement, the breakdown of community structures often instigates an increased rate of domestic violence and sexual assault. Women living under tarpaulins with little to no security are easy targets for predators besides being abused by the men they call their own. The men under the influence of substances and alcohol forget to respect the dignity of their women and treat them inhumanely. Given the fact and increased rate of reported violence, most places in this world are not safe for women, in general. The lack of privacy in these temporary shelters makes it tougher for women to protect themselves or their children. Moreover, the cultural stigma attached to victims of sexual violence often prevents them from seeking help or reporting crimes, perpetuating a cycle of carrying forward the legacy of social evils like child marriage and being married to one's rapist. The fear of retaliation or further ostracization forces many women to endure their plight in silence, making it difficult for authorities to address the issue effectively.

Girls in disaster-affected communities are unreasonably affected by child marriage, a serious societal scourge. Many families are pushed by economic hardship and insecurity to marry off their girls at an early age in the hopes of giving them better safety and financial stability. But for these young girls, child marriage frequently results in a life of ongoing abuse and poverty. They are compelled to drop out, cutting their chances to pursue higher education and better career prospects. Early pregnancy, which is linked to child marriage and poses serious health risks to young girls whose bodies are not yet ready for motherhood, is the primary cause of the high rates of mother and new-born mortality among girls and women in disaster-stricken areas, even though reaching gender equality is a major goal of the SDGs and controlling IMR and MMR are important goals of the MDGs. In addition to depriving girls of their youth, this practice feeds the vicious cycle of poverty and gender inequity.

Amongst the erosion hit community, the lack of education for girls is a critical issue that contributes to their vulnerability. Though it was reported to the authors from the respondents' side that there is a school in Birnagar, there is a huge lack of teachers. A teacher finds it almost impossible to manage a class as it becomes bigger. Additionally, cultural norms and patriarchal attitudes also play a significant role in perpetuating gender inequality in Birnagar. Women are often expected to conform to traditional gender roles from a very young age, which limits their opportunities for education, employment, and personal growth. These cultural expectations place a heavy burden on women, who are expected to manage household duties and childcare in addition to the economic contributions that they earn by binding Bidi (Indigenous Cigarette). The lack of support for women in balancing these responsibilities further entrenches their disadvantaged position. Challenging and changing these cultural norms is essential for creating a more equitable society where women have the freedom to pursue their aspirations and contribute fully to their communities. Also, the very few who attend school, remain uninformed of the vast career opportunities that exist after completing the 12th grade due to the unavailability of proper career counselling. Girls who do not have an education are denied the information and abilities necessary to seek paid work and stand out for their rights. Due to their lack of education, they are more susceptible to exploitation and rely on males for their financial well-being. To break the cycle of poverty and provide their children with the tools to live with dignity in this competitive world, they need to create better futures for themselves and their communities. It is imperative that all children, especially girls, have access to school.

Given the low rate of literacy amongst girls/women, Economic exploitation is one of the many significant issues faced by women in erosion-affected areas. With limited employment opportunities, women are often forced to accept low-paying and exploitative jobs. Employers take advantage of women's desperation to earn a living, paying them meagre wages for long hours of work and making them do exploitative work. This economic exploitation extends to the informal sector, where women working as domestic helpers or in hotels are often underpaid and overworked. The lack of legal protections and labour rights for these women leaves them vulnerable to further exploitation and abuse. In an attempt to improve their standard of living, many become victims of the vile and disruptive flesh trade. Women's economic empowerment and general well-being depend on giving them equal employment opportunities and guaranteeing the protection of their labour rights; these issues still need to be addressed, particularly in developing nations.

From the discussion with the respondent, it seemed to the authors that human trafficking is not an impossible practice in erosion-hit areas. Women and children are particularly vulnerable to human trafficking due to the economic instability and relocation these events bring. Traffickers entice them with false promises of improved living conditions and employment opportunities. After being trafficked, victims endure abuse, forced labour, and sexual exploitation. It is challenging for survivors of human trafficking to rebuild their lives and reintegrate into society because of the trauma and shame attached to the practice.

Another essential fact that came to notice by the authors is that the health of women in Birnagar is a major area of concern as there is hardly any health centre in the vicinity. The lack of nearby healthcare facilities means that women do not have access to essential medical services, including prenatal and postnatal care. Pregnant women and new mothers are particularly at risk, with many dying because they cannot be transported to distant hospitals in time. The absence of proper sanitation and hygiene in the temporary shelters further exacerbates health issues, leading to the spread of infectious diseases. Malnutrition is rampant among women and children, weakening their immune systems and making them more susceptible to illnesses. The physical toll of hard labour combined with poor living conditions significantly impacts women's health, reducing their ability to work and care for their families.

To sum up, the situation of women in the studied area after being uprooted due to Ganga erosion underscores the pressing requirement for all-encompassing measures to tackle the various facets of susceptibility they encounter amongst which only a few are addressed above. Social evils that perpetuate gender inequality disproportionately affect women in disaster-affected communities; these sins range from concerns about safety and privacy in temporary shelters to issues of economic exploitation, gender-based violence, and health. A complex strategy involving access to legal protections, economic opportunities, healthcare and education, as well as community-based support systems, is needed to address these concerns. To end the cycle of exploitation and create a more just and equitable society for all, we must empower women and advance gender equality, especially in geographical areas that are more vulnerable than others.

STRUGGLE AGAINST HUNGRY RIVER: RESILIENCE AND RESISTANCE

Despite these challenges, affected communities have demonstrated remarkable resilience in the face of adversity and violations of human rights. Grassroots organizations committed to protecting civil rights, such as, '*Jana Andalan* 'and the 'Ganga Bhangan Protirodh Action Nagarik Committee' have emerged as powerful advocates for community rights and environmental justice. Through collective action and advocacy, these groups have mobilized support, raised awareness, and pressured authorities to address the underlying causes of river bank erosion. In a group discussion, the informants from the erosion-affected Malda settlements revealed that there hasn't been any dredging in a while, which has caused continuous pressure to build up along the coasts, causing siltation and erosion to worsen. Despite the multifaceted challenges posed by river bank erosion, the affected communities in West Bengal have exhibited extraordinary resilience throughout history. Even if the needs of the survivors have hardly changed over time, they are still not acknowledged. The riparian community's top priority is for river erosion to be recognised as a disaster and to provide appropriate rehabilitation. The "*Ganga Bhangan Protirodh Action Nagarik Committee*" and "*Jana Andalan*," two grassroots organisations, members from which have shared grassroots information via interviews to develop this paper, have been instrumental in promoting environmental justice and community rights. These organisations have gathered support, increased public awareness, and put pressure on decision-makers to deal with erosion's underlying causes.

These grassroots movements are distinguished by their inclusive strategy and reliance on the community's combined strength. They involve a wide range of stakeholders, making sure that their opinions are heard during the decision-making process. The stakeholder list includes women, school-going children, youth etc. To address the intricate and interconnected vulnerabilities that these communities face, inclusion is essential. There have been cases of schoolchildren walking in rallies organised by the grassroots organizations in the area with similar demands, as well as news has also flashed regarding women-led protests calling for infrastructure improvements for the communities impacted by river erosion. There is hope for the future because schoolchildren are made aware of the problem at a young age.

One of the primary functions of grassroots organizations is to raise awareness about the issues affecting their communities. Through campaigns, public meetings, and media engagement, these civil action groups highlight the human and environmental impacts of river bank erosion. They documented cases of displacement, loss of livelihoods, and other social injustices, creating a compelling narrative that resonates with policymakers, media, and the broader public. Online and offline magazines are published by the members of grassroots organizations that document the day-to-day crises of the riparian communities. By framing river bank erosion as a matter of social and environmental justice, these organizations shift the discourse from one of inevitability to one of accountability and change. They challenge the narratives that marginalize affected communities and call for recognition of their rights and needs. This advocacy work is crucial for building a collective voice that can influence policy and practice.

Grassroots organizations play a critical role in holding authorities accountable for their actions and inactions. Through petitions, legal challenges, and direct action, they pressure local and national governments to take responsibility for the consequences of infrastructure projects like the Farakka Barrage. They demand transparent and inclusive decision-making processes that consider the long-term impacts on communities and ecosystems. Recently, both the civil rights organizations, that is Jana Andalan and Ganga Bhangan Protirodh Action Committee decided to collaborate to gather funds and build strong embankments utilising the Geosynthetic-tube method. These efforts often involve challenging powerful interests, including political bodies, sand mining syndicates, and other entities that benefit from the status quo. The resistance of grassroots organizations is therefore not just a response to environmental degradation but also a fight against systemic corruption and exploitation. Their bravery and tenacity in the face of these difficulties provide witness to their resilience.

ADAPTIVE STRATEGIES AND COPING MECHANISMS

The resilience of riverine communities is also evident in their adaptive strategies and coping mechanisms. Faced with the loss of land, homes, and livelihoods, these communities have developed various ways to survive and thrive despite the adversities. These strategies are often rooted in traditional knowledge and practices, which have been adapted to the changing environmental context.

Community-led initiatives play a vital role in this adaptive process. They provide support and assistance to those who have lost their livelihoods, helping them to rebuild their lives. These initiatives often include skills training, alternative livelihood programs, and mental health support, which help the affected individuals regain their economic independence and social stability.

Programmes for alternative livelihoods are also essential for fostering resilience. These programmes frequently centre on environmentally friendly and sustainable practices, like organic farming, aquaculture, or eco-tourism; however, as revealed in the discussion conducted with the members of the civil rights organization and survivors of the erosion, no such livelihood programmes are provided in the areas where the communities affected by erosion seek refuge. There has been no intervention by the government or non-governmental organisations in those areas to provide mental health support as well. The psychological impacts of displacement and loss being profound, affect individuals' ability to cope and rebuild their lives. The affected communities are supporting each other and proving their resilience.

To develop any type of resilience, social support is essential. It was disclosed in the interview by a member of Birnagar village who lost his home to Ganga river erosion, that there is no caste distinction among the impacted families. The community also feels that they are each other's support system because the crisis affects everyone who lost their home and other properties due to river erosion. These networks make it easier for people to share resources, knowledge, and encouragement, which aids people and families in overcoming obstacles. These bonds increase the community's overall resilience by fostering social cohesion. Even if anyone or a group of people are immoral, other members work together to try to correct the offending party. Though this might have advantage in most cases however it might be disadvantageous and lead to mob lynching in severe cases. Thus, the government and legal authorities must look into the community's needs.

A significant aspect of the resilience and resistance of these communities is the documentation and sharing of their stories. By capturing the experiences of those affected by river bank erosion, grassroots organizations create a powerful narrative that highlights the human dimension of environmental disasters. These stories are shared through various mediums, including reports, documentaries, social media, and public events.

It is clear from the conversation with members of *Jana Andalan* and *Ganga Bhangan Protirodh Action committee* that the organizations are working in alliances and trying further to build networks with other civil society groups. These partnerships would strengthen their capacity to advocate for change and provide more support to affected communities. By collaborating with like-minded organizations, they can share resources, expertise, and strategies, creating a more robust and coordinated response to the challenges they face. Such alliances would also help to bring the issue of river bank erosion to the attention of a wider audience. By connecting with regional, national, and international networks, grassroots organizations can elevate their cause and mobilize broader support. This can lead to increased funding, technical assistance, and policy attention, further strengthening their efforts.

The ultimate goal of grassroots organizations is to achieve systemic change that addresses the root causes of river bank erosion and promotes sustainable and just development. This involves advocating for policies that protect the rights of affected communities, promote environmental sustainability, and ensure equitable distribution of resources.

Environmental sustainability involves protecting and restoring natural ecosystems, promoting sustainable resource management practices, and mitigating the impacts of climate change. The use of Geo-Tubes as embankments is one of the many examples of sustainability. Social justice, on the other hand, requires addressing the inequalities and injustices that exacerbate vulnerability, ensuring that all individuals have access to resources, opportunities, and decision-making processes. The basic tenet of Human rights is to protect the dignity, rights, and freedoms of all individuals, particularly those who are marginalized and vulnerable. As the vulnerable riparian community stands far from living a dignified and private life, this article is a mere effort to spread the information on a global level seeking effective measures to support the needs of the community.

The resilience and resistance demonstrated by the riverine communities in West Bengal provide valuable lessons for disaster management and adaptation efforts globally. Community-led initiatives that prioritize local knowledge, inclusive governance, and social cohesion are essential for building resilience to environmental disasters.

Policymakers and practitioners can learn from these examples and incorporate community-led approaches into their strategies. This involves supporting grassroots organizations, investing in local capacities, and creating enabling environments for community-driven solutions. By doing so, they can promote more effective, equitable, and sustainable responses to environmental challenges.

POLICY SUGGESTIONS

Policy advocacy is a critical component of this work. As much as it is true that grassroots organizations mentioned in this work engage with policymakers at various levels, providing evidence, recommendations, and solutions based on their experiences and research. They call for policies that prioritize the needs and rights of marginalized communities, promote sustainable land and water management practices, and ensure accountability and transparency in decision-making processes. The writers are motivated by their on-field experiences to define additional policies to further define the scope of development.

In general, legal and institutional barriers hinder women's empowerment in disaster-affected areas. The lack of legal recognition and lack of property rights for women means that they are often excluded from decision-making processes and economic opportunities. Without legal protection, women cannot claim their rights or seek justice for the abuses they suffer. Strengthening legal frameworks to protect women's rights and ensuring their enforcement is critical for empowering women and promoting gender equality. This includes providing women with legal documentation, such as *patta* documents, which grant them land ownership and security.

Community-based interventions are vital for addressing the issues faced by women in Birnagar. Establishing women's groups and networks can provide a platform for women to share their experiences, support each other, and advocate for their rights. These groups can also play a crucial role in raising awareness about gender-based violence, health, and legal rights, and in providing education and vocational training for women. Community support systems can help build resilience and empower women to take collective action to improve their living conditions and protect their rights.

Government and non-governmental organizations (NGOs) have a critical role to play in addressing the plight of women in disaster-affected areas. Besides providing immediate relief, long-term support in the form of vocational training and job placements are essential for helping women rebuild their lives. This includes providing secure housing, healthcare services, education, and economic opportunities. Additionally, implementing programs that focus on women's empowerment and gender equality can help address the root causes of their vulnerability. Ensuring that women are included in disaster response and recovery planning is crucial for creating solutions that meet their specific needs and promote their well-being.

Education and awareness-raising campaigns are essential for changing attitudes and behaviours towards gender equality. Educating communities about the importance of gender equality and the rights of women and girls can help shift cultural norms and reduce discrimination and violence against women. These campaigns can also focus on the benefits of educating girls and the importance of women's economic participation, highlighting how these contribute to the overall development and prosperity of the community.

Promotion of livelihoods that are less dependent on vulnerable natural resources, these initiatives help to reduce the risk of future displacement and economic instability.

A holistic approach to resilience and resistance requires the integration of environmental sustainability, social justice, and human rights considerations. This means recognizing the interconnectedness of these issues and addressing them in a coordinated and comprehensive manner.

CONCLUSION: TOWARDS A JUST AND SUSTAINABLE FUTURE

Every human being has a fundamental right to development, which allows them to take part in, contribute to, and enjoy economic, social, cultural, and political advancement that fully realises their basic freedoms and all other human rights, regardless of their caste, class, gender, creed, race, or other special circumstances. As part of the community's development, grassroots organisations are working to balance the government's support and bring positive changes in the dire circumstances the riparian community faces. The riverine villages of West Bengal have demonstrated perseverance and resistance in the face of hardship, demonstrating their strength and drive. They have proven they can endure, prosper, and promote systemic change by grassroots organising, campaigning, and adaptable tactics. Their initiatives emphasise the value of locally driven solutions and the necessity of inclusive and equitable government. By adopting a holistic approach that integrates environmental sustainability, social justice, and human rights considerations, policymakers and practitioners can develop more effective strategies to address the broader consequences of ecological disasters. This requires a commitment to equitable and inclusive governance as well as investments in community-led initiatives and grassroots organizing.

Recognising the interdependence of environmental, social, and political challenges is ultimately necessary to build resilient communities. A more equitable and sustainable future for all can be built by empowering marginalised people to oppose injustice and build resilience.

The interaction with the survivors who lost their land and properties to river erosion revealed repeatedly that declaring river erosion as a disaster is one of their main demands. It can therefore be concluded that the survivors may have benefited from the services provided under the Disaster Management Act of 2005, the National Policy on Disaster Management of 2009, and the Land Acquization, Rehabilitation, and Resettlement Act of 2013 and from welfare schemes like The Prime Minister Awas Yojana, Pradhan Mantri Fasal Bima Yojana, National Social Assistance Program, MGNRGA, and ICDS services had they been registered as disaster affected. However, survivors hardly receive any assistance or recognition because river erosion is still not considered a national disaster.

ACKNOWLEDGMENT

Authors are sincerely thankful to Mr. Biplab Bhattacharya (Human Rights & Social Activist, *Jana Andalan*) and Abdullah Musleuddin (Ganga erosion survivors, social activist, and a member of *Ganga Bhangan Protirodh Action Nagarik Committee*) for their immense help in shaping this study. Without their constant support this study could have been an impossible task.

REFERENCES

Alkire, S., Meinzen-Dick, R., Peterman, A., Quisumbing, A., Seymour, G., & Vaz, A. (2013). The women's empowerment in agriculture index. *World Development*, 52, 71–91. DOI: 10.1016/j.worlddev.2013.06.007

Banerjee, M. (1999). A Report on the impact of Farakka Barrage on the Human Fabric: A study of the upstream and downstream areas of Farakka Barrage. On behalf of South Asian Network on Dams, River, and People (SANDRP), New Delhi.

Cornwall, A. (2003). Whose voices? Whose choices? Reflections on gender and participatory development. *World Development*, 31(8), 1325–1342.

Banerjee, S. N., & Chakraborty, P. (1983). Some observation on recent trends of shifting of Ganga between Rajmahal and Ahiron. *Journal of the Geographical Society of India*, 24, 318–332.

Das, J. D., Dutta, T., & Saraf, A. K. (2007). Remote sensing and GIS application in change detection of the Barak River channel, N.E. India. *Photonirvachak (Dehra Dun)*, 35(4), 301–312. DOI: 10.1007/BF02990786

Das, R., & Samanta, G. (2023). Impact of floods and river-bank erosion on the riverine people in Manikchak Block of Malda District, West Bengal. *Environment, Development and Sustainability*, 25(11), 13595–13617. DOI: 10.1007/s10668-022-02648-1 PMID: 36118734

Das, T., Haldar, K., Sarkar, S., Borderon, K., Kienberger, D., Das Gupta, M., Kundu, S., & Guha-Sapir, D. (2017). Impact of riverbank erosion: A case study. *Australasian Journal of Disaster and Trauma Studies*, 21(2), 73–81. https://trauma.massey.ac.nz/issues/2017-2/AJDTS_21_2_Das.pdf

Dewan, A., Corner, R., Saleem, A., Rahman, M. M., Haider, M. R., Rahman, M. M., & Sarker, M. H. (2017). Assessing channel changes of the Ganges-Padma River system in Bangladesh using Landsat and hydrological data. *Geomorphology*, 276, 257–279. DOI: 10.1016/j.geomorph.2016.10.017

Elahi, K. M., John, R. R., & Rogge, J. R. (1990). *Riverbank erosion, flood and population displacement in Bangladesh: a report on the riverbank erosion impact study. Riverbank Erosion Impact Study (REIS)*. Jahangirnagar University.

Gupta, D. N. S. (1951). *An outline of the river problems of West Bengal*. General Printers & Publishers.

Hassan, N. A., Ratna, S. J., Hassan, M., & Tamanna, S. (2017). Remote Sensing and GIS for the Spatio-temporal Change Analysis of the East and the West River Bank Erosion and Accretion of Jamuna River (1995-2015), Bangladesh. *Journal of Geoscience and Environment Protection*, 5(9), 79–92. DOI: 10.4236/gep.2017.59006

Howlader, N., & Rahman, M. A. (2016). Riverbank Erosion and Its Impact: A Study on Ganges River Affected Area. *Jagannath University Journal of Social Science*, 4(1-2), 73–85.

Islam, M. F., & Rashid, A. N. M. B. (2011). Riverbank Erosion Displaces in Bangladesh: Need for Institutional Responses and Policy Intervention. *Bangladesh Journal of Bioethics*, 2(2), 4–19. DOI: 10.3329/bioethics.v2i2.9540

Karim, A. H. M. Z. (n.d.). (2914). Flood and Riverbank Erosion Displacees: Their Indigenous Survival Strategies in Two Coastal Villages in Bangladesh. *Asian Social Science*, 10(4), 16–26.

Mandal, S. (2017). Assessing the Instability and Shifting Character of the River Bank Ganga in Manikchak Diara of Malda District, West Bengal using Bank Erosion Hazard Index (BEHI), RS & GS. *European Journal of Geography*, 8(4), 6.

Mirza, A. T. M., Rahman, T., Islam, S., & Rahman, S. H. (2014). Coping with flood and riverbank erosion caused by climate change using livelihood resources: A case study of Bangladesh. *Climate and Development*, 7(2), 185–191. DOI: 10.1080/17565529.2014.910163

Mirza, M. M. Q. (2004). The Ganges Water Diversion: Environmental Effects and Implications — An Introduction. In Mirza, M. M. Q. (Ed.), *The Ganges Water Diversion: Environmental Effects and Implications. Water Science and Technology Library* (Vol. 49). Springer., DOI: 10.1007/978-1-4020-2792-5_1

Mollah, T. H., & Ferdaush, J. (2015). Riverbank Erosion, Population Migration and Rural Vulnerability in Bangladesh (A case Study on Kazipur Upazila at Sirajgonj District). *Environment and Ecology Research*, 3(5), 125–131. DOI: 10.13189/eer.2015.030502

Mondal, M., & Satpati, L. N. (2012). Morphodynamic setting and nature of bank erosion of the ichamati river in swarupnagar and baduria blocks, 24 Parganasz (N), West Bengal. *Indian Journal of Spatial Science*, 3(2), 35–43.

Mukherjee, P. (2013, February 05). Cry of Help for Erosion Victims. Business Standard. Retrieved from https://www.business-standard.com/article/economy-policy/cry-of-help-for-erosion-victims-108021901020_1.html

Nazem, N. I., & Elahi, K. M. (1990). Impact of Riverbank Erosion on Land Settlements in Bangladesh. In A.Q. M. Mahtub (Ed.), *Proceedings of the Seminar on People and Environment in Bangladesh*, Dhaka, Bangladesh: UNDP and UNFPA

Pittman, A. (2010). Capturing change in women's realities a critical overview of current monitoring & evaluation frameworks and approaches.

Rabbi, H., Saifullah, A. S. M., Sheikh, M. S., Sarker, M. M., & Bhowmick, A. C. (2013). Recent study on river bank erosion and its impacts on land displaced people in Sirajgonj Riverine area of Bangladesh. *World journal of Applied Environmental chemistry, 2*(2), 36-43.

Rahman, M. (2013). Impact of Riverbank Erosion Hazard in the Jamuna Floodplain Areas in Bangladesh. *Journal of Science Foundation*, 8(1-2), 55–65. DOI: 10.3329/jsf.v8i1-2.14627

Sarif, M. N., Siddiqui, L., Islam, M. S., Parveen, N., & Saha, M. (2021). Evolution of river course and morphometric features of the River Ganga: A case study of up and downstream of Farakka Barrage. *International Soil and Water Conservation Research*, 9(4), 578–590. Advance online publication. DOI: 10.1016/j.iswcr.2021.01.006

Sarkar, S. D. (2022, May 18). Octogenarian MLA's riverbed vigil in Malda: Samar Mukherjee's relentless effort to prevent illegal mining of soil and sand. The Telegraph online. Retrieved from https://www.telegraphindia.com/west-bengal/octogenarian-mlas-riverbed-vigil-in-malda/cid/1865585

Sommer, M., Ferron, S., Cavill, S., & House, S. (2015). Violence, gender and WASH: Spurring action on a complex, under-documented and sensitive topic. *Environment and Urbanization*, 27(1), 105–116. DOI: 10.1177/0956247814564528

Tripathy, B., & Mondal, T. (2019). Impact of Riverbank erosion on Human life. *Thematics Journal of Geography*, 8(09), 53–56. DOI: 10.26643/tjg.v8i9.8101

Chapter 16

Women Labour Force Participation in India and Sustainable Development Goals: A Legal Framework

A. R. Vidya
Christ Academy Institute of Law, Bengaluru, India

Arpita Chakraborty
Christ University, India

ABSTRACT

Women, constituting nearly half of India's population, are crucial for the nation's economic growth. Gender equality as a core global agenda, is encapsulated in the United Nations' Sustainable Development Goals (SDGs), particularly SDG 5, which focuses on gender equality and the empowerment of all women and girls. Despite educational advancements, women in India continue to face significant barriers in other areas. Therefore, it is imperative to implement relevant legal instruments and policies to support environment-friendly agricultural practices, benefiting both ecological sustainability and women's economic participation. Judicial pronouncements related to women also contribute to sustainability and women's participation in labour.

DOI: 10.4018/979-8-3693-6069-9.ch016

INTRODUCTION

In India, women make up nearly half of the population. Their active participation in economic activities is crucial for the nation's growth. However, India's proportion of economically active women is the lowest globally, a pressing issue that needs immediate attention. The World Economic Forum's Global Gap Report 2022 further emphasises this urgency, ranking India 135th among 146 participant countries in terms of gender gap parity. In the economic participation and opportunity pillar, India ranks a staggering 143rd out of 146 countries.

In contrast, in the educational achievement pillar, India has closed the primary and secondary education gender gap. It is remarkably close to meeting the tertiary education gender gap, with scores of 0.961. Though educational achievement is an important pillar, women in India face other challenges, such as access to healthcare facilities, decent employment opportunities, etc. There is also a huge gender disparity in shouldering domestic and care work, too. Indian women spend significantly more time on unpaid caregiving and domestic work for family members than men (Mitra & Mehta, 2021).

Women's economic participation in India is of paramount importance, with the Female Labour Force Participation Rate (LFPR) serving as a crucial indicator of the country's development.

Women's empowerment through education is a global issue. To address such global issues, the UN Women was formed in 2010. It is the entity of the United Nations specifically for empowering women. Going towards gender equality is a core agenda not only in India but in many other countries as well. So, in September 2015, Member States of the United Nations Committed to working towards 17 Sustainable Development Goals (SDGs). SDG 5 is Gender Equality and Empowerment of all women and girls. Member countries of the United Nations set Millennium Development Goals (MDGs) in 2000. It aimed to achieve eight goals, which were decided to be accomplished by 2015. MDGs needed to be more effective in addressing all global issues, but these were more of a politically inclined agenda for developing countries (Pedersen, 2018). Therefore, the idea of Sustainable Development Goals (SDGs) came as progress to MDGs in 2015, where 193 member countries agreed to achieve the goals under SDGs by 2030. All the member countries of the United Nations organised a summit on Sustainable Development in September 2015 at the United Nations headquarters in New York. They decided on 17 sustainable goals to 'make the planet a better place'. SDGs are not only for developing countries; developed countries are also part of this ambitious project. Each country's government declared these 17 goals and 169 targets to achieve under these goals. Gender indicators to ensure gender equality under SDG 5 were declared in the Beijing Declaration and Platform for Action (BPFA) in 2015.

Empowering women by providing economic opportunities and eradicating gender-biased inequality are the goals of the policy implementation of all government entities, including the government of India. In India, Govt. is taking various steps towards empowering women economically by educating them. Still, according to ILO, in the past ten years, data show that there has been a drastic downward trend in female labour force participation in India. If it continues at this pace, ILO has projected that by 2030, India will continue to detract from achieving Sustainable Development Goal 5, i.e. eliminating gender inequality. The recent global pandemic has severely affected women's labour in India, especially those who were working in informal sectors, in terms of their quality of employment. Though the recent Periodic Labour Force Survey (PLFS) data show an upward trend in female labour force participation rate, the analysis shows no improvement in women's employment quality. Most of them are engaged as unpaid helpers in their family enterprises.

Sustainable development means fulfilling the needs of the current generation while not compromising the needs of future generations. Though India has a solid rural base in terms of economy, there is a growing trend towards industrialisation and urbanisation, which enlarges city size and pushes towards deforestation. The effects are environmental vulnerability and the impact of climate change. This will also prevent the country from achieving the Sustainable Development Goals (SDGs) because among all 17 SDGs, the 13th Goal is Climate Action.

Structural change means it is through industrialisation and urbanisation, which are the indicators of economic growth, but urbanisation and women's work participation are seen to have an inverse relation (Mitra, 2019a). However, more recent literature on urbanisation shows the complex relationship between urbanisation and women's labour force participation (Mitra, 2019b). Urbanisation and industrialisation are biased towards male-dominated factory production that replaces female-dominated home-based production (Goldin, 1986). Rural women are particularly affected by urbanisation and industrialisation, as the skill mismatch is serious (Boserup, 1970).

Despite some geographic disparities, women are the backbone of agri-food systems, representing about 36% of the global agricultural workforce. Their role is significant, as they typically carry out activities that include producing crops, tending animals, collecting fuel and water, processing food, caring for family members, and home maintenance. Their proximity to the farm sites gives them more flexibility to get involved in agricultural activities and unpaid domestic and care work.

The benefits of focusing on the growth of the agricultural sector are twofold: it contributes to better ecologically sustainable development and a broader scope for women to participate in productive activities. Agricultural activities, comparatively, have less harmful effects on the ecosystem, and India has a solid agricultural base. Proper balance in agricultural production through the modernisation of agriculture and implementation of the principles of sustainable development would enable the economy to grow faster. At the same time, the growth of a women-friendly agricultural sector would help more women to participate in the workforce. Therefore, there is a need to implement relevant legal instruments to protect environment-friendly agricultural production, which has the double advantage of achieving environment-friendly economic growth and the possibility of bringing more women into productive activities. Undoubtedly, this would push India towards achieving Sustainable Development Goals.

PERCENTAGE OF POPULATION ENGAGED IN AGRICULTURE AND FEMALE PARTICIPATION IN LABOUR MARKET

Empowering women by providing economic opportunities and eradicating gender-biased inequality are the goals of the policy implementation of all government entities, including the government of India. The government's commitment to this cause is evident in its efforts to promote economic empowerment, a crucial aspect of women's overall empowerment, enabling them to participate more fully in society and the economy. Women's education is another essential aspect of empowering them with knowledge and skills. Women empowerment is a process of marginalising power in women and girls so that they can play a significant role in society; they just need a little help to raise their voices against oppression, female foeticide, rape, dowry, gender inequality and more. Participation in the labour market is a crucial measure of economic empowerment. Unfortunately, in India, the female labour force participation rate (FLPR) has persistently declined over the past two decades, although the country is experiencing high economic growth.

The Periodic Labour Force Survey (PLFS) conducted by the National Statistical Office (NSO) provides comprehensive data on various labour market indicators in India, including female labour force participation rates (FLFPR). The data from 2011 to 2022, as per PLFS and NSSO Employment-Unemployment Survey, reveals a concerning trend. There was a significant drop in FLFPR in 2017-18, followed by a gradual recovery, underscoring the pressing need for immediate action.

Figure 1. FLPR from 2011-2022

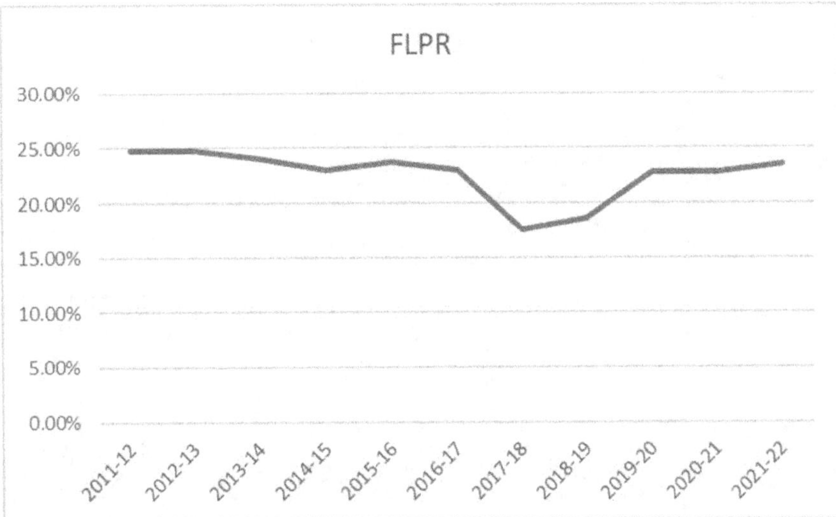

Source: PLFS and NSSO Employment-Unemployment Survey

Based on NSSO surveys, the early years indicate a relatively higher participation rate than the later years captured by PLFS data. The recent data shows an upward trend, which could be more significant.

Various socioeconomic factors influence women's participation in the labour force over this period. The participation rate of women in the labour force is highly dependent on economic development and the country's structural transformation. Much research works globally on the relationship between economic development and FLPR (Goldin, 1994; Kottis, 1990). Literature on structural transformation underlines how higher growth drives urbanisation, drawing people from rural to urban areas (Mitra, 2019). Urbanisation and industrialisation have inversely affected women in the labour market, as women lack adequate skills and education to cope with modern industrialisation.

As India undergoes a structural transformation, people are increasingly moving away from the agricultural sector and towards the industrial and service sectors. This shift is reflected in the gradual decline in the percentage of the population engaged in the agricultural sector.

The comprehensive data provided by the World Bank show agricultural employment as a percentage of total employment.

Figure 2. Percentage of population in agriculture

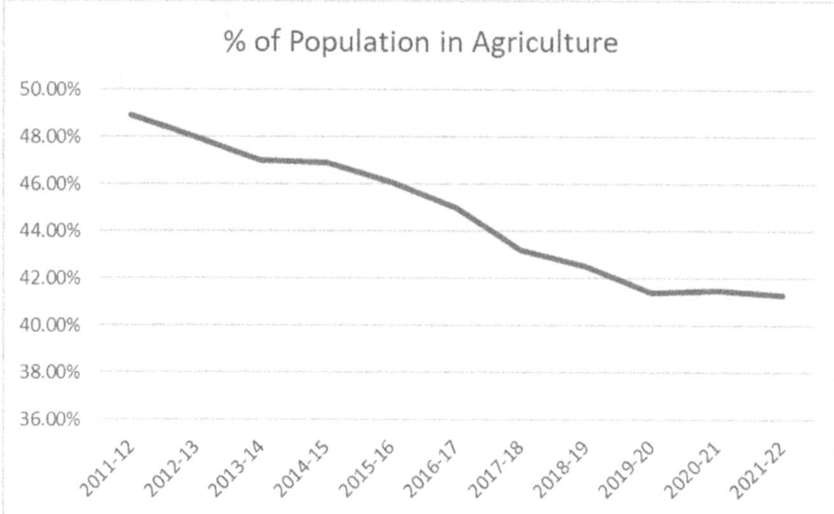

Source: The World Bank Report

The trend reflects a shift towards non-agricultural sectors and urbanisation, consistent with broader economic changes in India.

As the country undergoes structural transformation, we witness a decline in the agricultural sector's workforce and a rise in the industrial sector's contribution to GDP (Chakraborty & Sutradhar, 2023). This economic development and structural transformation manifest in rapid urbanisation, creating new job prospects and luring rural inhabitants to urban areas. While industrialisation and urbanisation present numerous opportunities, these benefits are unevenly distributed among the male and female labour forces. Urban, highly educated, and skilled women are better positioned in the job market, showcasing the potential for a more balanced workforce.

Conversely, at the grassroots level, rural unskilled women labourers encounter barriers that hinder their participation in the labour market. Traditionally, rural areas boasted higher FLFP rates than urban areas, primarily due to the prevalence of agricultural and informal sector work in these regions. However, the plight of these women, facing obstacles in accessing the job market, calls for a more inclusive and equitable approach to economic development.

Figure 3. FLPR and population in agriculture

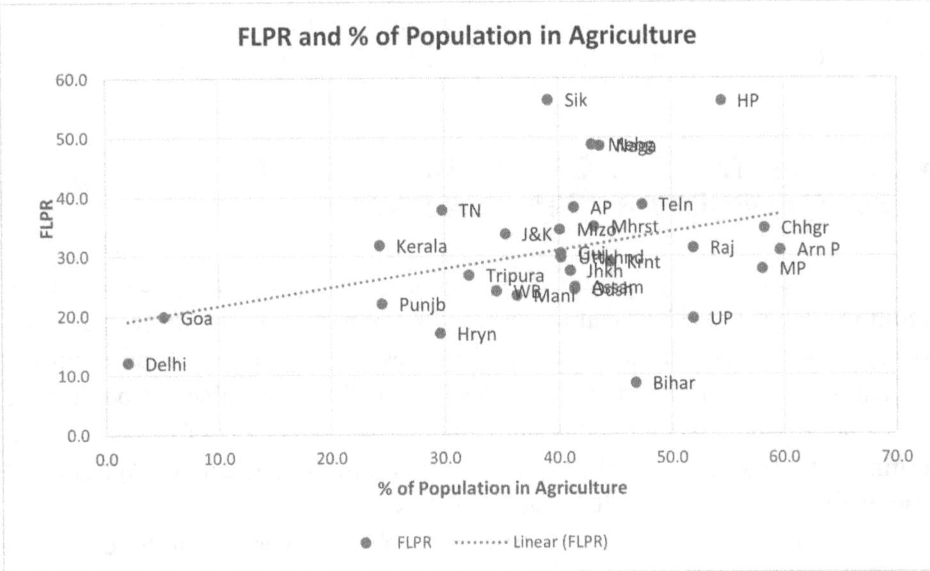

The above scatter plot shows the positive relation between the Female Labour force Participation rate and the percentage of the population engaged in agriculture across states in India, with some variation in Bihar and Uttar Pradesh. States like Sikkim, Himachal Pradesh, Meghalaya, and Nagaland have high FLPRs and a significant percentage of their populations are engaged in agriculture. Delhi and Goa have low FLPRs and a low percentage of their populations in agriculture. Kerala and Punjab show moderate FLPRs with lower percentages of their populations in agriculture compared to other states.

The positive relationship shows that keeping all other factors constant, if we increase the share of the population in agriculture, then obviously more women labour will be engaged in agriculture, and the female labour force participation rate will be higher. This might be because proximity to the farm sites gives women more flexibility to participate in agricultural activities and unpaid domestic and care work. We know that women in the family are mainly responsible for providing caregiving work and shoulder the burden of household chores. The feminisation of agriculture, which refers to the increasing share of women in agricultural employment, means that the sector is still the leading source of women's employment (Sundari, 2020). This has significant implications for gender equality and women's economic empowerment. Agriculture provides women with work near their homes, allowing them to switch between paid and unpaid work. With agriculture's declining importance in employment, women find it difficult to get jobs in other sectors due to skill mismatch (Chakraborty & Sutradhar, 2023).

So, it can be said that any drop in the percentage of the population working in agriculture due to industrialisation would drive women out of the labour force.

MODERNISATION OF AGRICULTURE AND FEMALE LABOUR FORCE PARTICIPATION RATE

Modernisation of agriculture is a process of transforming agriculture from traditional labour-based agriculture to technology-based agriculture (Kusz,2014). Machines are used in farming activities. To assess the extent of modernisation and mechanisation of agriculture, we have taken agricultural machine use per hectare of land as a proxy variable in this study. It measures the extent of machine use in farming activities, reflecting the level of mechanisation of agriculture. Agricultural machinery includes tractors, harvesters, seeders, etc.

For our research on the use of agricultural machinery per 1000 hectares of land in India, we have relied on the 'Agricultural Statistics at a Glance' reports. These reports, published by the Directorate of Economics and Statistics, Ministry of Agriculture & Farmers Welfare, Government of India, and also the Indian Council of Agricultural Research (ICAR), provide a comprehensive overview of the usage of various types of agricultural machinery across India, reflecting the growth in mechanisation over the years.

Figure 4.

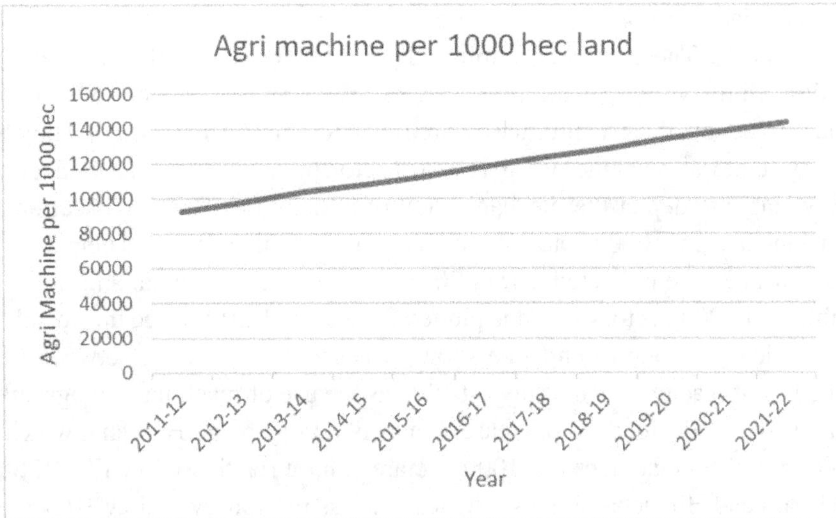

Source: Agricultural Statistics at a Glance 2022 report

To see the relationship between FLPR and the mechanisation of agriculture, we can plot the following scatter diagram

Figure 5.

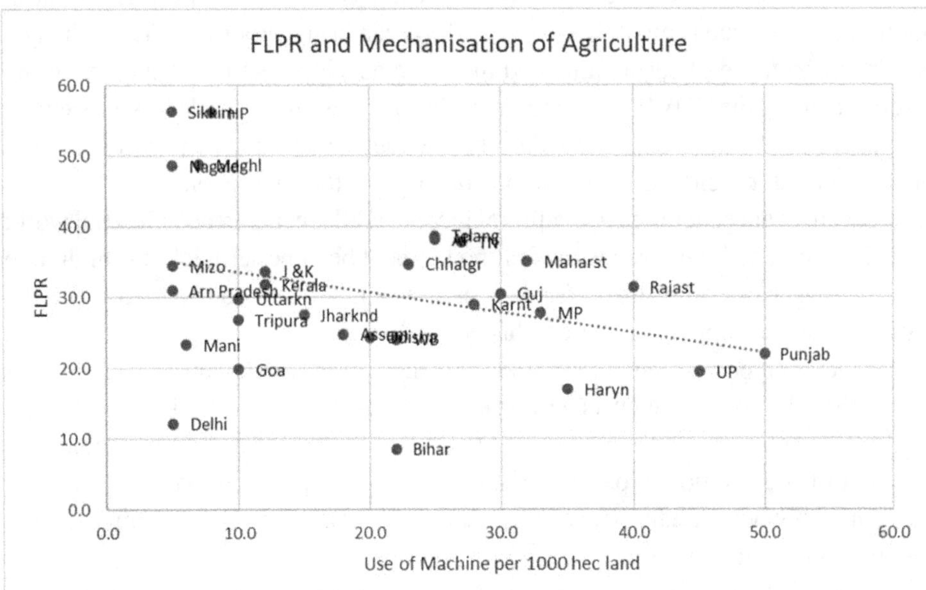

The scatter diagram depicts the relationship between the Female Labor Force Participation Rate (FLPR) and the use of machinery in agriculture across various states. The data for this analysis was collected from 'Agricultural Statistics at a Glance' reports. These reports, published by the Directorate of Economics and Statistics, Ministry of Agriculture & Farmers Welfare, Government of India, and the Indian Council of Agricultural Research (ICAR), and the methodology involved, where agricultural machine use per 1000 hector of cultivable land is taken as a proxy variable to depict the mechanisation of agriculture. The cross-section data on machine use per 1000 hector of land and female labour force participation rate (FLPR) of 2022 are collected across 30 important States in India, and the use of machines per 1000 hector of land is plotted against the FLPR to see the correlation between them. A dotted trend line shows a negative correlation between FLPR and the use of machines. This suggests that as the use of machinery in agriculture increases, the FLPR tends to decrease. States like Punjab and Haryana, which use machinery (40-50 machines per 1000 hectares), have relatively low FLPR. States like Sikkim and Himachal Pradesh, which use less machinery (below 10 per 1000 hectares), have high FLPR.

The data clearly show that agricultural mechanisation significantly influences the Female Labour Force Participation Rate. As states adopt more machinery in agriculture, manual labour is replaced by machines and as women are mainly involved in agriculture, their participation in the labour market declines. This trend underscores the broader impact of technological advancement on gender-specific employment patterns within the agricultural sector. It is a crucial area that requires the attention and action of policymakers, researchers, and gender equality advocates. Their role in addressing this issue is pivotal, and their efforts can bring about significant change.

The importance of sustainable and inclusive development in India came to the forefront during the COVID-19 pandemic when female agricultural workers stepped up significantly when men returned to their villages without industrial jobs; women worked in the fields to earn income and support their families. In this context, where women are generating agricultural income while men pursue industrial work, ensuring women's inclusiveness and empowerment becomes crucial. It is high time that challenges faced by female farmers should be taken into consideration and the legal framework supporting their rights strengthened.

Women's empowerment is essential for achieving the Sustainable Development Goals (SDGs) also. Women have a crucial role in all SDGs, with many targets evidently recognizing women's equality and empowerment as both the objective and part of the solution. Goal 5, dedicated to achieving gender equality and empowering all women and girls, stands as the stand-alone gender goal, emphasizing the importance of these aims.

Integrating gender equality into agricultural policy is crucial for the success of sustainable agriculture. In a country where women have the potential in farm sector, policies that recognize and support women's roles and rights in agriculture can lead to more effective and comprehensive approaches to food security. Enhancing financial services, women's access to land, education and technology will directly bolster sustainable agriculture efforts in our country.

This issue of inequality is prevalent among Indian women across various sectors, and agriculture is not an exception. Women often work nearly twice as many hours in the fields as men but are paid only seventy percent of what men earn, with payments frequently being irregular. A significant gender disparity exists in the agricultural sector, where one-third of women work as unpaid labourers on the farms of their parents, husbands, or in-laws, receiving neither recognition nor compensation for their efforts.

The legal status of women under agriculture-related legislation is determined by rights of women agricultural workers; rights to land and other natural resources; and rights concerning women's agricultural self-employment.

LEGAL FRAMEWORK GOVERNING FEMALE AGRICULTURAL LABOURERS

1. Constitutional Provisions

The Constitution of India not only grants equality to women but also empowers the State to adopt positive measures in favour of women for neutralizing the cumulative socio-economic, education and political disadvantages faced by them. Fundamental Rights, among others, ensures equality before the law, equal protection of law, prohibits discrimination against any citizen on grounds of religion, caste, sex or place of birth, and guarantees equality of opportunity to all citizens in matters relating to employment. Articles 14, 15, 15(3), 16, 39(a), 39(b), 39(c) and 42 of the Constitution, are of specific importance it this regard.

Constitutional Privileges (i) Equality before law for women (Article 14) (ii) The State not to discriminate against any citizen on grounds only of religion, race, caste, sex, place of birth or any of them {Article 15 (i)} (iii) The State to make any special provision in favour of women and children {Article 15 (3)} (iv) The State to direct its policy towards securing for men and women equally the right to an adequate means of livelihood [Article 39 (a)}; and equal pay for equal work for both men and women [Article 39 (d)}

v) The State to make provision for securing just and humane conditions of work and or maternity relief (Article 42) (vi) The State to raise the level of nutrition and the standard of living of its people and the improvement of public health (Article 47) (vii) To promote harmony and the spirit of common brotherhood amongst all the people of India and to renounce practices derogatory to the dignity of women {Article 51 (A)(e)}.

The constitutional background of rights of women agriculture workers is embedded in the Directive principles of state policy, part IV of the Constitution of India. The mandate on equal pay for equal work flows out of this provision. India uses the concept of "equal remuneration for the same work or work of a similar nature".

While Article 39(d) of the Constitution of India mentions about the principle of equal pay for equal work and Article 14 guarantees equality for all before the law. India ratified the Equal Remuneration Convention in 1958. "for the payment of equal remuneration to men and women workers and for the prevention of discrimination, on the ground of sex, against women in the matter of employment and for matters connected therewith or incidental thereto".

2. Legislative Provisions

To uphold the Constitutional mandate, the State has enacted various laws.

Under the Equal Remuneration Act 1976, employers are obliged to pay equal remuneration to workers of both sexes for the same work or work of a similar nature. The Act extends to the whole of India. The Act of 1976 was amended by the Equal Remuneration (Amendment) 1987. This Amendment extended the scope of the Act.

In India, the principle of equal pay means equal remuneration for "the same work or work of a similar nature" (Section 4), which in turn is defined in Section 2(h) of the Act as follows: "work in respect of which the skill, effort and responsibility required are the same, when performed under similar working conditions, by a man or a woman and the differences, if any, between the skill, effort and responsibility required of a man and those required of a woman are not of practical importance in relation to the terms and conditions of employment."

The Act is applicable to female farmers also and Section 4 of the act mandates that similar wages should be provided for a similar kind of work. Irrespective of the Act, there is a huge gap existing between the wages of male and female farmers and which should be looked into by the Government.

Another law, which is widely applicable to the vulnerable groups like agricultural women labourers is enforcement of minimum wages. The Minimum Wages Act, 1948 is primarily applicable to unorganized sector and empowers both Central and State Governments to fix/revise the minimum rates of wages in respect of scheduled employments under their respective jurisdiction. The minimum wages have been fixed at different levels by the different State Governments.

In addition to this, there are various laws enacted to ensure the provision of certain basic rights to female farmers. Section 25 of the Plantations Labour Act, 1951 prohibits the employment of women after 7 P.M. and before 6 A.M., except with the prior permission of the State Government. Maternity Benefit Act, 1961 applies to establishments that include plantation. Thus, all the maternity benefits provided in the statute are applicable to female farm labourers. The Court made it clear that grant of maternity leave is intended to encourage women to join and continue in the workplace.

Apart from this, agricultural labour, which constitutes the majority of the workforce, is exposed to many potential risks at the work location. The hospitals receive a large number of cases of accidents at work in the field. The health, hygiene and industrial safety set up, which has so far remained confined mainly to manufacturing sector should allocate a substantial part of its resources to providing services to agriculture sector.

The Governments of Gujarat, Kerala, Karnataka and Madhya Pradesh have insurance schemes for the landless agricultural labourers. This needs to be extended to the entire country. The Welfare Boards for Mine Workers, Beedi and Cigar Workers etc. are financed by the cess levied on these commodities. A Cashew Workers Welfare Board is set up in Kerala.

What is lacking in India is a gender-based agricultural law to support female farmers in the struggles they face every day. Even the Economic Survey Report 2017-18 has suggested adopting gender-specific interventions. Undoubtedly, sanction needs to be created to protect the rights of women. Recognising the same, Women Farmer's Entitlement Bill, 2011 was introduced by Prof. M.S. Swaminathan (former member of Rajya Sabha) in the year 2021. Unfortunately, the bill lapsed and no action was then taken. The bill gave recognition to female farmers in a real sense, considering them to be a farmer irrespective of their landholdings or marital status. The bill could have solved some major issues such as land rights, water rights, access to credit, etc prevailing in society. It even provided for penalties in case of non-compliance and formation of a board for redressal of grievances. .

Code on Social Security, 2020

The Code on Social Security, 2020 framed by the Government aims to extend social security to all employees and workers either in the organized or unorganized or any other sectors and for matters connected therewith or incidental thereto. Separate Social Security Funds will be set up by the Central and State Governments for unorganized workers including women agriculture.workers. Registration provisions are also mentioned. A National Social Security Board is envisaged to recommend and monitor schemes for these categories of workers.

Rights To Land and Other Natural Resources

The Land Acquisition (Right to Fair Rehabilitation and Resettlement) Act, 2013, (LARR) and the Forest Rights Act, 2006, (FRA) are two integral laws in this realm. While the former involves the acquisition of land from citizens, the latter confers land rights upon them. The LARR repealed the older Land Acquisition Act of 1984, which has been described as a draconian piece of legislation as it inflicted huge injustices on farmers in the name of development. The LARR instead made numerous progressive provisions, such as the need to obtain consent from affected families and a higher compensation rate.

The FRA is geared towards forest-dwelling Adivasi communities and recognises their right over forest land and resources, which they depend on for their livelihood, habitation, and other socio-cultural needs. These communities have been historically wronged by parties seeking to exploit the resource-rich forests that they rely on. Therefore, the FRA was designed to safeguard them.

These laws promote women's land rights in a variety of ways.

The Land Acquisition, Rehabilitation, and Resettlement (LARR) Act, 2013, is a significant piece of legislation in India aimed at ensuring fair compensation and rehabilitation for those affected by land acquisition. Here are the key points regarding the inclusion of women and land rights:

The Act recognizes both landowners and those who depend on the land for their livelihood as affected families. This inclusion benefits a large percentage of landless women who earn their livelihood through activities like farming and livestock rearing. As a result, these women are entitled to rehabilitation under the Act. The LARR Act recognizes widows, divorcees, and deserted women as separate families. This acknowledgment advances women's land rights by recognizing the rights of single women to compensation and rehabilitation.

The Act facilitates joint ownership of compensatory land by allowing it to be titled in the names of both the husband and the wife of the affected family. This provision promotes gender equity by ensuring that women have a stake in the compensatory land, even if the original land was solely in the husband's name. Overall, the LARR Act, 2013, is progressive in its approach to women's rights, particularly in the context of land acquisition and rehabilitation.

The Forest Rights Act, 2006, is instrumental in promoting gender equality in land rights. Here are its key provisions related to women's rights: The Act recognizes women's rights to land as equal to men's by granting individual forest rights through joint titles, ensuring both husband and wife are named. It provides individual forest rights to unmarried women, deserted women, and widows, ensuring they are not excluded from land ownership. The Act mandates that women constitute at least one-third of the members in forest rights committees. These committees assist in preparing claims for forest rights and advise the gram sabha in verifying and approving these claims. Overall, the Forest Rights Act, 2006, strengthens women's land rights and ensures their active participation in the decision-making processes related to forest lands.

Convention on the Elimination of All Forms of Discrimination against Women (CEDAW) has been ratified by India in July 1993. Article 14(2)(g) of the convention directs the signatory states to ensure that women have equal access to marketing facilities, agricultural credit and loan, appropriate technology and most importantly, equal treatment in land and agrarian reforms and land settlement schemes.

Judicial Pronouncements

The Supreme Court urged the Central government to consider amending Section 2(2) of the Hindu Succession Act, which specifically excludes female members of Scheduled Tribes from the purview of the Act. The Supreme Court in its judgment acknowledged the discriminatory nature of the Travancore Christian Succession Act, 1916, which governed the inheritance rights of Syrian Christian women in Kerala. The Act, based on customary law, granted women only a one-fourth share of the property that their brothers inherited. and ruled that the one-fourth share for women was unjust and unconstitutional, emphasizing the need for gender-neutral laws in matters of inheritance.

The Supreme Court held that every woman in a domestic relationship has a right to reside in the shared household of her husband even after his death.

Under India's current system, different inheritance laws apply to each of the nation's various religious communities in a bid to uphold religious freedom. The majority of Indian women fall within the ambit of the Hindu Succession Act, 1956, (HSA) that governs Hindus, Sikhs, Buddhists, and Jains.

In an important case, the Supreme Court held that even in cases prior to the Hindu Succession Act, 1956, if a property of a male Hindu dying intestate is a self-acquired property or obtained in the partition of a coparcenary or a family property, the same would devolve by inheritance and not by survivorship, and a daughter of such a male Hindu would be entitled to inherit such property in preference to other collaterals.

The Court noted that ancient text as also the Smritis, the commentaries written by various renowned learned persons and even judicial pronouncements "have recognized the rights of several female heirs, the wives and the daughter's being the foremost of them."

CONCLUSION

As India looks forward to a future of sustainable development, recognizing and empowering women in its agricultural sector is inevitable. The untapped potential of these invincible contributors is the key to unlock the true transformative power of rural India. It is time to shed light on their roles, address the challenges they face, and implement comprehensive initiatives that empower and reward their invaluable contributions to the Food Producing Sector of the country.Conclusion

Our analysis highlights a significant issue of gender inequality in the Indian labour market. By concentrating policy interventions on the agricultural sector, we can drive substantial positive changes in both the labor market and the environment. Rapid industrialization and urbanization have led to a declining share of the agricultural population, which disproportionately impacts female labor compared to male labor. Unemployed women often face challenges transitioning to other sectors due to skill mismatches and inadequate qualifications. This disparity emphasizes the need for legal measures that ensure fair opportunities for all.

Structural transformation, characterized by increased industrialization and a reduced share of the population engaged in agriculture, can accelerate economic growth. However, data indicates that this economic growth is not benefiting the female labor force, as their participation consistently decreases with industrialization, despite its contribution to the country's overall economic progress.

Moreover, the second part of our analysis shows a growing trend of using machinery may drive economic growth, but throwing female labour out of the workforce because one-way machines are replacing manual labour, where women are predominant and on the other way, this heavy machinery (tractors, harvesters etc.) can not be run by women for apparent reason.

Therefore, women are kept away from the labour force despite economic growth.

When examining the environmental aspect, it becomes clear that economic growth driven by the industrial sector may not be environmentally sustainable. Rapid industrialization negatively impacts the environment in two significant ways: accelerated extraction of non-renewable natural resources and increased pollution due to aggressive industrial growth. This underscores the urgent need for sustainable development, with a greater focus on developing the agricultural sector.

The agricultural sector demands more attention from lawmakers for several reasons. Firstly, agriculture has been a cornerstone of India's economy since time immemorial, and this heritage should be preserved and advanced. Secondly, women are generally more comfortable engaging in the agricultural sector for various reasons, as discussed earlier in this study. Therefore, increased focus on agriculture will boost female participation in productive activities. Lastly, and most importantly, enhancing the contribution of the agricultural sector to economic growth will lead to more sustainable and inclusive development for the country.

Therefore, it is imperative to develop the agricultural sector in a gender-inclusive manner, ensuring it keeps pace with the development of other sectors and the appropriate implementation of agricultural machinery. The goal is to achieve sustainable development, which encompasses economic growth and promotes inclusive social and ecological progress. While this objective is challenging, it is crucial and requires effective policy implementation and legal instruments to protect and advance our agricultural sector. India's strong agricultural heritage should serve as a foundation for this endeavor, ensuring that the sector grows in tandem with others. By modernizing agriculture appropriately, we can enhance female participation in the labor force and protect the environmental balance, ultimately leading to sustainable and inclusive development.

REFERENCES

Abraham, R., Lahoti, R., & Swaminathan, H. (2021). *Childbirth and women's labour market transitions in India* (WIDER Working Paper No. 2021/128). UNU-WIDER.

Boserup, E. (1970). *Women's role in economic development*. St. Martin's Press.

Chakraborty, A., & Sutradhar, R. (2023). Unpaid Work by Women: A State-level Analysis of Evidence from Time-use Survey. *Indian Journal of Human Development*, 17(3), 461–476. DOI: 10.1177/09737030231218147

Goldin, C. (1986). The economic status of women in the early republic: Quantitative evidence. *The Journal of Interdisciplinary History*, 16(3), 375–404. DOI: 10.2307/204496

Goldin, C. (1994). The U-shaped female labour force functions in economic development and economic history.

Arunachala Gounder (Dead) by LRs v Punnusamy & Ors (2011) Civil Appeal No. 6659/2011

https://sustainabledevelopment.un.org/content/documents/2322UN%20Women%20Analysis%20on%20Women%20and%20SDGs.pdf

Kamla Neti (Dead) Through Legal Representatives(Appellant) vs. Special Land Acquisition Officer and Ors. (Respondents), (2023) 3 SCC 528

Kottis, A. P. (1990). Shifts over time and regional variation in women's labour force participation rates in a developing economy: The case of Greece. *Journal of Development Economics*, 33(1), 117–132. DOI: 10.1016/0304-3878(90)90009-Z

Kusz, D. (2014). Modernisation of agriculture vs sustainable agriculture.

Mary Roy, V. (1986). The State Of Kerala 1986 AIR 1011. *SCR*, (1), 371.

Mehrotra, S., & Sinha, S. (2017). Explaining falling female employment during a high growth period. *Economic and Political Weekly*, 52(39), 54–62.

Mitra, A. (2019a). Women's work in response to urbanisation: Evidence from Odisha. *ANTYAJAA: Indian Journal of Women and Social Change, 4*(1), 92–106. https://journals.sagepub.com/doi/full/10.1177/2455632719836804

Mitra, S. (2019b). Women and unpaid work in India: A macroeconomic overview. https://slidetodoc.com/ women-and-unpaid-work-in-India-a-macroeconomic/

Mitra, S., & Mehta, S. K. (2021). It is time to compensate women's unpaid labour. *India Development Review.*https://idronline.org/its-time-to-compensate-womens-unpaid-labour/

Pedersen, C. S. (2018). The UN Sustainable Development Goals (SDGs) are a gift to business. *25th CIRP Life Cycle Engineering (LCE) Conference* (pp. 21-24). Copenhagen: Elsevier. doi:DOI: 10.1016/j.procir.2018.01.003

Perelli, C., Cacchiarelli, L., Peveri, V., & Branca, G. (2024). Gender equality and sustainable development: A cross-country study on women's contribution to adopting climate-smart agriculture in Sub-Saharan Africa. *Ecological Economics*, 219, 108145. DOI: 10.1016/j.ecolecon.2024.108145

Prabha Tyagi v. Kamlesh Tyagi (2022) 2022 LiveLaw (SC) 474 | CrA 511 OF 2022

https://timesofindia.indiatimes.com/blogs/developing-contemporary-india/womens-land-rights-in-india-whats-missing-from-our-land-laws/

Sundari, S. (2020). Structural changes and quality of women's labour in India. *The Indian Journal of Labour Economics : the Quarterly Journal of the Indian Society of Labour Economics*, 63(3), 689–717. DOI: 10.1007/s41027-020-00245-2 PMID: 33046952

Vineeta Sharma v. Rakesh Sharma & Ors.[(2020) 9 SCC 1]

Chapter 17
Childhood at Risk:
Examining the Effectiveness of the Framework in Protecting Children From Sexual Offenses

Saurabh Chandra
https://orcid.org/0000-0003-4172-9968
Bennett University, India

Megha Gupta
https://orcid.org/0009-0005-1864-7827
Independent Researcher, India

ABSTRACT

Child sexual abuse is a social evil that is frequently around the globe. They are becoming more numerous every day, which makes it difficult to evaluate the situation. The chance of negative psychosocial and mental health effects associated with child sexual abuse is unquestionably increased, but even stability processes identified a number of protective factors, including as family support. The perpetrators of this crime take advantage of their position of vulnerability, trust, and power in order to commit child sexual abuse or attempt to do so for financial gain or other purposes. Given how similar the terms "child sexual exploitation" and "child sexual abuse" are, it can be challenging to distinguish between them. Children who experience child sexual abuse suffer grave consequences. It serves as a huge roadblock to growth and advancement. Many health behaviours brought on by child sexual abuse result in negative mental, physical, and social outcomes throughout life. There are numerous circumstances that can contribute to child sexual abuse.

DOI: 10.4018/979-8-3693-6069-9.ch017

LITERATURE REVIEW

The effectiveness of legal frameworks in protecting children from sexual offenses is a complex issue, with a range of factors influencing their success. Niriella (2018) and Billis (2011) both emphasize the importance of strong legal frameworks in preventing and addressing sexual violence against children, with Niriella specifically focusing on the need for prevention and enforcement mechanisms. Wunneburger (2008) and Manurung (2023) highlight the role of specific laws and measures, such as sex offender registries and the Child Protection Law, in providing protection. However, the implementation and enforcement of these laws are crucial, as Manurung (2023) notes the need for legal education and appropriate sanctions to address and prevent sexual crimes.

The legal framework for protecting children from sexual offenses in India is a complex and evolving system. Seth (2015) and Mathew (2016) both highlight the need for a more effective and coordinated approach, with Mathew specifically emphasizing the role of NGOs in providing services and improving trial procedures. The Protection of Children from Sexual Offences Act, 2012 (POCSO) is a significant step forward, but it also has some unintended consequences, such as issues with age determination and mandatory reporting (Belur, 2015). Pandey (2019) further underscores the importance of understanding the risk factors, outcomes, and protective factors of child sexual abuse in India, suggesting a conceptual framework for addressing these issues.

A range of studies have explored the issue of child sexual abuse in India and the legal framework in place to protect children. Sharma (2023) proposed a conceptual framework for understanding the risk factors, outcomes, and protective factors of child sexual abuse in India. Manral (2022) discussed the provisions of the Protection of Children from Sexual Offences Act and its application in the medical and psychological fields. Belur (2015) highlighted the benefits and unintended consequences of the Act, particularly in relation to the criminalization of sexual behavior under 18 years of age. CasimirRaj (2021) provided an overview of the different types of child abuse, including sexual abuse, and the legal framework for protecting teenagers. These studies collectively underscore the importance of a comprehensive approach to addressing child sexual abuse in India, including the need for effective implementation of the legal framework and consideration of unintended consequences.

The Protection of Children from Sexual Offences Act, 2012 (POCSO Act) is a crucial legislation in India, aimed at safeguarding children from sexual assault, harassment, and pornography. It establishes special courts for the trial of such offenses and incorporates child-friendly mechanisms for reporting, evidence recording, and investigation (Mohanty 2021, Acharya 2020). The Act is comprehensive and applies to all, regardless of caste, creed, gender, or religion, with a focus on punishing offenders in both rural and urban areas (Acharya 2020). The judiciary plays a vital role in enforcing the Act and ensuring a swift and secure legal process for child victims (Kumar 2024). However, further research is needed to explore the intricacies and practical implementation of the Act (S. 2022).

The Protection of Children from Sexual Offences Act, 2012 (POCSO Act) is a significant piece of legislation aimed at safeguarding children from sexual abuse and exploitation. It was introduced to address the inadequacies of the Indian Penal Code and to comply with the Constitution of India and the United Nations Conventions on the Rights of the Child. The Act covers a range of sexual offences against children and provides for the establishment of special courts for their trial. However, there are concerns about the Act's mandatory reporting requirement, which could lead to unintended consequences in some cases (Mohanty 2021, Acharya 2020, Andrade 2013).

The Protection of Children from Sexual Offences (POCSO) Act was implemented in India in 2012 to address the high prevalence of child sexual abuse (CSA) and protect children's rights (Aslammoideen Chekkara & Sony Kunjappan, 2022; J. Ganesha, 2019). While the Act serves as a legislative deterrent, its implementation faces challenges due to corruption and outdated legal proceedings (J. Ganesha, 2019). Similar frameworks exist in other countries, such as Vietnam's legal system for protecting children against online sexual abuse (Pham Thi Thanh Thuy, 2023). The UK justice system also grapples with child-on-child sexual abuse cases, highlighting the need for appropriate interventions (Elaine Craig, 2020). To enhance the effectiveness of these frameworks, experts suggest a holistic governance approach, including prevention strategies and community development (Aslammoideen Chekkara & Sony Kunjappan, 2022). Additionally, addressing gaps in legal provisions, improving support systems, and providing education on sexual abuse prevention are crucial for protecting children from sexual offenses (J. Ganesha, 2019; Pham Thi Thanh Thuy, 2023).

Research in India has highlighted the alarming prevalence of sexual offences against children, with a significant increase in reported cases in recent years (Picess, 2018; Kethineni, 2020). The Protection of Children from Sexual Offences (POCSO) Act, 2012, was introduced to address this issue, but its implementation has faced challenges (Renu, 2019). These offences have a particularly devastating impact on female children, who are often unaware of the abuse they are experiencing (Singh, 2022). The need for more effective laws and measures to protect children from such crimes is evident.

Recent research highlights the urgent need for effective frameworks to protect children from sexual offenses. In India, where child sexual abuse (CSA) is prevalent, the POCSO Act serves as legislative deterrence, but a holistic governance approach is recommended for prevention (Aslammoideen Chekkara & Sony Kunjappan, 2022).

Global efforts to combat sexual violence against children emphasize evidence-based interventions, including school-based programs and community mobilization (D. Ligiero et al., 2019). The UK justice system faces challenges in addressing child-on-child sexual abuse, with technological advances blurring lines between normal curiosity and harmful behavior (Craig, 2020). Risk assessment frameworks for sexual offending have identified four key domains: sexual interests, distorted attitudes, socioaffective functioning, and self-management. Studies show that combining static and dynamic risk factors improves prediction of sexual reconviction, suggesting a need for comprehensive risk assessment procedures (Thornton, 2002).

INTRODUCTION

Out of all the population, half population is covered with the children. In our society children constitutes an essential number of population but ironically it is the children only who are least aware about their rights, or we can say that only a few ratios are aware about their rights in the society. Because of this lack of awareness, it results into exploitation of the children. In our society children are being considered as sensitive and delicate as flower. Hence, they deserve to get necessary protection, care and love for their proper development and not exploitation. One of the prominent reasons behind the child exploitation is the Innocence of the Child. It is the innocence which makes a child an easy target to exploit and adds them to the vulnerable category. It is the innocence of the child which makes a child attractive and due to which converts a child into a victim of exploitation, since due to the innocence only a child is unable to understand and interpret the true meaning of a relationship.

Children are viewed as national assets; it is the responsibility of the state to ensure that their personalities develop appropriately. Children are essential to a nation's progress. Many aspects of their growth should be considered, including their cognitive and emotional development as well as the development of their love, affections, and security, among other things. A civilised society can be recognised by how it treats its children. Governments and society have a duty to recognise that children need to be protected and cared for.

As per World Health Organization (WHO) sexual abuse is the involvement of a child in sexual activity that he or she does not fully comprehend, is unable to give informed consent to, or for which the child is not developmentally prepared and cannot give consent, or that violates the laws or social taboos of society. Child sexual abuse is evidenced by this activity between a child and an adult or another child who by age or development is in a relationship of responsibility, trust or power, the activity being intended to gratify or satisfy the needs of the other person. This may include but is not limited to the inducement or coercion of a child to engage in any unlawful sexual activity; the exploitative use of a child in prostitution or other unlawful sexual practices; the exploitative use of children in pornographic performance and materials.

Child sexual abuse is a social evil that is frequently portrayed in myths and stereotypes around the globe. They are becoming more numerous every day, which makes it difficult to evaluate the situation. Technological improvements are unquestionably helpful to the investigation process, but they can also help us enhance our resources, especially in areas with limited resources. The chance of negative psychosocial and mental health effects associated with child sexual abuse is unquestionably increased, but even stability processes identified a number of protective factors, including as family support.

The perpetrators of this crime take advantage of their position of vulnerability, trust, and power in order to commit child sexual abuse or attempt to do so for financial gain or other purposes. Given how similar the terms "child sexual exploitation" and "child sexual abuse" are, it can be challenging to distinguish between them. Children who experience child sexual abuse suffer grave consequences. It serves as a huge roadblock to growth and advancement. Many health behaviours brought on by child sexual abuse result in negative mental, physical, and social outcomes throughout life.

REASONS FOR CHILD SEXUAL ABUSE

There are numerous circumstances that can contribute to child sexual abuse. The underlying causes run deep within the nation's social structure. Children of various societies are raised in a society where it is expected of them to respect and obey adults, and that they should not question those adults' decisions. Extreme sensitivities that believe questioning or resisting authority figures is a sign of poor upbringing which eventually results in a society that supports sexual predators. Children who are silent and obey without inquiry may become obedient targets as a result. The downside of this situation is that instances of sexual abuse of minors are marked by shame and in isolation. Gender violence is most often unseen and is shrouded in a culture of silence.

The causes and factors of violence against women include entrenched unequal power equations between men and women that foster violence and its acceptability, aggravated by cultural and social norms, economic dependence, poverty and alcohol consumption, etc.

a) Disability

People with disabilities are frequently kept apart from the general population, which restricts their ability to participate in everyday social situations. This seclusion heightens a person's reliance on service suppliers. Because of this, the abusers exploit it to continue to have influence over the victims and to sexually assault them. Due to their insufficient education and general ignorance about healthy relationships, legal rights, and individual rights, people with disabilities are more likely to be sexually abused.

b) Poverty

Scholars from all over the world have looked into the connection between child sexual abuse and poverty. The consequence of child sexual abuse is unfavourable in poverty. The majority of sexual abuse is committed by underprivileged families. Not all children who are sexually abused are from low-income homes; others are from middle-class and wealthy families. Adult perpetrators prey on poor and vulnerable children, claiming to support them but ultimately taking advantage of them.

c) Domestic Violence

Child sexual abuse is directly tied to domestic violence in a household. Domestic violence is highly influenced by ongoing disputes and disturbances in families, which subsequently cause family division and divorce. These familial dysfunctions frequently serve as the setting for child sexual abuse.

All children, regardless of age or gender, are susceptible to child sexual abuse, but girls are more at risk. As has already been said, one of the main causes of child sexual abuse is poverty, but it is not the only one.

FORMS OF CHILD SEXUAL ABUSE

a) Institutional Child Sexual Abuse

Not all sexual assaults are attempted by unidentified individuals. There are many instances of institutionalised child sexual abuse where the perpetrators are people in positions of power who are familiar to the victim. They take place in trustworthy organisations like schools, ashrams, orphanages, etc. These Atrocities are more serious because they are actually hard to find. Due to a fear of stigma, the majority of them are not reported, and even when they are, relatively few of them result in conviction.

b) Child Sexual Abuse in Cyber Space

The issue of child pornography demands our attention as another crucial topic. The prevalence of child sexual abuse increased during the COVID-19 epidemic. According to research by the Indian Child Protection Fund (ICPF) on the consumption of child pornography in India, there has been a dramatic rise of about 95%.

c) Child Trafficking and Sexual Abuse

Child trafficking is the act of enlisting, transferring, transporting, or housing a child for the purpose of exploitation. For the goal of commercial sexual exploitation, girls are trafficked. Transgender people are also trafficked for sexual exploitation, in addition to girls. Boys are trafficked for use in agricultural, mining, armed labour, etc.

Nobody is new to the practise of trafficking kids and then using them for sexual exploitation. Although there is no official definition of child trafficking in India, the Immoral Traffic (Prevention) Act makes provision for it. a missing kid will be treated in accordance with the JJ Act. The term "commercial sexual exploitation of children" (CESC) encompasses both outdoor and indoor sexual activity, such as that which takes place in brothels, massage parlours, etc.

Equal amounts of males, females, and transgender people are trafficked. Here, the victim is not just a girl child. According to research by the National Center for Missing and Exploited Children (NCMEC), trafficking for the purpose of child sexual abuse affects males and transgender people equally. Children are trafficked for a variety of reasons, including child tourism and prostitution, in order to sexually abuse them.

Hence, child sexual abuse is the least acknowledged form of the abuse, since our criminal justice system is more of accused oriented as well as in comparison with the child sexual abuse it is also more of adult-dominated. Therefore, in order to protect the rights of the accused it is often neglected that there is a victim also who needs the same protection of rights.

CONCEPTUAL FRAMEWORK FOR CHILD SEXUAL ABUSE

During the past times, the growth of children wasn't given any specific attention. Eventually, as things progressed, many countries all over the globe began to pay special attention to children's development. Children's growth must be given top priority in every civilized society because they are a country's greatest resource. Children require particular care and therapy since they have unique physical and mental traits. Every state's government is required to adopt policies that promote children's growth.

As per the general interpretation of the society a child is being considered as someone who is immature and unable to understand the consequences of a decision and hence is considered that a child is not under the capacity to decide for himself. However, as per the legal interpretation the definition of child varies from every other enactment.

The Indian Constitution doesn't define the word "child." "A child implies any human being below the age of eighteen years," states Article 1 of the UNHRC (United Nations Convention on the Rights of the Child,1989), "unless under the legislation applicable to the child, majority is acquired earlier." The legal definition of a child frequently changes depending on the objective. In India, there are various laws that define the term "Child" depending on the situation. According to the, "Indian Majority Act, 1875" the age of majority is 18 years old. According to the "Child Labour (Prohibition and Regulations) Act,1986", a child is defined as a person who has not reached the age of 14. According to the, "Child Marriage Restriction Act, 1926", a child is defined as a person who, if a male, has not reached the age of 21 and, if a female, has not reached the age of 18. A person who has not reached the age of eighteen is considered a juvenile or child under the, "Juvenile Justice (Care and Protection) Act, 2000."

WHAT IS CHILD SEXUAL ABUSE?

Child abuse is a serious offense against human dignity and a negative childhood memory. It is also a societal ill that has been in civilization since the beginning of existence. Any intentional harm or mistreatment of a kid under the age of eighteen is considered child abuse. This problem encompasses all forms of abuse and maltreatment of a kid by a parent, caretaker, or someone else acting in a parental capacity, such as a trainer, instructor, friend, family, or a visitor. Child abuse can have an effect on a child's development on a physical, psychological, and social level as well as on their sexual behaviour.

Child sexual abuse (CSA) is a global issue with serious long-term effects. CSA is defined by the World Health Organization (WHO) as "the involvement of a child in sexual activity that he or she does not fully comprehend and is unable to give informed consent to, or for which the child is not developmentally prepared, or else that violate the laws or social taboos of society." "Sexual activity, attempted intercourse, oral-genital contact, exhibitionism or exposing kids to adult sexual activity or pornography, and the use of the child for prostitution or pornography" are just a few of the behaviours that fall under the umbrella of CSA (Putnam, F. W. 2003).

IMPACT OF CHILD SEXUAL ABUSE

Many physical as well as mental consequences of child sexual abuse can have both short- and long-term repercussions. Early detection and successful intervention can lessen the immediate and long-lasting impacts of child sexual abuse and aid in the victims' healing. Some typical outcomes that might follow from child sexual abuse are listed below.

a) Psychological Effect of CSA

Distress, shame, anxiousness, embarrassment, self-blame, avoidance, poor self - esteem, physiological concerns, repression, traumatic patterns, sexual problems, and interpersonal issues are all strongly connected with childhood sexual assault. One typical long-term consequence among victims is depression. "Therapists typically deal with the more general psychological problems before treating the sexual problems of survivors. Victims may find it difficult to explain the abuse, thus thinking adversely about themselves. This is because physical relationships are delicate and exposed. By dealing with their thoughts regarding the abuse and developing skills in areas like independence, self-love, and self-awareness, survivors are more likely to succeed in relationships. It is crucial that research into the long-term repercussions of childhood sexual abuse continues. It is well recognized how serious this issue is and the additional major effects it has on victims' lives." "The creation of special courts is one technique to get around the problems presented by adversarial legal systems; essential features of such courts include case management procedures for early case resolution and specialisation at the prosecutorial and judicial levels.

Persistent pain: It is a typical symptom of child sexual abuse and can appear in various body parts, including the head, neck, and back.
Sleep disturbances: Children who have experienced sexual abuse may lead to fatigue, flashbacks, or anxiety attacks.

b) Mental Health Effect of CSA

PTSD: Survivors of child sexual abuse may get PTSD 'post-traumatic stress disorder,' which can bring on extreme panic attacks and anxiousness as well as flashbacks to the horrific event.
Depression: Those who have experienced CSA may develop low self - esteem, which can lead to feelings of despair and hopelessness as well as a loss of interest in formerly pleasurable activities.
Anxiety: Victims of child sexual abuse may have anxiety, which can lead to uneasy, fearful, and worried sensations that interfere with day-to-day activities.
Drug abuse: As a stress reliever for the trauma children have undergone into the habit of alcohol or illicit substance.
Eating disorders: As a way to regain control over their bodies, survivors of child sexual abuse may develop eating disorders such anorexia or bulimia.

c) Sexual Health Effect of CSA

Sexual dysfunction: Individual who have experienced child sexual abuse may develop long-term sexual dysfunction, which may manifest as discomfort during sexual contact or a lack of interest in sexual activity. To prevent themselves from being triggered by sexual activity, survivors of child sexual abuse may completely refuse sexual interactions or intimacy.
Gynaecological issues: Women who were sexually abused as children may be more likely to develop gynaecologic issues include pelvic pain, menstruation issues, and STDs.
Urination issues: Victims of child sexual abuse may experience incontinence or urinary tract infections.
Hypersexuality: As a coping mechanism for the trauma they endured, some survivors may engage in compulsive sexual behaviour. Risky sexual activities may result from this, which may have unfavourable effects.
Confusion over sexual orientation: As a result of the abuse they endured, survivors may face issues with their sexual identity and orientation.
Body image problems: Survivors may feel shame and humiliation for their bodies, which can affect their sexual life.

d). Relationship Effect of CSA

Problems with trusting others: it includes such as relationships, friends, and family members, might result from CSA. Those who have experienced abuse may find it difficult to trust others to act in their best interests and may be on continual alert for clear signals of betrayal.

Issues with intimacy: Victims of child sexual abuse may experience difficulties with emotional and physical intimacy. They might have trouble forming close connections, expressing their emotions, or having sexual relationship in future, without being nervous or upset.

Personality Issues: Sexual abuse of children can cause emotions of meaninglessness, unworthiness and shame. Survivors may struggle to establish healthy relationships because they may feel shattered or unlovable.

Setting up boundaries: Survivors of child sexual abuse often have trouble establishing and upholding limits in their relationships. They might be too sympathetic or have trouble saying no, which can result in relationships that are toxic or even violent.

Issues with communication: Those who have experienced child sexual abuse may find it difficult to express their needs and wishes in their relationships. They may have developed the habit of keeping things private or of hiding their genuine emotions, which can cause miscommunication and trouble.

e). Educational Effect

Individuals who have been sexually abused may be more likely to face scholastic challenges, such as poor attendance, poor grades, and greater rates of school dropout. Also, they may experience behaviour problems like hostility, withdrawal, or trouble relating to classmates, which might hinder their academic success. Children who have been sexually abused may also be more likely to face developmental delays, learning problems, and speech and language issues. It is crucial that teachers and other school personnel recognise the warning symptoms of CSA and offer victimized children a secure and encouraging environment. This can entail offering counselling, making allowances for a individual behavioural or academic issues, and making sure the youngster won't experience more abuses or trauma while attending school. Overall, CSA may have a substantial and complex impact on schooling, so it's critical for teachers and school personnel to be aware of these issues and to offer the right kind of assistance and resources to help kids cope with any difficulties they might encounter.

INDIAN STATUTORY FRAMEWORK FOR CHILD SEXUAL ABUSE

Until 2012, provisions of the Indian Penal Code, 1860 (IPC), Immoral Traffic (Prevention) Act, 1956 (ITPA), and Juvenile Justice (Care and Protection of Children) Act 2000 (JJ Act) applied to sexual offences against children. To deal with child sexual abuse cases, the Government has brought in a special law, namely, The Protection of Children from Sexual Offences (POCSO) Act, 2012. The Act has come into force with effect from 14th November 2012 along with the Rules framed thereunder. Being a signatory to the convention on the rights of children adopted by the UNO, our country has brought this legislation into force. The act in question aims at ensuring the physical, emotional, intellectual and social development of the child.

Being a special penal legislation providing for special provisions for children as per Articles 15(3) of the Constitution of India, the intention of the special legislature will have to be kept in mind while applying the provisions of this act. For effective implementation of the provisions of this act, a coordinated effort is required by the Investigation Agency, Prosecuting Agency and the Criminal Courts dealing with the matter. The act differs substantially from the criminal cases arising out of violation of various provisions of Indian Penal Code.

Some of the salient features of this act are in respect:

The POCSO Act, 2012 is a comprehensive law to provide for the protection of children from the offences of sexual assault, sexual harassment and pornography, while safeguarding the interests of the child at every stage of the judicial process by incorporating child-friendly mechanisms for reporting, recording of evidence, investigation and speedy trial of offences through designated Special Courts.

The said Act defines a child as any person below eighteen years of age, and defines different forms of sexual abuse, including penetrative and non-penetrative assault, as well as sexual harassment and pornography, and deems a sexual assault to be "aggravated" under certain circumstances, such as when the abused child is mentally ill or when the abuse is committed by a person in a position of trust or authority vis-à-vis the child, like a family member, police officer, teacher, or doctor. People who traffic children for sexual purposes are also punishable under the provisions relating to abetment in the said Act. The said Act prescribes stringent punishment graded as per the gravity of the offence, with a maximum term of rigorous imprisonment for life, and fine. The POCSO Act was enacted to provide more stringent punishments for the offences of child abuse of various kinds and that is why minimum punishments have been prescribed in Sections 4, 6, 8 and 10 of the POCSO Act for various categories of sexual assaults on children.

POCSO is gender neutral, meaning that crimes of this nature committed against children will be handled by this act regardless of the gender of the child. This Act sets a burden of proof of guilty until proven innocent|| rather than the general rule of ——innocent until proven guilty. The

flagship Indian legislation that prevents offences against Children has several important clauses

to keep our children (both male and female) safe and secure. The Act defines Penetrative Sexual

Assault (Section 3), Aggravated penetrative sexual assault (Section 5), Sexual Assault (Section 7), Punishment for penetrative Sexual Assault being not less than 7-years (Section 4), Punishment for Sexual Assault (Section 8), Sexual Harassment (Section 11), Child Pornographic (Section 13), Mandatory reporting of Child Abuse cases (Section 19(1)), the burden of proof on Accused (section 29), A Special Public Prosecutor being appointed (Section 31) and many more clauses that would deter anyone from committing such offences.

In keeping with the best international child protection standards, the said Act also provides for mandatory reporting of sexual offences. This casts a legal duty upon a person who has knowledge that a child has been sexually abused to report the offence; if he fails to do so, he may be punished with six months" imprisonment and/ or a fine.

The said Act also casts the police in the role of child protectors during the investigative process. Thus, the police personnel receiving a report of sexual abuse of a child are given the responsibility of making urgent arrangements for the care and protection of the child, such as obtaining emergency medical treatment for the child and placing the child in a shelter home, should the need arise. The police are also required to bring the matter to the attention of the

Child Welfare Committee (CWC) within 24 hours of receiving the report, so the CWC may then proceed where required to make further arrangements for the safety and security of the child.

The said Act makes provisions for the medical examination of the child in a manner designed to cause as little distress as possible. The examination is to be carried out in the presence of the parent or other person whom the child trusts, and in the case of a female child, by a female doctor.

The said Act provides for Special Courts that conduct the trial in-camera and without revealing the identity of the child, in a child-friendly manner. Hence, the child may have a parent or other trusted person present at the time of testifying and can call for assistance from an interpreter, special educator, or other professional while giving evidence; further, the child is not to be called repeatedly to testify in court and may testify through video-link rather than in a courtroom. Above all, the said Act stipulates that a case of child sexual abuse must be disposed of within one year from the date the offence is reported. It also provides for the Special Court to determine the amount of compensation to be paid to a child who has been sexually abused, so that this money can then be used for the child''s medical treatment and rehabilitation.

The said Act recognises almost every known form of sexual abuse against children as punishable offences, and makes the different agencies of the State, such as the police, judiciary and child protection machinery, collaborators in securing justice for a sexually abused child.

Further, by providing for a child-friendly judicial process, the said Act encourages children who have been victims of sexual abuse to report the offence and seek redress for their suffering, as well as to obtain assistance in overcoming their trauma. In time, the said Act will provide a means not only to report and punish those who abuse and exploit the innocence of children, but also prove an effective deterrent in curbing the occurrence of these offences.

The said Act is to be implemented with the active participation of the State Governments. Under Section 39 of the said Act, the State Government is required to frame guidelines for the use of persons including non-governmental organisations, professionals and experts or persons trained in and having knowledge of psychology, social work, physical health, mental health and child development to assist the child at the trial and pre-trial stage. The following guidelines are Model Guidelines formulated by the Central Government, based on which the State Governments can then frame more extensive and specific guidelines as per their specific needs.

Some of the salient features of this act are in respect:

(a) Reporting of cases and registration of the cases and the obligation of the police to report to the Special Court and the child welfare committee.
(b) Recording the statements of the victim by the investigating officer and the precautions to be taken by the investigating officer.
(c) Not allowing the advocates of the accused by the magistrate at the time of recording the statement of witnesses u/s 164 of Cr.P.C.
(d) Provision for audio-video recording of the victim during the investigation.
(e) Provision for establishment of Special Courts to be presided over by a sessions judge and prosecuting the case by a public prosecutor and the jurisdiction of the Special Court to try even the offence punishable u/s 67-B of The Information Technology Act 2000.
(f) Mode of recording the evidence of abused child and time schedule to record such evidence and the steps to be taken for creating comfortable atmosphere to the child.
(g) Power to take the assistance or guidance of experts or persons having knowledge about the child welfare.

(h) Provision for awarding adequate compensation and interim compensation to the abused child.
(i) Mandatory presumption in respect of the offences u/ss 3, 5, 7 and 9 of this Act the moment case is prosecuted.
(j) Provision for alternative punishment to the accused and punishment for false reporting or lodging false complaints.
(k) Obligation of government in creating awareness about the act and monitoring the implementation of the Act and imparting training to the concerned.
(l) Power of Special Court to decide as to whether the accused alleged of the offence is a child or not and effect of such an order.

It necessitates specific handling of instances involving child sexual abuse, including the establishment of special courts, special prosecutors, and support groups for young victims.

The Act's stated objective indicates that India joined the United Nations Convention on the Rights of the Child in 1992, which commands States to protect children from:

-Enticing a minor to engage in any illegal sexual behaviour.
-Taking advantage of children by forcing them to engage in prostitution and other sexual acts.
-Using kids in obscene content and actions.

The Act emphasises a child's overall development, including their physical, emotional, intellectual, and social growth. The fact that this Act includes child-friendly processes from the time a case is reported until it is resolved ensures that children are protected at all stages of the judicial process. It is crucial to notice that the term "rape" is not used in the Act in order to avoid any stigmatisation.

CLASSIFICATION OF OFFENCES UNDER POCSO

Penetrative Sexual Assault

It covers actions where an individual inserts his penis into a child's mouth, vagina, urethra, or anus, or either forces the child to do the same through him or someone else. This term includes the inserting of an object or part of the body other than the penis into a kid's vagina, urethra, or anus, as well as any actions that cause the child to do so with him or someone else. Sexual assault includes applying his mouth to a child's penis, vagina, urethra, anus, or any other part of the body and either making the child do it with him or someone else. Manipulating a child's body part to penetrate the vagina, urethra, anus, or any other part of the body is also considered sexual assault.

Aggravated Penetrative Sexual Assault

The definition of aggravated penetrative sexual assault includes a number of sorts of incidents. If an individual who works for the government or the police, is a soldier or member of the armed forces or other security services, or is on remand, if any of the aforementioned people perpetrate the crime of penetrative sexual assault in their houses, even observation homes, or any educational institution, it will be considered an aggravated form. In addition, employing a lethal instrument to commit a penetrative sexual assault, setting something on fire, inflicting severe bodily harm, or physically incapacitating a child all constitute aggravated penetrative sexual assault. If an individual who is connected to the child by blood, marriage, adoption, or foster care, or who is linked to the parent of the child through a domestic connection, or if they live in the same residence, they are considered to have perpetrated a penetrating sexual assault. This type of abuse is often known as "institutional" child sex abuse. whoever commits penetrative sexual assault on a child below twelve years, is guilty of committing aggravated penetrative sexual assault.

Sexual Assault- Non-Penetrative

If an individual touches the child's vagina, breast, penis, or anus with the aim of abusing them sexually, or if they force the child to do the same with them or anybody else. In addition, sexual assault occurs when any physical act is carried out that involves contact with the skin other than penetration.

Aggravated Sexual Assault- Non-Penetrative

If an individual of trust sexually assaults a child, if the child is in that person's care, or if the two people are related through blood, marriage, adoption, or another relationship, the sexual assault is said to be aggravated.

Sexual Harassment

Sexual harassment can take many different forms, such as:

(a) exposing his body or a portion of it with the goal for a child to see it, or
(b) forcing the child to see pornographic material.

(c) Attempts to follow the kid or attempts to get in touch with the child through electronic or digital platforms.
(d) Attempts to extort or threatens to utilise any content—whether true or made up—of any part of the child's body in digital or electronic form.
(e) Attempts to seduce the youngster in order to produce material that is pornographic.

This is significant because it demonstrates that identical grounds were also added to the IPC (Indian Penal Code) S. 354 in 2013 by the Criminal (Amendment) Act.

Child Pornography

This type of sexual abuse includes using the youngster for pornography or keeping pornographic materials on them. It also covers situations in which a youngster is utilised for sexual enjoyment through social channels. Additionally, the usage of pornographic content for commercial purposes is a violation of this legislation.

GENDER NEUTRAL PROVISIONS

Since the Act is gender neutral, both male and female children are covered by it.

Before this Act, a person who committed child sexual abuse was punished under the IPC (Indian Penal Code). The concept was not, however, gender neutral. But following the start, this Act provides equal safeguards against sexual abuse of both male and female children. The definition of sexual assault has been expanded to include peno-urethral, peno-oral, and peno-anal acts as well, so it is no longer solely limited to peno-vaginal actions.

CHILD FRIENDLY PROVISIONS

There need to be female police officer present who are not wearing their uniforms when the statement is being recorded. This recording must be made with a child's trusted adult present. There has to be every effort made to ensure that the requirements of S. 157 CrPC are being followed. Rather it be the child's identity, family, school, or other information, it should never be made public. However, it can only be approved after a Special Court order that is supported by reasons, and even then, only when it is in the child's best interests. In order to protect the victim from any influence or fear, the law has also made sure that they do not have to interact with the accused person during the course of the trial. The one who was assaulted doesn't have to recite her words again and again because the entire testimony is videotaped using a video conference system.

SPEEDY TRIAL

According to CrPC S. 173(1A), the investigation must be finished within two months. Additionally, the whole trial must be finished in a year to avoid keeping cases on hold for too long. Long-pending issues could have an effect on the mental well-being of children. This time constraint is regarded as another crucial component of this Act. The POCSO Act 2012 provides for establishment of Special Courts for the purpose of ensuring speedy trial. Further, as per the Act, the evidence of the child shall be recorded within a period of thirty days of the Special Court taking cognizance of the offence and the Special Court shall complete the trial, as far as possible, within a period of one year from the date of taking cognizance of the offence. The Act was amended in 2019 to introduce more stringent punishment including death penalty for committing sexual crimes on children, with a view to deter the perpetrators & prevent such crimes against children.

MEDICAL EXAMINATION OF THE ACCUSED AND THE VICTIM

Only female doctors should perform the medically necessary examination of the victim child in accordance with S. 27 of the POCSO Act of 2012 and S. 164 A of the CrPC. The availability of the parent of the child or another trustworthy person must be guaranteed while the check-up is being conducted. They must express their consent for such medical examinations and the collection of samples for forensic examination in writing. In a similar manner, S. 53 A of the CrPC would be followed, which states that DNA samples of the suspect would be collected by a police officer of at least sub inspector rank. Here, the accused's consent is not required.

MANDATORY REPORTING

Anyone who works with children and young people in the education, social, religious, or health fields is required by POCSO to report CSA (child sexual abuse) as per S. 19. If an individual neglects or doesn't report then it may lead to legal repercussions like, A maximum of six months in jail and/or fines are possible penalties, all of which are intended to promote law abiding behaviour.

CONCLUSION

Child sexual abuse is a global issue that affects millions of children every year, with devastating and long-lasting consequences. While all countries have laws and policies in place to address this issue, there are significant variations in the legal frameworks, cultural norms, and societal attitudes towards child sexual abuse across different countries. A comparative analysis of child sexual abuse laws and policies across different countries can provide valuable insights into the strengths and weaknesses of various approaches, as well as the challenges and opportunities for addressing this issue effectively. By examining the similarities and differences in legal frameworks, prevention strategies, and victim support services, we can gain a better understanding of the most effective ways to protect children from sexual abuse and support survivors of abuse.

It is important for parents, educators, and communities to be vigilant about online safety, provide guidance and supervision to children, and report any suspected cases of cyber child sexual abuse to the appropriate authorities. Children should be educated about online risks, safe internet practices, and encouraged to report any uncomfortable or inappropriate online experiences. Collaborative efforts among government, civil society, technology companies, and individuals are essential to combat cyber child sexual abuse and protect the well-being of children in India and worldwide.

Child sexual abuse is a horrific crime affecting millions of youngsters around the world. It is a form of exploitation that can leave the victim with severe and long-term distress. Despite attempts to prevent and address child sexual abuse, it is a widespread and complex issue with numerous problems. These issues include a lack of knowledge and understanding of the problem, difficulties in reporting and investigating occurrences, and the need for a comprehensive preventative strategy. In this response, the author goes through these issues in further depth and emphasise how important it is to solve them in order to keep children safe. Child sexual abuse is a widespread and devastating issue that can have serious and long-term effects for the victim. To address this issue, effective techniques for preventing abuse, identifying and responding to cases of abuse, and assisting victims in their recovery must be developed and implemented.

REFERENCES

Ashok, S., Ramaswamy, S., & Seshadri, S. (2022). A Transdisciplinary Perspective on the Adolescent Consent-Abuse Binary. *Journal of Indian Association for Child and Adolescent Mental Health*, 18(3), 210–213. DOI: 10.1177/09731342221143854

Belur, J., & Singh, B. (2015). Child sexual abuse and the law in India: A commentary. *Crime Science*, 4(1), 26. Advance online publication. DOI: 10.1186/s40163-015-0037-2

Bhullar, D., Mittal, P., & Singla, A. (2015). Child sexual abuse in India: A review. *International Journal of Ethics*, 1(2), 38–44. Advance online publication. DOI: 10.18099/ijetv.v1i2.6816

Carson, D. K., Foster, J. M., & Tripathi, N. (2013). Child sexual abuse in India: Current issues and research. *Psychological Studies*, 58(3), 318–325. DOI: 10.1007/s12646-013-0198-6

CasimirRaj, M., Director, D.B., & Nepolian, A. (2021). CHILD SEXUAL ABUSE IN INDIA. *International Journal of Approximate Reasoning*, 9, 353–360.

Chaudhary, B., & Kr, R. (2018). Medical examination of child of sexual assault under the protection of children from sexual offences (POSCO), act 2012., 2, 50-53. https://doi.org/.DOI: 10.18231/j.ijfmts.2017.016

Choudhary, V., Satapathy, S., & Sagar, R. (2019). Qualitative study on the impact of child sexual abuse: Perspectives of children, caregivers, and professionals in Indian context. *Journal of Child Sexual Abuse*, 28(4), 489–510. DOI: 10.1080/10538712.2018.1563262 PMID: 30681937

Choudhry, V., Dayal, R., Pillai, D., Kalokhe, A. S., Beier, K., & Patel, V. (2018). Child sexual abuse in India: A systematic review. *PLoS One*, 13(10), e0205086. DOI: 10.1371/journal.pone.0205086 PMID: 30300379

Craig, E. (2020). Child's play or sexual abuse? Reviewing the efficacy of the justice framework in dealing with child on child sexual abuse in the United Kingdom. *Journal of Child Sexual Abuse*, 29(6), 734–748. DOI: 10.1080/10538712.2020.1719448 PMID: 32286188

Kumar, A., Pathak, A., Kumar, S., Rastogi, P., & Rastogi, P. (2012). The problem of child sexual abuse in India laws, legal lacuna and the bill–PCSOB-2011. *Journal of the Indian Academy of Forensic Medicine*, 34(2), 169–174.

Kumar, D.S., & Kumar Paswan, P. (2024). Role and importance of judiciary in protecting children from sexual offences. *Prayagraj Law Review.*

Manral, I., Panjrattan, C., Saini, R., Roy, S., & Singh, A. K. (2022). A multi-speciality approach to the protection of children from sexual offences act: A review. *Industrial Psychiatry Journal*, 32(1), 4–8. DOI: 10.4103/ipj.ipj_169_21 PMID: 37274576

Mathew, L. A. (2016). A Multidisciplinary Approach to Child Protection for Sexual Abuse in India: The Law.

Kulkarni, U. K., Kulkarni, K. V., Kokre, R. N., & Daundkar, B. B. (2016). Forensic study on child sexual abuse under POCSO act. *European Journal of Biomedical and Pharmaceutical Sciences*, 3(7), XX–XX.

Mathews, B., Lee, X., & Norman, R. (2016). Impact of a New Mandatory Reporting Law on Reporting and Identification of Child Sexual Abuse: A Seven Year Time Trend Analysis. Queensland University of Technology (QUT) - Law & Justice Legal Studies Research Paper Series. https://doi.org/DOI: 10.1016/j.chiabu.2016.04.009

Mohanty, H., & Banerjee, D. (2021). An Analysis of Protection of Children from Sexual Offences Act, 2012 (POCSO ACT). SSRN Electronic Journal. https://doi.org/DOI: 10.2139/ssrn.3768096

Mohanty, H., & Banerjee, D. (2021). An Analysis of Protection of Children from Sexual Offences Act, 2012 (POCSO ACT). SSRN *Electronic Journal*. DOI: 10.2139/ssrn.3768096

Mota, S. A. (2002). The U.S. Supreme Court Addresses the Child Pornography Prevention Act and Child Online Protection Act in Ashcroft v. Free Speech Coalition and Ashcroft v. American Civil Liberties Union. *Federal Communications Law Journal*, 55, 4.

Pandey, S., & Reddy, S. (2019). Article: Risk Factors, Outcomes, and Protective Factors of Child Sexual Abuse in India: A Conceptual Framework.

Picess, A. R. (2018). Child sexual abuse in India: Law with lacunae and loopholes. *Supremo Amicus*, 5, 76.

Renu, R., & Chopra, G. (2019). Child sexual abuse in India and the Protection of Children from Sexual Offences (POCSO) Act 2012: A research review. *Integrated Journal of Social Sciences*, 6(2), 49–56.

Seth, K., & Sharma, R. (2015). Overview of Laws against online child sex abuse in India, U.K and U.S. *International Journal of Research*, 2, 1169–1177.

Seth, R., & Srivastava, R. (2017). Child Sexual Abuse: Management and prevention, and protection of children from Sexual Offences (POCSO). *Indian Pediatrics*, 54(11), 949–953. DOI: 10.1007/s13312-017-1189-9 PMID: 29217802

Sharma, D.P., & Kumar, D.A. (2023). Child Sexual Abuse Risk Factors, Outcomes, and Protective Factors in India: A Conceptual and framework model. *Innovative Research Thoughts*.

Manurung, W., & Dame Panjaitan, J. (2023). Implementation of Legal Protection of Children and Women as Victims of Sexual Crimes. *International Journal of Social Research*.

Sharma, G., Kumar, S., & Banga, H. (2022). protection of children from sexual offences with special reference to POSCO Act, 2012. *International Journal of Health Sciences*, 6706–6713. Advance online publication. DOI: 10.53730/ijhs.v6nS4.10317

Singh, M. M., Parsekar, S. S., & Nair, S. N. (2014). An epidemiological overview of child sexual abuse. *Journal of Family Medicine and Primary Care*, 3(4), 430. DOI: 10.4103/2249-4863.148139 PMID: 25657958

Singh, S., & Kumar, D.R. (2022). Sexual Offences Against Female Children in India: A Critical Study. *legal researchd development an international refereed e Journal*.

Sinha, I., & Singh, R. (2023). Critical Analysis of POCSO: In Light of the Rights of Adolescents. International Journal For Multidisciplinary Research. https://doi.org/DOI: 10.36948/ijfmr.2023.v05i06.10289

Topping, K., & Barron, I. (2009). School-Based Child Sexual Abuse Prevention Programs: A Review of Effectiveness. *Review of Educational Research*, 79(1), 431–463. DOI: 10.3102/0034654308325582

Tyagi, S., & Karande, S. (2021). Child sexual abuse in India: A wake-up call. *Journal of Postgraduate Medicine*, 67(3), 125–129. DOI: 10.4103/jpgm.JPGM_264_21 PMID: 34380802

Varghese, D. P., & John, S. (2021). The protection of children from sexual offences act, 2012: Strong provisions, weak implementation? *Iustitia (Bloomington, Ind.)*, 12(1), 95–123.

Wahlsten, D. (2004). Leilani Muir versus the Philosopher King: Eugenics on trial in Alberta. *Genetica*, 99(2-3), 185–198. DOI: 10.1007/BF02259522 PMID: 9463073

Wunneburger, D.F., Olivares, M., & Maghelal, P.K. (2008). *Internal Security for Communities: A Spatial Analysis of the Effectiveness of Sex Offender Laws*.

Chapter 18
The Unequal Burden: Climate Change and Vulnerable Populations

Sini John
Christ Academy Institute of Law, Bengaluru, India

Surima Singh
Christ Academy Institute of Law, Bengaluru, India

ABSTRACT

Climate change disproportionately affects vulnerable people, exacerbates inequalities and threatens human rights. In this chapter the author has explores the complex interplay between climate change and human rights, focusing on rights to life, health, food, water, shelter and culture highlighting the challenges faced by Indigenous peoples, women, children and poor meet on climate change adaptation emphasis. The chapter discusses the role of international instruments such as the UNFCCC and human rights treaties in addressing these issues and explores the need to create stronger legal safeguards, greater public participation and focus on stronger groups to build a just and equitable future

INTRODUCTION

Evidently, climate change – the phenomenon widespread worldwide – goes far beyond solely the environmental issue affecting human societies and their basic rights. When temperatures go up, seas rise and the weather goes as haywire the ramifications seen on human beings become diverse and extensive. These climatic

DOI: 10.4018/979-8-3693-6069-9.ch018

changes distort people's existence, uproot them, and increase existing gaps in the society since different groups are affected differently.

It should be noted that there has been improvement in the last couple of years with acknowledge of climate change as a crucial human rights issue. Climate change and its impact are against human rights where people's rights to life, health, food, water, housing, and an adequate standard of living are interfered with. The sociology between climate change and human rights points more to this that there is need to address climate change holistically with human rights consideration.[1]

Despite growing recognition of the human rights dimensions of climate change, translating awareness into concrete action faces significant hurdles. The international climate change regime, while encompassing numerous conventions, can be slow and cumbersome. Reaching consensus on ambitious mitigation strategies often gets bogged down in political negotiations and competing national interests. This lack of swift action allows the climate crisis to worsen, further jeopardizing human rights.

Adding to this challenge is the issue of accountability. The human rights framework, while robust in outlining obligations, often lacks clear mechanisms to enforce them. While states have a legal duty to protect their citizens' rights, holding them accountable for inaction on climate change remains difficult. This lack of clear enforcement mechanisms creates a situation where lip service is paid to human rights, but concrete action lags.

However, there are strong advocates for a human rights-based approach to climate action. Human rights bodies at the international level, environmental NGOs, and the climate activism community all support this approach. It focuses on states and other obligation bearers' roles and responsibilities in fulfilling human rights in relation to climate change. This shift in perspective emphasizes the need for concrete action and strengthens arguments for holding states accountable for inaction.

That is why this chapter is intended to discover more nuances of the connection between climate change and human rights from the international perspective. In this paper, using the principles of international law, relevant conventions, and analyzed cases, it is planned to reveal how the global community addresses this crucial interrelation.[2] The chapter will also explore the obstacles and prospects of improving the use of human rights paradigm to climate change mitigation, population shelters, and their sustainable developments with stronger arguments for more suitable and just measures for vulnerable groups.[3]

BACKGROUND OF THE CHAPTER

In the context of the reality of climate change, various aspects of human rights are significantly transformed, inasmuch as climate change influences many aspects of human life. Extremely linked to the undesired outcomes of climate change are not only an immediate and definite effect on the natural environment but also an indirect effect on the economy and the society. This section explains how climate change affects the most basic human rights such as; the right to life, health, food, water, shelter, and culture.[4]

The right to life is among the stem rights that are inherent and inalienable in every human being. Global warming endangers this right by raising the chances of adversities like hurricanes, floods, and heatwaves. Such occurrences cause damages, resulting to loss of lives, injuries and evictions especially the affected vulnerable groups in the coastal and low-lying areas. Catastrophic storms and gradual increase in the sea level poses a great threat to the existence as well as safety of people.

Extremes of climate have an effect on the health of the people. Higher temperatures enhance the spread of diseases transmitted by vectors as well as malaria and dengue fever, heatwaves on the other hand cause cardiovascular and respiratory diseases. Thus, adding to the accrued pollution and frequent incidences of wildfires exacerbates health conditions affecting the respiratory system. Moreover, Climate change vehemently related with disasters involving psychological impacts that may pose long-term effects on the mental health status of the individuals including children and the elderly in particular.

The right to food is violated by climate change via its impacts on the yields of agriculture production.[5] This results in food deficit, food inflation, and even higher food vulnerability mainly to countries who are rain fed agriculture based. Most of these difficulties affect the lives of smallholder farmers and other rural people who are already considered disadvantaged since they are most affected in aspects concerning their food supply and feeding.[6]

Recent analysis knowledge shows that clean water is one of the highest human necessities that is experiencing emerging risks due to climate change. Variations in the sources and quality of water include precipitation, melting of glaciers and changes in river flows. Severe and frequent droughts and reduced snow melt water affect availability of water used for human consumption, cleaning and agriculture. This scarcity is worse felt by minorities, thus worsening the inequality aspect in already strained societies and even provoking water resource related conflicts.

The general human right to access shelter is violated by climate change as storms as well as flooding makes lives and properties uninhabitable. Every year natural disasters such as floods, hurricanes, and typhoons make people homeless displacing millions of people and leaving them in the search for shelters. Coastal erosion and storm surges endanger structures; this results in the removal of inhabitable land. This leads to humanitarian problems like shelter of CLIP affected people, social issues, and basic needs and assistance of the displaced population.[7]

Education as a component of cultural rights is also threatened by climate change affected indigenous people and local communities who have their practices, traditions and self-identities intertwined with climate. The shifts in the ecosystem and loss of biodiversity have impacts on the traditional practices for instance fishing, hunting, and farming. Climatic changes including the continuous rise in the sea level is posing a significant danger to many important cultural practices and religious shrines. Communities' displacement eliminates social relations and undermines cultural sustainability, which makes it possible to preserve ethnicity.

Global warming impact is more on the most vulnerable groups and escalates cross- cutting inequalities and fabricates fresh barriers to odds already confronted by deprived groups. First Nations, second-class women, children and poor people are more vulnerable as they depend most of their time on the climatic influenced environment and they cannot easily fund ways of minimizing the effects of climate on their lives.[8]

Whereas Indigenous people bear the brunt of climate change impacts because of their focus to the land and dependent on natural resources for cultural, social, and economic activities.[9] The alteration of ecosystems affects their way of living through hunting, fishing and farming because of the changing occurrence of biodiverse species. natural effects such as high water levels and natural disasters endanger their territories and holy groove thus expelling them from their home. The Indigenous peoples exist in the rural settings and, therefore, they are not affluent in most cases and cannot easily rebound from the effects of climate related disasters.

As it has been presented, women are also in a vulnerable position because climate change affects them as individuals in the context of their gendered roles, responsibilities, and restrictions in the access to resources and decision-making power. In most cultures, it is women who go to the market to buy food, water and energy which becomes more difficult when climate change affects the supplies. When there are climate related disasters like flooding and hurricanes women are more vulnerable to being physically or sexually violated including rape, especially by their husbands and male relatives. Also, their poor representation in local, national and international decision-making forums has retarded gender sensitive climate policy and undertakings.[10]

Because of the children's physical and developmental status, they are considered the most sensitive to climate change. They lose their school and class time through adverse weather conditions like floods and hurricanes besides facing ill-health. Food scarcity and fresh water become scarce due to climate change and malnutrition and water borne diseases are common and their effects are manifested in children for instance. Furthermore, climate-related disasters impose psychological stresses and traumatic experience that hinder children's mental health and development productivity in future.

Climate change hits the poor because they are usually disadvantaged in terms of access to resources, services, and physical structures. These are communities which are disadvantaged and they usually live in areas with worst impacts of climate change for instance flood prone zones, slum areas, and regions. These populations do not have the resources to purchase 'rebuild ready' structures, insurance, or health care, that would better prepare them to 'roll with the punches' that are increasingly occurring more frequently due to climate change. Thus, they suffer higher impacts and are slower in rehabilitation from the impacts of climate-related disasters.[11] Both poverty and climate vulnerability deepen socio-economic inequalities making many people stay in a poverty-environment nexus.

Global warming endangers the life of every inhabitant; however, weaknesses, such as indigenous people, women, children, and people in poverty face bigger challenges. These groups are more at risk because of their position socially, economically, culturally, and because they have little control over many of the decisions affecting them. That is why to enhance resilience and fairness, it is necessary to focus on their particular weaknesses concerning the climate crisis.

LEGAL FRAMEWORKS AND MECHANISMS FOR PROTECTING HUMAN RIGHTS IN CLIMATE ACTION

The connection between human rights and climate change has fostered the creation of numerous International legal instruments and system of addressing protection of human rights in the context of climate change processes. These frameworks stress on awareness and understanding of climate policies and actions that are human rights compliant and affirmative for all human beings, especially the marginalized ones. In this regard there are several important international instruments and mechanisms that operates in this context.

United Nations Framework Convention on Climate Change (UNFCCC)

The most important international legal act devoted to climate change is the United Nations Framework Convention on Climate Change signed in 1992.[12] However, the UNFCCC per se does not mention human rights as such; yet, in the post-UNFCCC regime, there has been a growing acknowledgement with regard to human rights' assimilation into climate change policy. The Paris Agreement signed under the

UNFCCC in 2015 has been symbolic in terms of attaining such agreement. In Paris Agreement, the preamble affirms Parties' commitment to Human Rights principles, as recognized in international Human Rights Treaties and individuals and communities' rights to be free from human rights abuses in the process of combating Climate Change. Procedurally, it mentions indigenous peoples and local communities right, rights of migrants, children, persons with disabilities, and individuals in vulnerable circumstances.

Human Rights Council and Special Procedures

Evidently the United Nations Human Rights Council has been very active in the bid to ensure that the human rights aspect of climate change is well addressed.[13] Among those adopted, the Council has highlighted that climate change undermines human rights and that climate change should be addressed through a human rights perspective, and principle. The Council has also decided on special procedures such as the Special Rapporteur on human rights and the environment and the Special Rapporteur on the rights of the indigenous people for addressing the effects of climate change on human rights. Each of these special procedures gives useful information, contributes for the development of consciousness and presents some conclusions of how human rights can be incorporated into climate policies.

International Human Rights Treaties

Threat and effects of climate change also have legal ground in the form of human rights in the international human rights treaties. The cultural rights are outlined in the International Covenant on Civil and Political Rights and International Covenant on Economic, Social, Cultural Rights limbed by climatic change are the rights to life, health, food, water and shelter.[14] All international agreements on climate change impose a legal duty on states parties to take steps towards the mitigation and eradication of the impact of climate change on the aforementioned rights. Also, the CEDAW and the CRC put the need to safeguard rights of women and the children on the realization of human rights amidst effects of climate change.

Regional Human Rights Mechanisms

Another set of actors is the regional human rights mechanisms who are also involved in responding to the effects of climate changes on human rights. For its part, the Inter-American Commission on Human Rights (IACHR), the European Court of Human Rights and the African Commission on Human and Peoples' Rights have all addressed situations and issued decisions concerning protection of the environment and rights. These regional bodies offer other channels, through which people and communities can claim climate induced human rights abuses, and ensure compliance by states with regional human rights instruments.

Climate Litigation

Climate litigation has therefore been portrayed as one of the effective ways of promoting and protecting human rights in as much as climate change is concerned. Modern people and groups of people, civil society actors have again and again sought justice and remedies in national and international courts for climate change affecting their rights.[15] Landsmark case for example the Urgenda case in the Netherlands have set important milestones in assertion of the state's responsibility to protect human rights from the effects of climate change. All these legal actions have led to more regard to impacts of climate policies on human rights as well as enhanced more climate change measures.

CASE LAWS AND EXAMPLES

Examining attitudes of the case laws from different jurisdictions C thus gives an insight of how climate change impacts human rights and the responses to such impacts. This section also presents several historical and recent cases that reveal the multifaceted effects of climate change on the rights of people and the growing acknowledgment by courts of the necessity of climate change mitigation to defend these rights.

Uganda Foundation v. The State of the Netherlands (Netherlands)[16]

Human rights are thus revealed to be closely linked to climate action through the monumental Urgenda case. In 2015 the District Court of The Hague supported the Urgenda Foundation and 940 Dutch citizens and instructed the Dutch government to decrease the greenhouse gas emissions at least to the level of 25% from that of 1990 till the year 2020. It was concluded that similar to the decision in the AER case, the sections 6 and 17 duty on the government to protect the rights contained in the ECHR were breached because the common law domestic tort of negligence in mitigating climate change was not properly applied. This was affirmed by the Dutch Supreme Court in its verdict in 2019 where it asserted that scopes of actions by states can be legally binding objectives of the climate actions required to uphold human rights.

Leghari v. Federation of Pakistan (Pakistan)[17]

In 2015, the Lahore High Court in Pakistan endeavored into the known case Leghari v. Federation of Pakistan. Farmer Ashgar Leghari sued the Pakistani government for not properly executing NCCP and FIT CCP for the tenure of 2014-2030. In Leghari's favor, the court found out the adverse effects of climate change in relation to water, food and energy which consequently are part of the individual's rights to life and dignity. The court compelled the government to form a Climate Change Commission for policy implementation, showcasing activism as pertinent for forcing climate change action for humanity's sake.

Juliana v. United States (United States)[18]

21 youth plaintiffs filed a complaint against the USA federal government as its policies supporting fossil fuel would infringe youth's rights to life, liberty, and property, and also neglect public trust resources. While the Ninth Circuit Court of Appeals dismissed the case in 2020, in part due to the technical argument of standing, the court recognized the literature and the seriousness of climate crisis and the government's part in it. The case also drew lessons on the significance of engaging in climate litigation by young persons in order to demand protection of human rights in climate change circumstances.

Human Rights Committee's View on Teitiota v. New Zealand (International)[19]

In 2020 the UNHRC delivered its much a waited decision in the case of Teitiota v New Zealand. Muslim refugees forced to flee their homes due to persecution, based their case for asylum on legal rights to seek refuge, citing as grounds, impending dangers posed by climate change; promulgated by global warming and climatic shifts, which made his homeland of Kiribati uninhabitable. While the Committee denied refugee status to Teitiota it acknowledged the fact that climate change is a real threat to the right to life, thus, the countries hearing the asylum seekers should consider it. This decision made an international human rights body see climate change as a possible ground for asylum to be applied, an aspect that connects human rights with climate displacement.

Torres Strait Islanders v. Australia (Australia)[20]

It is in 2019 that a group of the Torres Strait Islanders lodged a complaint to the United Nations Human Rights Committee against the Australian government for its positive inaction on climate change as violating the rights to culture, family, and life under ICCPR. They also claimed that sea levels were rising and weather conditions worsening leading to losses of homes food produce and even their identity. Such climatic changes have adversely affected indigenous peoples and this case exemplifies the effects of climate change in indigenous regions due to international inaction thus proving the need for the protection of the indigenous people through international human rights mechanisms as a way of making governments stick to their climate change duties.

ROLE OF OTHER ORGANIZATIONS

The fight against climate change involves a combination of different tasks, which need to be solved by international organizations, governments, civil society, and people. These actors have essential functions in the realization of human rights in relation to climate change adaptation and mitigation, guaranteeing the climate actions are sustainable, rights-based processes.

Governments

Safety of human rights is basically a government's responsibility and governments are agents of change in implementing climate change measures relating to these rights. Within national policy and law, national governments can mainstream human rights on climate so that adaptation and mitigation interventions reduce further vulnerabilities. It can embrace the national climate action plans which focus on the most exposed groups of citizens like indigenous people, women, children and the poor.[21] In addition, governments can participate in the international climate talks, call for more ambitious climate policies that incorporate human rights frameworks. Another important level, therefore, is the local government because they are responsible for implementing and modifying policies to suit the local population as well as enhancing the strength and continuance of development initiatives at the bases.[22]

Civil Society

CSOs have a critical role in the advocacy of human rights-based approach to climate change. They inform the public about the effects of climate change on human rights issues, demand fairness and justice in climate change policies and ensure governments and corporations are responsive to their actions or decisions. Thus, CSOs give a face to the voiceless and vulnerable subjects in the climate change decision-making arena. They also practice climate litigation where they take cases to the national and international Court to enforce human rights on climate change. Working as a means of convening a diverse group of stakeholders, CSOs are able to multiply the outcomes of their influence and initiate collective action on climate justice.

Individuals

Citizens, activists, and researchers remain critical for developing human rights-based approach on climate change. International celebrities like Greta Thunberg and the Intergenerational anthem for the climate movement of the youth, the Fridays for the future have awakened the world to climate change crisis and urging the leaders to act with humanity. Scholars help in an effort by offering research findings on climate change consequences on human rights and the suitability of various policy measures.[23] Climate participation is possible in two ways: by individuals having Climate Refusal Rates regarding their actions, decisions, and choices in their daily lives as well as engaging in Climate Positive Actions like embracing eco-friendly policies, practices, and supporting top-notch environmental projects. Using their rights & responsibilities, people can inform society's sentiments and its rules, thereby directing the transformation of its world into a better one.

Any positive change towards the adoption of human rights approaches to climate change adaptation and mitigation has to involve the collaboration of IOs, governments, CSOs, and people. These stakeholders are required to jointly contribute to the climate actions with specific considerations for equity, non-discrimination, and the protection of human rights for every human being or more so the vulnerable persons.[24] The inclusion of human rights in climate policies and activities makes it possible for world communities to mitigate the impacts of the climate crisis and uphold justice in the related development processes.

CHALLENGES AND GAPS IN ADDRESSING HUMAN RIGHTS IN CLIMATE CHANGE POLICIES

Implementing human rights within climate change has various difficulties and shortcomings which discourage the protection and enforcement of human rights. Another aspect is that human rights are not protected as strictly in the field of climate change violation.[25] Even if such global treaties as the Paris agreement refer to human rights protection, there are frequently no robust enforcement measures in place to back up these-commitments. Most states are still deficient in the legal and institutional requirements that are required in order to implement the international human rights standards. Such a difference between the international level practices and national level practices hinders rights for human beings most impacted by climate changes.

Another significant issue is that research for the climate policies is insufficient when it comes to addressing vulnerable groups. The adverse effects of climate change are felt by the vulnerable group of people including the indigenous peoples, women, children and persons living in poverty and they are the ones who are generally left out when it comes to policy making and policy enforcement.[26] These populations are often excluded from participating in decision making processes hence policies which are put in place do not consider their plight and the struggles they go through thus continuing to worsen existing inequalities. For climate policies to be effective, it is necessary to take proactive measures toward engagement of the above-mentioned groups in climatic policies formulation and implementation.[27]

Other challenges make it even more difficult to extend the sphere of human rights to climate fields and incorporate them into climate policies. To be meaningful, participation must mean having the means to obtain information, being able to use the information in discuss and engage the decision-makers and having chances to influence the decisions made. Nevertheless, most groups considered by society as outliers struggle to gain equal access to information and be involved in climate change regulation. Language problems, illiteracy, and limited access to communication technology that may pass information on and involve such communities can be a challenge. Secondly, socio-political barriers including discrimination, political marginalization also limits their stand in the decisions made on climatic change issues.[28]

Another massive challenge to meeting human rights in climate change policies is the rivalry of objectives. Decision makers in the form of governments and policymakers are usually presented with certain dilemmas– growth, prosperity, and sustainable use of resources among others. When these priorities are harmonized, then the result is achievable at the compromise of human rights. For example, construction of massive hydraulic facilities to counter climate change like dams, bio fuel plantations and so on means that the rights of people are violated, often a number of individuals and their means of subsistence are eliminated. Whereas, the balanced approach is the solution to these conflicting objectives, the protection of human rights will be given top priority while addressing the climate objectives.[29]

Moreover, global inequalities intensify the difficulties for human rights preservation in the context of climate change. Climate-affected countries, especially the developing nations, have the least capacity and resources to initiate and effectively implement its policies to counter climate change impacts concerning human rights. Some of these countries may find it hard to embrace human rights in relation to climate change policies because they are battling with social-economic challenges and are closely attached to donors' funding. In addition, the UN Framework Convention on Climate Change, namely that developed countries have emitted the lion's share of GHG, brings into turn principles of climate justice to whom and how resources and funds should be allocated to finance adaptation and mitigation measures.

These measures show that human rights are not well protected in climate change policies mainly due to weak enforcement mechanisms at both the national and international levels. Despite the fact that such policies are developed, the problem can be found in insufficient monitoring, reporting, and accountability which results in the poor implementation of the measures and inadequate observation of human rights.[30] Furthermore, in so many situations, there are no discernible sanctions or remedial actions for non-implementation of the human rights obligations in climate actions. Enhancing these enforcement mechanisms is critical to entrench accountability in states and other actors as well as human rights in climate policies.[31]

Some of the problems that arise when human rights are incorporated in climate changes goals are the lack of legal teeth, the failure to take into consideration vulnerable groups, lack of participation, failure to address dual Mandates and the failure to address equity, weak accountability among others. To eliminate these problems, it is crucial to establish an effective cooperation of all subjects in the context of the creation of inclusive, fair, and enforceable climate policies with a focus on the protection of human rights for all, including marginalized communities.

RECOMMENDATIONS FOR ENHANCING HUMAN RIGHTS-BASED APPROACHES TO CLIMATE ACTION

The attempt to mainstream human rights into climate action has to be holistic and precise, broad and specific suggestions are required. These recommendations mainly aim at enhancing the legal annexes, inclusiveness, the consideration of disadvantaged groups, the prioritization, and enforcement interventions. Therefore, through the implementation of these measures, the stakeholders can encourage climate just transitions for climate change mitigation that respects human rights.

Strengthening Legal Frameworks

One of the intervention measures is to enhance the legal instruments on international and national levels to promote the human rights in climate change processes. Global conventions and/or regulations should ought to contain provisions for human rights and there ought to be stringent ways of monitoring and implementation.[32] At the national level, it should be required for each country to accept general liability for human rights protection within climate legal systems and national pertinent policies corresponding to international norms. This can be done by enacting of laws, formulation of national strategic frameworks on climate change that include respect, promotion, protection and fulfilment of HHRs.

Ensuring Inclusive Participation

Mainstreaming concerns of the marginalized is essential in formulation climate policies because these people require protection of their rights and freedoms. Governments and international organisation should design the ways and means to ensure that these groups can be effectively involved in climate decision making.[33] These activities encompass both the dissemination of information and data, the development of human capital, and the social spaces for dialogue and consultation. By providing the stakeholders, especially the indigenous people, women, children and other vulnerable groups a voice and including them as policy participants better climate policies will be formulated.

To ensure meaningful participation, governments and international organizations must prioritize the protection of rights and freedoms for these groups. This requires a concerted effort to create inclusive spaces for dialogue, consultation, and decision-making. By amplifying marginalized voices, policymakers can gain a deeper understanding of the challenges faced by these communities and develop policies that address their specific needs and priorities.

Addressing The Needs of Vulnerable Populations

To address climate change, the underprivileged group suffering from climate change impacts should be prioritized into climate change measures. Governments should undertake the analysis of these apparatuses in order to establish the vulnerability and sensitivity of the above-stated groups and then proceed to formulate adaptation and / or mitigation strategies and policies. For instance, there is the possibility of the introduction of social protection programs to protect vulnerable population that may be affected by frequent floods, rising sea levels or other similar factors. Also, the energy should be made to pay attention to protect indigenous peoples' cultural identity and intellectual property rights concerning sustainable environment utilization.[34]

Balancing Priorities

Government and the policymakers have to reconcile between the two competing objectives through the equal management of climate change goals and human rights. This entails undertaking of analysis in order to assess the COUNCIL human rights implications of climate projects and policies.

Likely at the top of this list is climate action and human rights, each with their own system-level claims on policymakers. Trying to balance these two important goals is an extremely delicate, and comprehensive mission. Climate policies, therefore, should be carefully reviewed and implemented in a human rights-based approach with special consideration on vulnerable groups. This requires a full-fledged evaluation of the social, economic and ecological impacts of climate actions. If states manage the trade-offs in respect of these rights and maintain a balanced mix between climate change mitigation, adaptation measures on one hand, and human rights protection on the other - they will not put at risk fundamental freedoms as well as equalities.

This balance requires too that participation in the decision-making process involved not only civil society and indigenous communities but other marginalized groups as well. They are essential to realize process.

Enhancing Enforcement Mechanisms

There is a necessity of effective compliance measures that would ensure level of responsibility of states and other players regarding their commitments in the sphere of human rights within climate change process. Global organizations like the United Nations should improve the methods of supervision and documentation to assess the observance of human rights conventions in climate policies.[35]

There must also be independent monitoring bodies to investigate and enforce human rights violations related directly or indirectly to climate change. These bodies can serve to deter abuses and impose accountability. Strong enforcement provisions need to be built into international climate agreements as well. That consists of specifying what acts can breach the agreement, penalties for breaching and how claims will be solved. Enforcing human rights in the face of climate change: capacity development for international law enforcement

Promoting Climate Justice

Climate justice consists of aspects such as admitting and responding to the roles of developed countries in climate change issues and supporting developing countries in their climate actions. Industrialized countries must for honor their obligations to support financially and technologically the developing countries in order to put into practice adequate and human rights' friendly climate policies.

Climate justice is based on the understanding that adapting to changes in our atmosphere caused by humans calls for more than simply mitigation of greenhouse gas emissions; it also calls for equity and fairness as we confront future difficulties. It recognizes that people who are most affected by a crisis have frequently had the least influence in its inception. It is morally and ethically required by heavy industrialized nations—who historically had the most of the blame for greenhouse gas emissions—to help developing countries make the transition to low-carbon economy and strengthen their ability to withstand climate-related calamities.

There must be support for capacity building, technological transfer, and funding. Furthermore, climate justice calls for compensation for the losses and harm that vulnerable nations have endured as a result of extreme weather conditions linked to climate change. Developed nations have the ability to create a just and sustainable future for the global climate by honoring their commitment to support, at least in theory, an emissions trajectory that is compatible with the 1.5 degree target—that is, until they give the necessary financial assistance!

Fostering Collaboration and Partnerships

It became evident that cooperation and engagement of governments, IGOs, NGOs, and business entities are vital for the promotion of HRBA to climate change. It is recognized that multi-stakeholder initiatives can combine resources, experience, and work on creating new strategies to respond to climate-related issues. Non-governmental organizations should remain active in general, fighting for human rights, and pressurizing governments and large companies into embracing humane practices.

In effectively taking climate action, the approach would be multi-stakeholder, where the different strengths of governments, international organizations, civil society, and the private sector are better harnessed. Strong partnerships are called for in the sharing of knowledge, resources, and expertise in the solution of the complex challenges of climate change. Collaborative platforms may support such actions through the development of innovative solutions, knowledge exchange, and building capacity for each stakeholder to contribute to climate action.

Furthermore, partnerships can act as a key bridge between policy making and implementation to ensure that climate policies are responsive to the needs of grassroots communities. Partnerships at all levels can foster collaboration among governments, businesses, and civil society to achieve systemic transformation. Collaborative efforts can succeed only if there is mutual trust, transparency, and accountability among partners.

CONCLUSION

Human rights are a crucial factor in climate change, and therefore, climate change issues should be addressed concerning total and equal integration. Each pillar of human rights is affected by climate change; right to life, health, food, and culture the effects being more devastating on vulnerable groups including indigenous peoples, women and children, and those living in absolute poverty.[36]

The UNFCCC and the Paris Agreement as well as the climate litigation foster the concept of human rights in responding to climate change. Legal actions to enforce human rights obligations in relation to climate change are illustrated by Urgenda Foundation v. The State of the Netherlands. Nevertheless, the problems remain: legal Recourse unadequate protection of vulnerable subjects, access issues, RAE conflict of interests, and globalization.[37]

To surmount these challenges there is need to enhance the legal protection, promote the people's inclusion and care about the vulnerable groups. Something as simple as this shows just how important and complex the balancing act is of competing interests and the strengthening of enforcement of climate polices. Climate justice advocacy initiatives, cooperation and awareness raising of climate change human rights can enable people / societies to record meaningful engagement.[38]

When adopting the proposed strategies, the stakeholders will be in a position to foster climate policies that will uphold, recognize, and guarantee people's rights. Still, with progressive endeavors, it becomes possible that climate action will maintain justice and equity with wide-spread resilience contributing immensely to the development of a just world in the future.

REFERENCES

Atapattu, S. (2015). *Human rights approaches to climate change: challenges and opportunities*. Routledge.

Bunn, I. D. (2012). *The right to development and international economic law: legal and moral dimensions*. Bloomsbury Publishing.

Albuquerque, C. D. (2010). Climate Change and the Human Rights to Water and Sanitation. Position Paper by the Independent expert on the human rights to safe drinking water.

Human Rights Committee's View on Teitiota v. New Zealand, CCPR/C/127/D/2728/2016.

Human Rights Council. Res. 7/23, U.N. Doc. A/HRC/RES/7/23 (Mar. 28, 2008).

Human Rights Watch. World Report 2022, https://www.hrw.org/world-report/2022 (last visited July 8, 2024).

Intergovernmental Panel on Climate Change. Climate Change 2022: Impacts, Adaptation and Vulnerability, Summary for Policymakers (2022), https://www.ipcc.ch/report/ar6/wg2/ (last visited July 8, 2024).

Intergovernmental Panel on Climate Change. Climate Change 2022: Impacts, Adaptation and Vulnerability, Summary for Policymakers (2022), https://www.ipcc.ch/report/ar6/wg2/ (last visited July 8, 2024).

International Council on Human Rights Policy. Climate Change and Human Rights: A Rough Guide (2008), https://www.ichrp.org/files/reports/45/136_report.pdf (last visited July 8, 2024).

Juliana v. United States, 217 F. Supp. 3d 1224 (D. Or. 2016).

Knox, J. H. (2020). Constructing the human right to a healthy environment. *Annual Review of Law and Social Science*, 16(1), 79–95.

Lavanya, R. (2010). The Role of Human Rights in Climate Change Adaptation, 22 J. *Envtl.*, L, 229.

Limon, M. (2009). Climate Change, Human Rights and the UN: A New Role for the Office of the High Commissioner for Human Rights? 38 Brighton J. *Int'l*, L, 37.

Olawuyi, D. S., & Justice, C.Right to Development. (2015). A Perspective from Sub-Saharan Africa, 42 J. *Energy & Nat. Res.*, L, 16.

Quirico, O., & Boumghar, M. (Eds.). (2015). *Climate change and human rights: an international and comparative law perspective*. Routledge.

Ranjan, P. (2020). The Right to Development in a Climate-Constrained World: The Green Climate Fund as an Opportunity for Transformative Development, 22 Indian J. *Int'l*, L, 183.

Res, G. A. 70/1, U.N. Doc. A/RES/70/1 (Sept. 25, 2015).

Torres Strait Islanders v. Australia, UN Doc CCPR/C/135/D/3624/2019.

United Nations Environment Programme, Climate Change and Human Rights, https://www.unep.org/resources/report/climate-change-and-human-rights (last visited July 8, 2024).

United Nations Framework Convention on Climate Change, Climate Change: Impacts, Vulnerabilities, and Adaptation in Developing Countries, https://unfccc.int/resource/docs/publications/impacts.pdf (last visited July 8, 2024).

United Nations Office of the High Commissioner for Human Rights. Mapping Human Rights Obligations Relating to the Enjoyment of a Safe, Clean, Healthy and Sustainable Environment, https://www.ohchr.org/Documents/Issues/Environment/MappingReport/MappingHRObligationsRelationClimateChange.pdf (last visited July 8, 2024).

United Nations Office of the High Commissioner for Human Rights. Mapping Human Rights Obligations Relating to the Enjoyment of a Safe, Clean, Healthy and Sustainable Environment, https://www.ohchr.org/Documents/Issues/Environment/MappingReport/MappingHRObligationsRelationClimateChange.pdf (last visited July 8, 2024).

United Nations Special Rapporteur on Human Rights and the Environment, Human Rights Obligations Relating to the Enjoyment of a Safe, Clean, Healthy and Sustainable Environment. (2018), https://www.ohchr.org/Documents/Issues/Environment/SREnvironment/A_HRC_37_59.pdf (last visited July 8, 2024).

ENDNOTES

1 Intergovernmental Panel on Climate Change, *Climate Change 2022: Impacts, Adaptation and Vulnerability*, Summary for Policymakers (2022), https://www.ipcc.ch/report/ar6/wg2/ (last visited July 8, 2024).

2 United Nations Framework Convention on Climate Change, *Climate Change: Impacts, Vulnerabilities, and Adaptation in Developing Countries*, https://unfccc.int/resource/docs/publications/impacts.pdf (last visited July 8, 2024).

3 United Nations Special Rapporteur on Human Rights and the Environment, *Human Rights Obligations Relating to the Enjoyment of a Safe, Clean, Healthy and Sustainable Environment* (2018), https://www.ohchr.org/Documents/Issues/Environment/SREnvironment/A_HRC_37_59.pdf (last visited July 8, 2024).

4 Human Rights Watch, *World Report 2022*, https://www.hrw.org/world-report/2022 (last visited July 8, 2024).

5 Sumudu Atapattu, *Human Rights Approaches to Climate Change Mitigation and Adaptation*, 2 Carbon & Climate L. Rev. 26 (2014).

6 United Nations Office of the High Commissioner for Human Rights, *Mapping Human Rights Obligations Relating to the Enjoyment of a Safe, Clean, Healthy and Sustainable Environment*, https://www.ohchr.org/Documents/Issues/Environment/MappingReport/MappingHRObligationsRelationClimateChange.pdf (last visited July 8, 2024).

7 Sarah Mead, *Human Rights and Climate Change Litigation: Trends and Insights*, 26 Geo. Envtl. L. Rev. 23 (2018).

8 G.A. Res. 70/1, U.N. Doc. A/RES/70/1 (Sept. 25, 2015).

9 Rajamani Lavanya, *The Role of Human Rights in Climate Change Adaptation*, 22 J. Envtl. L. 229 (2010).

10 Intergovernmental Panel on Climate Change, *Climate Change 2022: Impacts, Adaptation and Vulnerability*, Summary for Policymakers (2022), https://www.ipcc.ch/report/ar6/wg2/ (last visited July 8, 2024).

11 Catarina de Albuquerque, *Climate Change and the Human Rights to Water and Sanitation*, 21 Waterlines 1 (2014).

12 United Nations Environment Programme, *Climate Change and Human Rights*, https://www.unep.org/resources/report/climate-change-and-human-rights (last visited July 8, 2024).

13 Paris Agreement, Dec. 12, 2015, T.I.A.S. No. 16-1104.

14 Damilola S. Olawuyi, *Climate Justice and the Right to Development: A Perspective from Sub-Saharan Africa*, 42 J. Energy & Nat. Res. L. 16 (2015).

15 Human Rights Council, Res. 7/23, U.N. Doc. A/HRC/RES/7/23 (Mar. 28, 2008).

16 Urgenda Foundation v. The State of the Netherlands (Netherlands), [2015] HAZA C/09/00456689

17 Ashgar Leghari v. Federation of Pakistan, (2015) W.P. No. 25501/2015.

18 Juliana v. United States, 217 F. Supp. 3d 1224 (D. Or. 2016).

19 Human Rights Committee's View on Teitiota v. New Zealand, CCPR/C/127/D/2728/2016.

20 Torres Strait Islanders v. Australia, UN Doc CCPR/C/135/D/3624/2019.

21 Miller, David. "The responsibility to protect human rights." *Legitimacy, justice and public international law* 48 (2007).

22 McClean, Emma. "The responsibility to protect: The role of international human rights law." *Journal of Conflict and Security Law* 13.1 (2008): 123-152.

23 Native Village of Kivalina v. ExxonMobil Corp., 696 F.3d 849 (9th Cir. 2012)

24 Nickel, James W. "How human rights generate duties to protect and provide." *Hum. Rts. Q.* 15 (1993): 77.

25 Limon, Marc. "Human rights and climate change: Constructing a case for political action." *Harv. Envtl. L. Rev.* 33 (2009): 439.

26 Setzer, Joana, and Lisa Benjamin. "Climate change litigation in the global South: Filling in gaps." (2020): 56-60.

27 McInerney-Lankford, Siobhan, Mac Darrow, and Lavanya Rajamani. "Human rights and climate change: a review of the international legal dimensions." (2011).

28 Ferguson, Laura, et al. "Non-communicable diseases and human rights: Global synergies, gaps and opportunities." *Global Public Health* 12.10 (2017): 1200-1227.

29 Marc Limon, *Climate Change, Human Rights and the UN: A New Role for the Office of the High Commissioner for Human Rights?*, 38 Brighton J. Int'l L. 37 (2009).

30 *Supra* Note, 11.

31 United Nations Office of the High Commissioner for Human Rights, *Mapping Human Rights Obligations Relating to the Enjoyment of a Safe, Clean, Healthy and Sustainable Environment*, https://www.ohchr.org/Documents/Issues/Environment/MappingReport/MappingHRObligationsRelationClimateChange.pdf (last visited July 8, 2024).

32 Isabella D. Bunn, *The Right to Development and International Economic Law: Legal and Moral Dimensions* (2012).

33 Siobhán McInerney-Lankford, *Climate Change and Human Rights: The Role of the International Human Rights System*, 35 Harvard Envtl. L. Rev. 345 (2011).

34 John H. Knox, *The Human Right to a Healthy Environment: Procedural Aspects* in *The Human Right to a Healthy Environment* 152 (John H. Knox & Ramin Pejan eds., 2018).

35 International Council on Human Rights Policy, *Climate Change and Human Rights: A Rough Guide* (2008), https://www.ichrp.org/files/reports/45/136_report.pdf (last visited July 8, 2024).

36 Prabhash Ranjan, *The Right to Development in a Climate-Constrained World: The Green Climate Fund as an Opportunity for Transformative Development*, 22 Indian J. Int'l L. 183 (2020).

37 *Supra* Note, 27.

38 Ottavio Quirico & Mouloud Boumghar eds., *Climate Change and Human Rights: An International and Comparative Law Perspective* (2015).

Chapter 19

Climate Change, Women's Rights, and the Way Forward: Notes on the Indian Perspective

Mayank Kamboj
Amity University, Noida, India

Shilpy Verma
https://orcid.org/0000-0001-8138-3450
Amity University, Noida, India

ABSTRACT

The particular focus of this study is on bringing to light the differential nature of climate change effects across the population, as seen through the lens of Amartya Sen's crucial thesis on 'missing women.' We argue that there is a need for economists to map how climate change makes invisible women, particularly in the labour force. The precarious accessibility women have, and the freedom women enjoy by virtue of gaining access to education, housing, credit and so on, is doubly at risk from the effects of climate change. It is essential to bring to light the susceptibility of the labor market at large, and the susceptibility of women workers in particular, to the vagaries of nature. Be it in the form of heat stress, or global warming disrupting large scale supply chains, or even wiping out small industries, the impact of climate change on the labor market is substantially large and deserving of closer study, especially through a fruitful application of Sen's essential ideas on human development and gender parity.

DOI: 10.4018/979-8-3693-6069-9.ch019

INTRODUCTION

The sight of Delhi ravaged by continuous heat spells (Mitra & Berlinger, 2024) has brought climate change back to the forefront of public discourse. Fainting spells amongst students and workers, closed shops and deserted streets seem an all-too-common sight in the wake of new extremes this summer's heat has brought with it. Social media users are aghast at images of birds who have passed away in the heat. While the summers seem increasingly dystopian, the myopic focus on the effects of climate change-induced erratic climatic patterns on the country's urban middle-class rings alarm bells: how is the countryside faring in the face of hitherto unseen extremities of heat?

Existing research indicates that the impact of climate change is largely differentiated along caste, class and gender. A recent study, for instance, found that the burden of heat stress disproportionately impacted workers from marginalized castes (Shah et al., 2024) Yet scholarship on the gendered burden of climate change remains scarce. Climate action plans tend to forego women's specific needs or dedicate inadequate attention and infrastructural support to them. Even as poverty and food insecurity are expected to disproportionately rise amongst women due to climate change, little is done to truly address the gendered impact of global warming.

From an economic perspective, it is crucial to assess the level and intensity of impact that climate change is expected to wreak in order to effectively devise policies to remedy the same. In particular, the labor market is expected to undergo major changes over the years as it adjusts itself to increasingly hostile conditions imposed my climate change. The beginnings of this transformation within the labor market are already visible, and this project will highlight the ways in which women laborers occupy a unique doubly disadvantaged position in the face of climate change. A truly intensive study of the distribution of the burden of climate change along the lines of gender is necessary for devising sustainable solutions to mitigate any crises that will surely arise in due course of time.

One of the most stellar contributions to providing a theory for the imbalanced sex ratios in India and China comes from Amartya Sen's work where he argues that nearly 100 million women were 'missing,' in the sense that they didn't exist, contrary to projections of their share in the population by experts (Sen, 1990). Most fascinating was that such a phenomenon seemed to be restricted largely to Asian countries, and more specifically to India and China, where patriarchal power relations continue to operate in insidious ways. Sen attributed the 'missing women' to a variety of causes: female feticide, women's precarious access to resources, and women's exposure to risk of gender-based violence. In recent times, research has shown how lower wages, that is the gender pay gap, may also have contributed to imbalance in the sex ratio in China (Qian, 2008). Having access to social security nets significantly boosts one's chances of living a decent and safe life, so it is not a surprise that the absence of economic liberties and rights for women in Asian countries may have contributed to skewing the sex ratio.

The purpose of this paper is to suggest to researchers an intervention in studying the differential impact of climate change in Asian countries; through the framework of Sen's 'missing women' argument. If climate change affects women and gender minorities in ways unique to their social positionalities, and in ways that manifest in the form of economic indicators like financial stability and access to resources, then it is essential to study, and use statistical techniques to quantify, the exact impact that climate change has had on contributing to the 'missing women' phenomenon. Throughout this paper, we try to show how climate change may contribute to 'missing women' on two fronts; first, Sen's argument that almost 100 million women are missing throughout the world by patriarchal forces that literally pose a danger to their lives, and second, by using an economic interpretation of the 'missing women' phenomenon, and studying how climate change may push women out of the labour force. We wish to emphasise on the need for researchers to study how climate change influences women's decision to enter the labour force, and further, how existing legal barriers may further compound the problem by not granting women liberties equivalent to those granted to their male counterparts.

A proper analysis of the reason for this gendered difference is rooted in an understanding of the ways in which the patriarchy in India functions: first, social stigma and economic control over women's lives either entirely prevents women from pursuing any job other than household chores (the hidden economy) or pushes women into jobs with little to no security and mobility. A good portion of women in rural areas are engaged in menial agricultural work, or running small cottage-level industries from within the confines of their homes. Second, lack of freedom of movement greatly inhibits the level and extent of work women can pursue. Small businesses are more susceptible to supply-chain disruptions than larger ones. Third, women's specific needs with respect to sanitation facilities, for example, are met with a dismissive attitude by policymakers at large. Since climate action plans are designed keeping men in mind, there is poor planning to address women's needs. The combination of such social factors at large increases the vulnerability of women to climate change. There is clearly a need to get a better understanding of the ways in which women stand at risk and the ways in which policies must be designed by foregoing the "one-size-fits-all" principle.

To better understand the urgency with which we must address labor market disruptions due to climate change, we refer to studies which indicates the number of workers in selected workplaces across India who were exposed to higher-than-recommended threshold temperatures (Venugopal et al., 2016) and demonstrate that the combined proportion of workers exposed to harsh temperatures is unusually high. Using statistical tools, it is observed that a staggering 57% workers reported a decline in productivity owing to rising temperatures. This decline registered in the form of failure to attain targets, absenteeism and heat stress-induced fatigue and sickness. The study drives home the need to perceive climate change as a negative externality that impedes economic efficiency. By hampering the productivity of workers, climate change induces a downward pressure on national output, away from the socially optimal quantity. From a legal point of view, large-scale barriers exist for women which influence their decision to enter the labour force. These legal restrictions range from women's rights to inheritance and property and security of tenure, and legal barriers that prevent women from exercising their economic freedoms (Gonzalez et al., 2015). We shall explore how climate change and these legal impediments together adversely affect women's economic participation.

Let us look at the impact of climate change on agricultural activity. We peruse data from the Economic Survey of 2018 from India which studies the percentage decline in Kharif and Rabi crops on unirrigated and irrigated lands in response to extreme temperature and rainfall shocks, and predicts that the impact would be most adverse for farmers with unirrigated lands, which comprise about 52% of India's total agricultural land. On an average, the survey predicted, the erratic climatic patterns may drastically reduce farmer incomes by around 15-18%. A reduction in incomes would have far-reaching macroeconomic impacts: it would affect the consumption levels in the economy, and reduce savings and investment.

How are women affected by climate change? The gendered dimensions to the climate crisis cannot be ignored. Rehabilitation of women displaced due to climate change-induced disasters is far more difficult owing to a lack of resources being released for the same. While men find it easy to migrate in search of non-farm work, societal stigma keeps women chained to their households. In particular, a lack of freedom to pursue alternative jobs pushes women in agriculture into economic crises when extreme temperatures affect crop yields. The role of women in agriculture is essential, and yet climate change threatens to usurp their only means of subsistence and relative financial freedom. From a legal perspective, women's ownership and security of tenure over land is precarious. Women constitute only 14% of landowners in India, even though more than three-fourths of women are engaged in agriculture. Even after the 2005 amendment to the Hindu Succession Act, there is little parity amongst men and women when it comes to land ownership. Without owning land, women are left out from social security nets and government schemes that could help insulate them from the effects of climate change (IDR 2024). In Bangladesh, landless women who engage in agricultural activities often find themselves stuck in endless bureaucratic processes to gain the security of title that would give them greater flexibility and control over land (Md. Golam Sarwar et al., 2007). On the one hand, adequate laws find little practice on-ground because imbalanced power relations between men and women have not been remedied through an accompanying social revolution, and on the other, improper or poor implementation of laws that secure women's rights to own land also prove to be a major impediment for women farmers.

Since women farmers tend to focus more on growing subsistence crops rather than cash crops, they are important for crop diversification as well. It is essential to center women in climate-adaptive policies and recognize them as a vulnerable group in the face of the climate crises. Although research on the effects of climate change on women workers is scarce, this paper collates the major findings up till now and charts a way forward by suggesting policies that can help protect vulnerable women workers in agriculture and manufacturing from sliding into poverty and food insecurity at the behest of climate change, especially, but not limited to, legal remedies that recognize women as equal citizens and recognize, and protect, their rights and liberties. To do so would require an infrastructural overhaul, and to challenge the apathy of the bureaucratic class in resolving the plight of those most gravely affected.

Broadly, the paper has three major objectives. First, it aims to collate the major findings in studying the gendered effects of climate change on workers in India. Since research on this topic is scarce, the paper acts as a tool for the reader in reviewing existing literature. Second, the paper aims to identify major research and knowledge gaps in our understanding of the ways in which climate change impact women workers. It helps in charting a path forward for researchers to identify unanswered questions or unexplored areas. Third, the paper conducts an analysis of existing legal loopholes which compound women's vulnerability to climate change, especially in the context of South Asian countries.

Exploring the impact of climate change through the lens of Amartya Sen's study on "missing women" represents a crucial advancement in understanding environmental justice. Sen's work highlights systemic gender inequalities that result in a significant disparity between the expected and actual number of women in many societies. By applying this framework to climate change, we can better grasp how environmental disruptions exacerbate existing gender biases and contribute to further marginalization of women. This new direction in research is vital because it integrates the gendered dimensions of climate vulnerability into broader environmental studies, revealing how climate change disproportionately affects women, particularly in low-income and marginalized communities. It emphasizes the need for climate policies that address these inequalities and ensure that women's voices and needs are adequately represented in climate action. By focusing on this intersection, the research can drive more inclusive and equitable climate strategies, promote gender-sensitive adaptation measures, and advocate for legal reforms that protect the most vulnerable populations. Ultimately, this approach can lead to a more nuanced understanding of climate change impacts and foster comprehensive solutions that advance both environmental and gender justice.

ANALYZING TRENDS

2.1 Literature Review

Existing data on the gendered dimensions of climate change throws insight onto the gravity of the crises we face. Major research studying the effects of climate change on worker productivity in the Indian context found a decline of around 4-9% in the total national output per degree when wet bulb temperatures exceeded the threshold of 27 degree Celsius (Somanathan et al., 2021). The study further concluded a rise of 1-2% in absenteeism amongst workers on days of elevated temperatures. The study noted that this decline in productivity and rise in absentees wasn't observed in casual and daily wage workers, where single-day absences are heavily penalized. Another important study focused on the labor market in India in the agricultural sector. The study collected monthly labor supply data across eight agricultural zones in India across four years and found that the decline in labor force participation for women due to draughts is far larger than decline observed amongst men (Afridi et al., 2021). In the event of a draught, women spend more days on average seeking employment than men, even though their engagement in non-farm alternatives remains relatively stagnant, indicating a failure to diversify into non-farm sectors during seasons of poor crop yield. Further, women's real wages fall by as much as 29% due to draughts alone. The absence of laws regulating women's wages in the informal economy in India compound the problem gravely, and provide little recourse for women workers except collective bargaining through unions to protest for their cause.

A possible explanation for this phenomenon is the high social cost women face that severely inhibits their career mobility. These costs are a result of two identifiable factors; rigid social norms that disproportionately relegate household chores to women and dissuade women from searching for alternative avenues of employment beyond their immediate physical vicinity, and laws, or the lack thereof, to ensure a safe and decent working environment for women. Women who are married and have children in rural areas, the study found, were, on an average, less likely to diversify to non-farm jobs than women who were unmarried.

Other works by notable economists like Amartya Sen have indicated the disadvantaged position women in India find themselves in. Not only are rigid gender norms strictly enforced by threat and coercion, but women also enjoy lesser access to resources that their male counterparts either control completely or tend to distribute preferentially amongst men. These resources include education, healthcare and credit especially. Building upon Sen's capability approach that stresses on improving accessibility to the tools for a healthier life, studies have argued that the impact of climate change on women laborers in rural areas in the form of poverty and job insecurity can be greatly mitigated by ensuring women's equal access to the job market at large, and to non-farm activities in particular (Roy & Venema, 2002). However, this is not to suggest that women in other sectors do not face economic difficulties during climate crises.

Of women employed in non-farm work, a staggering 34% work in the manufacturing sector. The growth of women's employment in the manufacturing sector is primarily driven by a rise in small-scale cottage industries, instead of any job creation as such (Klasen & Pieters, 2013). This suggests that skilled employment opportunities for women in the secondary sector has not seen any growth over the past few years. This is further aggravated by the abysmally low ratio of migrant workers who are women, indicating structural barriers for women to migrate in search of better opportunities. We have already established that workers suffer from a stark decline in productivity due to the heat stress induced by climate change. This situation is worse for women who face greater levels of job insecurity and penalties for failure to deliver targets than their male counterparts. With respect to entrepreneurship opportunities, boosting women's security of land tenure through stringent laws could help in increasing women's ownership of land, and thereby helping them in using this land as collateral while accessing institutional credit (Bin Humam et al., 2023).

Against this backdrop of legal loopholes is the heated debate that rages presently in India over the implementation of a uniform civil code, which its proponents argue would bring an end to community-specific family laws, and help standardize matters pertaining to succession, marriage, divorce and other family matters across the vast myriad communities that form the heterogenous fabric of India. Whether such a uniform civil code could help in building climate resilience amongst women and other vulnerable groups by providing them legal recognition of their rights remains to be seen. Climate change, as it is, finds little mention in Indian laws, especially from a gendered perspective. It would be a crucial challenge for makers of this uniform set of laws to incorporate a climate-conscious worldview within it.

How is all of this connected to Sen's 'missing women'? In recent times, studies show women's economic participation in India has declined to levels seen never before since the Independence (Nikore et al., 2021). The researchers trace the causes of this decline to a variety of factors; disparities in skill training, wage gaps and the concentration of women in sectors registering low growth, with little opportunity to migrate, to name a few. During the lockdown, nearly 13.4 million workers dropped out of the workforce. These were women who were employed in various vocations, and had to drop out, leaving a vacuum in their wake. These form the 'missing women' in India's labour force, and similar trends may be observed before the pandemic was imposed, and after, once adequate data is studied by researchers in this regard. Throughout this paper, our aim is to show how a lack of gender-sensitive laws may bolster the 'missing women' phenomenon by intensifying the effects of climate change on women.

2.2 Methodology Employed

Secondary analysis has been employed to study major interventions in this field till date. This study used comparative analysis for an investigation into the role of climate change-induced factors in impacting productivity and women workers' lives. Secondary data analysis has been harnessed for this purpose. This data is available from institutions like Government of India, State Governments, United Nations and the International Labor Organization. Finally, policy analysis has been done to analyze the pros and cons of various policy suggestions in remedying the problem at hand. Both qualitative and quantitative data is used to gauge various indicators of economic well-being such as productivity, wages, absenteeism from workplace etc.

2.3 Variables Used

1. Measures of national output, such as GDP, NDP, NNP, GNP are used in the project to study the changes in output (and consequently, income) due to global warming. The formulae for calculating these aggregates are based on the rules set down by the NSSO. GDP is the total monetary value of all goods and services produced within a country's borders over a specific period, usually a year or a quarter. It measures the economic performance of a country by capturing the market value of all final goods and services produced. GDP does not account for the depreciation of capital goods or the depletion of natural resources.
2. Employment ratio for men and women is calculated as the fraction of men and women employed in some capacity out of the total number of men and women looking for, and able to, work. Other related variables, calculated as per rules set by the NSSO, include the proportion of skilled and unskilled labor and the proportion of male and female migrant workers.
3. Crop yield is calculated as the amount of crop produced in per unit area of land (either irrigated or otherwise) with a constant level of other factors except temperature and rainfall. The aim is to study fluctuations in crop yield due to erratic climactic patterns identified by extreme temperature or rainfall shocks.
4. Real wages are used to study the difference in male and female workers' incomes, adjusted for inflation.

2.4 Tracing Women's Rights and 'Missing Women'

In the case titled *M K Ranjitsinh & Ors versus Union of India & Ors,* the Supreme Court of India ruled that the right to an environment free from the adverse effects of climate change is an essential component of Article 21 in the Indian Constitution, which deals with the protection of rights and personal liberties for all (Down to Earth). In their ruling, the Chief Justice of India, and other Justices involved in the ruling, highlighted the lack of a legislation that singularly talks about climate change, especially in a country like India whose geographical location makes it far more vulnerable to the effects of climate change than other countries. Given the lack of any law that explicitly talks about climate change, it is not surprising that there is paid little attention to the gendered dimensions of climate change, and how the lack of proper laws further creates legal barriers for women that intensify the effects of climate change on them.

International organizations engaged in climate action have long recognized that climate change is a human rights issue, and a women's rights issue specifically because of its differential impact on women and gender minorities. A rights-based approach to climate justice is one which recognizes advancements in women's rights, both in terms of recognizing them legally and ensuring they are translated into a change in power relations between men and women on-ground, to ensure gender-responsive infrastructure in dealing with climate change. In particular, law and policymakers must make use of the gender integration continuum, a tool devised to gauge the gender-responsiveness of policies and programs, to decide whether the law or policy they are formulating can materially impact women, through such avenues as participatory decision-making, leadership building, economic participation, social changes and so on. Incorporating a gender-transformative approach to lawmaking can perhaps help refine existing legal frameworks to better adapt to the impending crisis of climate change.

In particular, let us examine three basic rights which are generally recognized the world over, and also in India, with respect to women. First, the right to life and liberty, which is a fundamental right granted to all Indians as per the Constitution of India. A gender-transformative audit of the violations of this law should focus on how women's right to life is violated at a large-scale in the form of female feticide and other forms of gender-based violence that threaten their lives. In particular, to incorporate the dimension of climate change within our analysis, we must be mindful of the fact that climate change tends to have a multiplier effect on social conflict. Areas which have experienced erratic draughts or floods because of climate change have also faced increased instances of violence against women, such as in the form of child marriage or domestic abuse. In addition, climate change also poses serious questions for the sexual and reproductive health of women and other gender minorities. Those women who live in poverty are often unable to seek medical help because of their inability to travel, or pay for, the long routes to the doctor's clinic. How can a gender-transformative approach help mend these situations? A gender-transformative policy or program should ideally use gender-disaggregated data to guide the implementation of its program and aim to transform the material conditions of women on the ground by granting them access to decision-making positions, leadership roles, trusteeship roles and so on.

The second right is the right to equality, and dignity, and to a life free from discrimination and harassment. As noted above, climate change acts as a multiplier for gender-based violence, and a feminist evaluation of the laws that grant such rights must take into account the climate change dimension within it as well. In addition, research shows that women's wage gap may widen due to climate change-induced disasters, and that women may be at greater risk of their right to livelihood being threatened at the behest of climate change than their male counterparts.

Perhaps the most central to this discussion, however, are the laws that govern women's right to property and land, especially. Women were not recognized as coparceners under the Hindu law initially. The two forms of property that Hindu law recognized women could hold were the 'Stridhana' which refers to property granted to women at the time of marriage, and women's estate. The former, women could dispose of as they wished. The latter, however, posed unique limitations on the possibility of disposal, and could only be done insofar as it helped in religious duties or legal necessity. Women were seen as unfit to be vested with the control and disposal of property, and as such were not admitted in the coparcenary. It was in 1956 that the Hindu Succession Act was passed, which posed a significant loophole within it; women were not granted the right to inherit ancestral property by birth right. In 1975, in response to feminist critiques of the act, Kerala became the first state to pass a legislation that fully abolished men's right by birth to property, bringing an end to the Hindu joint family system. Several states followed suit by amending the act at a state level to declare that daughters, too, could be coparceners in joint family property. In 2005, the amendment to the act at a federal level granted women equal rights to property by birthright as their male counterparts.

Land ownership has been one of the most essential legal barriers that compound women's vulnerability to climate change. When women do not possess ownership right over land, they are not granted access to social security schemes in the event of climate change-induced natural disasters which could insulate from the economic losses sustained in the aftermath of the disaster. In addition, studies show that women are more likely to adapt sustainable agricultural techniques if they are assured of the security of their tenure over their land. In times when poor agricultural techniques greatly expose farmers to the disastrous effects of climate change, it is all the more important to recognize women's land ownership rights to help pave the way for a sustainable future. Yet as highlighted earlier, women's land ownership patterns in India continue to remain dismal, owing to societal stigma, patriarchal conditioning and other such reasons.

But despite the existence of these laws which grant women rights that help safeguard against the consequences of climate change, we observe, as discussed earlier, that a larger proportion of women are 'missing' from the labour force than would be the case under normal conditions. Where, then, is the law going wrong, such that it is leading to millions of 'missing women,' as discussed by Amartya Sen years ago? We submit that it is not the law leading to the 'missing women,' and since the law is not the problem, stricter implementation of existing laws will probably bring little effects, especially when it comes to bringing about a material transformation in the lives of women. Rather, we believe it is the lack of the ability of the law to transform power dynamics on-ground that is the problem. In the context of India, we believe the implementation of the law, and women's access to institutional sources of justice, get more and more precarious as they are filtered through the multiple social locations women occupy. Given the prevalence of corruption and bribery in the country, the law becomes a distant figure in the lives of those it seeks to transform. This effect is especially pronounced in rural areas where alternative forms of justice, such as Khap Panchayats, take the role of dispute resolution and arbitration amongst villagers, and often tend to brush women's issues aside, or engage in victim-blaming and disparaging women. Given the patriarchal conditioning and people's belief that corruption and bribery are so deeply entrenched in our society as to be considered practically infallible enough to protect them from the law, it become difficult for the law to truly shift the power differential that exists between men and women in the country.

This is how women go 'missing.' It is not so much that our laws are inadequate, but that our social mindsets remain stagnant, stuck in a past that the world has moved on from. We remain rigid in our prejudices and continue to repose our faith in patriarchal systems, even as we see first-and the violence such systems perpetrate. Many of us even benefit from this violence. Women, despite being equal citizens under the eye of the law, remain 'missing'. They 'vanish' from spheres of social life, from politics to the economy. On the one hand, women are disproportionately the victims of premature deaths, owing to gender-based violence, which as we have seen climate change tends to exacerbate. On the other hand, women are forced, by virtue of their vulnerability to climate change, to drop out from the workforce, leading to the phenomenon of women 'missing' from the economy as well. This two-fold disappearance of women does not augur well for the country.

2.5 Comparative Analysis of Developing Countries

We conduct a comparative analysis between other lesser developed countries such as Bangladesh and South Africa. Using a study conducted by the United Nations in Bangladesh (UN Women, 2022), we understand that the adverse impact of climate change on women workers is similar across developing economies.

In Bangladesh, women's role in agriculture is systematically suppressed, resulting in greater job insecurity and lower wages. Women further have lower decision-making powers owing to limited land ownership rights. The restricted mobility of women often places them in vulnerable positions and the possibility of gender-based violence. For example, women find it difficult to gain social sanction to migrate in search of employment in the event of floods or draughts. A study by the ILO further found that the vulnerability of women to climate change due to institutional loopholes further aggravates the decent work deficit amongst women (ILO, 2017). It compounds the problem of women being primarily employed in small enterprises and the informal economy.

Different forms of pollution, research suggests, also have a gendered dimension. Air pollution reduces productivity and increases workload. All forms of pollution, the ILO report suggests, have a multiplier effect on social conflict. There is sufficient research to indicate that climate change could lead to a drastic increase in gendered violence over time. Researchers have driven home this point in their insightful study on the rise in gender violence during and after climate change-induced disasters (Desai & Mandal, 2021). Climate change leads to an acute food shortage, and in Malawi, the study discusses, minor girls are often forced into marriages to supplement family incomes enough to survive during famines. During prolonged dry spells, women in Uganda had to face an uptick in domestic violence, even though little was done to punish the male perpetrators.

Climate change is known to destroy jobs, and statistics show that a major fall in national output may be anticipated due to internal displacement and a drastic decline in labor productivity. However, the coverage for unemployment benefits remains limited for women in Eastern and Southern Asia, and Northern Africa. The figure below indicates the unemployment benefit coverage in a set of countries (ILO 2018).

Studies conducted in South Africa find that a huge proportion of women are employed in the low-skilled sector, and suffer from heat exposure (Shayegh & Dasgupta, 2022). This in turn decreases the relative availability of women employed in the sector under study. Women's optimal threshold temperature is significantly lower at 26.7 degree Celsius than that of their male counterparts, beyond which their supply of labor begins to decline. This indicates that women are at greater risk to heat stress than male workers.

Will laws be effective in regulating the activities of the informal economy? On the one hand, this is true, since legislation can safeguard women by providing them job security, fair wages, and gender-responsive sanitation facilities. On the other, there remains the question of shifting power dynamics within the informal economy. For example, even if women are paid fair wages, do they have the power to collectivise and bargain for better working conditions from their employers? Although the right to unionize has been granted to all workers in India, efforts to create unions are systematically suppressed often, either by villainizing unions as nuisance-creating elements or by union-busting rackets. Against this context, how can the law function as a tool to evade such union-busting, and ensure that everyone has the right to unionize and bargain?

2.6 Analysing the Gendered Dimensions of Climate Change: An Overview

The purpose of this section is to emphasise on those points which have been raised, and will continue to be raised, for the entirety of this paper. We acknowledge that climate change differentially impacts women. An assessment of women's heightened vulnerability to climate change can be more effective by using UNFCC's tool on the three-pronged approach to measuring vulnerability to climate change. These factors are firstly, the exposure of people to climate change, which refers to how exposed they are to global warming because of their geographical locations. For example, countries like India which lie in the tropical region are more directly affected than European countries, as evinced by news of the brutal heatwaves that ravaged India's capital, Delhi, only a few months ago. The other two factors, which are harder to measure, refer to sensitivity and adaptability. Both of these are measured at an individual as well as a systemic level, highlighting the fact that climate action must simultaneously consist of building people power and bringing about institutional changes through advocacy and organizing, for example.

Sensitivity and adaptability to climate change are also gendered, reflecting deep-rooted social and economic inequalities that influence how different gender minorities may experience and respond to environmental changes, especially climate change. Women, particularly in developing regions, are often more vulnerable to the adverse effects of climate change due to their socioeconomic roles and responsibilities. In many societies, women are primarily responsible for securing food, water, and energy for their families, and these tasks become increasingly difficult as climate-related issues such as droughts, floods, and extreme weather events become more frequent and severe. This vulnerability is compounded by limited access to resources, such as financial capital, technology, and education, which hampers women's ability to adapt to changing environmental conditions effectively. Moreover, traditional gender norms and expectations often restrict women's participation in decision-making processes related to climate adaptation and resilience. This lack of representation means that policies and interventions may not fully address their needs or leverage their knowledge and experiences.

Up till now, we have discussed broadly the various studies conducted to investigate the gendered dimension of climate change on employment. We summarize our findings in this section. Broadly, the impact of climate change on women's employment can be measured along four major observations. First, regions hit by climate change-induced natural disasters find an uptick in violence against women. Child marriages, the practice of dowry and domestic violence are prevalent problems deeply entrenched in a patriarchal society. Such forms of abuse seek to maintain complete control over women's lives, and the threat of women victims seeking rehabilitation or escaping their abusive conditions leads the perpetrators of the crime to dissuade women, often by coercion, from joining the labor market and getting jobs. In this context, generating sustainable livelihoods for women is imperative for their economic freedom. However, the uptick in violence against women associated with climate change makes it difficult for women to access the labor market. Often, women have to compensate by joining the informal economy, where working conditions are dangerous and ruthless.

Second, climate change reduces women workers' productivity, and is associated with health-related issues caused by heat stress. There is little to no provision for women's sanitation facilities in the informal sector. Climate change-induced disasters are known to cause often irreversible health issues amongst women. For example, pregnant women are at greater risk of preeclampsia from extremely saline water in areas hit by floods. Given women's poor access to healthcare facilities, the health risks associated with jobs may often prove to be too huge a cost for women to incur. It is important to increase women's access to both the skilled and unskilled job market to ensure job security.

Third, climate change reduces the incomes of women-led agricultural households. It has already been shown that studies conducted by researchers in rural areas indicate that women find it harder to switch to non-farm jobs in the case of a poor harvest season, unlike their male counterparts. This also rings true for women helming small-scale cottage industries. Climate change destroys raw materials and disrupts transportation services. Often, women need to borrow in order to buy enough capital to restart their business. However, women's access to formal credit institutions in rural areas is unfortunately precarious. The void has been filled by women's self-help groups, which have provided a veritable opportunity for women to gain financial independence and begin small-scale enterprises. However, the threat of climate change disrupting women's jobs and depressing their incomes is always a huge possibility.

Fourth, climate change-associated changes may act as a poverty trap that can push huge proportions of women throughout the country into poverty. This may be due to depressed wages, poor sanitation, an increase in women's household labor (time poverty) and a worsened access to major infrastructure including education and healthcare. Studies indicate that during times of relative financial struggles, families tend to drop girls out of schools instead of sons. Climate change, combined with regressive social norms codified into the lives of millions across the country, threaten to usurp the gains made in recent times by feminist advocates and NGOs. This is a highly concerning situation that merits proper attention by policymakers.

There is sufficient evidence to indicate the role of patriarchal forces in inhibiting women's lives. Traditional stereotypical notions of women bearing the burden of household chores prevents them from accessing the job market at large. Women's access to education and healthcare is seen expendable, and it is encouraged for women to sacrifice basic necessities such as food and water for their families. As a result, women are disproportionately hired in the informal economy. Their financial freedom remains precarious at best, and women are often subjected to economic abuse and their right over their hard-earned finances considered suspect by men. As a result, women have relatively lesser capital, and further face stigma in accessing credit as well. These factors compound women workers' vulnerability to climate change.

Employers are reluctant to promote gender-inclusive policies that, at a basic level, cater to women's sanitation and hygiene requirements. Heat stress further exacerbates this problem, and causes immense pain, physical and psychological, to women. This further discourages women from continuing work, and explains their drop from the labor market at lower temperatures than their male counterparts.

Very little countries accept the role of the gender dimension as a compounding factor in distributing the burden of climate change among workers. There is a need for policies that are specifically designed keeping the gender differential in mind. These are outlined in the next section.

At this stage, it is essential for us to highlight Amartya Sen's capability approach as a framework that can help devise gender-transformative and climate-resilient policies and programs. Throughout this paper we have noted how merely being provided rights may prove incapable of bringing about transformation socially, especially in terms of shifting power dynamics in favor of oppressed communities. Amartya Sen and Martha Nussbaum's significant contribution to the study of welfare economics is this normative framework that asks policymakers to not just ensure the existence of tools that grant a fulfilling and dignified life to people, but to also ensure that they are able to access the same. Chief amongst the factors influencing accessibility are education, healthcare and skill-formation, which form the human development approach Sen is seen to be the champion of. The use of the capability approach in climate action is widely acknowledged by social scientists. Climate change has a differential impact on disadvantaged and disempowered communities, and as such, it is essential that everyone is equally protected, or alternatively, equally at risk, to climate change. This is known often as achieving a climate-just future, through the equitable distribution of the burden of climate change among all stakeholders.

How is the capability approach relevant to our discussion of women's rights and 'missing women'? Women's rights and the recognition thereof form a baseline for developing indicators of women's resilience against climate change, but a baseline is merely a foundation from which to begin. It is through the capability approach that we can determine whether the law has been truly successful in materially impacting the lives of those it wishes to safeguard from violence, discrimination and so on. For example, we have already noted how the grant of land ownership rights in India has not materially impacted land ownership patterns as recently as the last few years. This situation is further compounded by the existence of social identities like caste and class markers. Oppressed-caste women, for example, face more barriers and greater discrimination that may hamper their security of tenure or their right to hold titles. The capability approach helps spell out for us a way through which we can gauge the phenomenon of 'missing women.' The fact that women are 'missing,' as highlighted by Sen and several researchers after him as well, can be attributed to causes that go beyond direct threats to women's lives. In other words, women can be 'missing' because of those factors which may not at first seem to explain skewed sex ratios in many countries, such as lower wages, absence of social security nets or disproportionate access to healthcare facilities, education and so on. However, what Sen's 'missing women' essay has done is that it has remarkably explained how the sex ratio depends also on these factors which directly influence women's capability to cope with, recover from and deal with violence.

All of this discussion is merely qualitative in nature. As the authors of this paper, we have not used statistical techniques to estimate the proportion of women who have gone 'missing' from the workforce because of climate change-related factors. The purpose of this paper has been to advance the study of the gendered dimensions of climate change by pointing to researchers a possible direction deserving of closer study, through Sen's 'missing women,' essay. We hope that this paper serves as an inspiration to researchers in this field and helps advance the dialogue surrounding the intersection of gender, environment, economics, and law.

Finally, a note to researchers who may embark on estimating the extent of the 'missing women' phenomenon through this paper. It is essential for us to keep in mind the multiple other social identities that exist alongside gender, and compound the effects of climate change further. We have already briefly mentioned a study that discusses the inequitable distribution of the heat burden amongst those belonging to the oppressed castes. It is crucial that we keep in mind to disaggregate data by relevant social identities to gain useful inferences from our dataset. Climate change, after all, is anything but a simple, unidimensional issue. It is a multifaceted social evil that affects all of us differently, and the only way we can devise policies and solutions that can help mitigate the effects of climate change and adapt to a rapidly climate-changing world, is through a recognition of the diverse array of social dignities which in turn manufacture unique experiences and require unique solutions as well.

ANALYSIS OF RESULTS AND POSSIBLE SOLUTIONS

Amazon was in the news in recent times for adopting a controversial policy in its warehouses in India: it made workers take an oath that prevented them from accessing restrooms and even water until they worked on their targets (Sanzgiri, 2024). Given the lack of heat adaptive infrastructure, Amazon workers and huge swathes of gig workers throughout the country, were suffering from heat stress caused by continuous heatwaves during the periods of April-June. The first, and most rudimentary, policy advise would thus be to forego such apathetic managerial decision, and to adopt an empathetic framework for effective policymaking.

One essential policy suggestion recognized in the ILO 2018 report on employment under climate change recognizes the role of ecosystem services in promoting employment. Natural rainfall, for example, and fertile land enable agricultural activity for vast portions of the rural segments. Women have traditionally been the most involved in conservation activities in indigenous societies, and recognizing their roles as trustees of natural resources is essential for a humanistic and inclusive approach to climate change adaptability.

The report further highlights that several green energy and sustainability sectors that have arisen in response to climate change and provided jobs see lesser participation from women. This is essential to remedy in light of a report (WEF, 2023) that suggests that hiring for new green market roles such as sustainability evaluation have consistently outpaced overall hiring rates in the market. It is crucial to improve women's participation in this market, given their historical role as trustees of natural resources worldwide. The figure shows the proportion of women employed in renewable energy sectors out of total number of workers in that sector. The data is obtained from select states in India and was studied by the United Nations (UNDP, 2021).

In Senegal, in collaboration with the UNDP, vocational training for green jobs is especially targeted at underrepresented groups such as women. This has helped enable women to switch to non-farm work and immensely improved their financial stability.

Certain mitigation strategies such as extended unemployment benefit coverage and direct cash transfers have helped in face of the brutal climate disasters caused by the El Nino in recent times. The beneficiaries continue to be disproportionately skewed toward men, but policymakers are new increasingly aware of this gender disparity and working on the same.

A wider network of public employment programs has proven to be beneficial in encouraging women's employment (ILO 2018) especially in environmentally relevant jobs such as ecological conservation. Evidence from the implementation of the MGNREGA act indicate an increase in women's employment and household autonomy. A similar employment-generation scheme called the Working for Water in South Africa has harnessed around 60% of women workers to help in conservation of water and prevention of soil erosion.

The involvement of women has further strengthened ecotourism in regions such as the Western Ghat. The novel 'Moco-Moco" program in Guyana has made use of untapped entrepreneurial potential amongst women workers for disaster management and rehabilitation assistance.

Women can thus be protected from the adverse effects of climate change with policies that can be broadly categorized as follows: first, policies that aim at improving women's participation in decent and organized work. This is done through employment generation schemes and vocational training, along with conscientious attempts to boost women's employment in the green market sector. Second, policies must create strong and extensive social security nets that provide women with assistance and rehabilitation in the case of displacement or disruption of job due to climate change. Third, policies must encourage public-private partnerships that collaborate to ensure climate-sustainable plans, such as ensuring carbon neutrality and encouraging women's representation on the job market. A combination of these factors will surely help in mitigating the impact of climate change on women workers.

Studies indicate that women-led households are expected to incur losses of around 8% more of their incomes than those led by men, owing to discrimination against women and disparities in wages, in the case of heat stress alone (OWSA, 2024). This quantification of losses along gendered lines due to climate change has enabled us to assess the gravity of the problem at hand. If left unchecked, climate change could have major macroeconomic implications for a nation. An estimated loss of around 37 billion amongst women-led households, for example, could greatly contract savings, affect investments and drastically alter consumption patterns. Climate adaptive policies are the need of the hour.

A comprehensive analysis of policies which must be implemented in India must not only seek experience from other developing economies, but also be aware of the dynamic nature of the rural-urban divide in the country. Ensuring social and financial security for women in light of climate change requires different areas of focus depending on whether the women are a part of the rural economy, the urban economy, or the peri-urban economy which is known to be the most socio-economically dynamic of all regional groups. Infrastructural and legal loopholes have to be remedied; this involves a more stringent push for combating corruption and dealing with misogyny. Recognizing women's manifold vulnerability due to their gender and class is crucial for crafting empathetic and inclusive policies that truly address their needs.

CONCLUSION

Very little is done to amend the gender disparities that exist in who share the burden of climate change in India. This poses a unique set of challenges from the perspective of employment particularly. On the one hand, women already suffer from wage gap and restrictions from accessing the formal economy. On the other, climate change is not only expected to compound this problem, but also undo several strides made in recent times in promoting women's participation in the labor force. Women face uniquely different problems than male workers due to climate change. This is especially because of restrictions placed on women's mobility and economic freedom. As such, women require social security nets that protect their jobs and businesses. However, a one-size-fits-all approach has proven to be limited ins cope and coverage. The need for a comprehensive policy that specifically addresses these gender disparities has never been more pronounced. Such policies can help mitigate the adverse effects of climate change among women workers if planned and implemented properly, and in thorough consultation with all stakeholders.

We find that women are consistently exposed to higher proportions of heat stress and greater levels of social and economic insecurity due to climate change. These occur on various levels, such as an increased exposure to gendered violence, the failure of adequate social security nets to protect their businesses and the greater levels of exploitation women workers endure in the labor market. There is a need to provide a comprehensive climate-adaptive solution to this problem. Further research must focus on incorporating this gender dimension into policymaking. Climate change has been widely recognized to be a driver of compounding existing social inequalities. Interventions in studying the impact of climate change have increasingly suggested the use of intersectional and gender-sensitive frameworks that help policymakers in more effectively allocating resources and mitigating risks of their programs having unintended consequences. In particular, the differential in power dynamic between men and women have led to women being more vulnerable to climate change than their male counterparts.

Vulnerability is generally assessed using a three-pronged approach, namely exposure, sensitivity and adaptability. On the latter two fronts, patriarchal conditioning, societal stigma and prejudice have disproportionately impacted women and thereby increased their vulnerability to climate change. With regards to sensitivity, women are more likely to slide into abject poverty, food insecurity and subsequently malnutrition in the event of climate change-induced natural disasters. This is because women are seen as less deserving of an equal access to resources such as food, healthcare, education and so on in comparison to male members of the family. This in turn also affects women's adaptability to climate change, because the lack of social security nets and further, the lack of adequate funding to uniquely empower women in the fight against climate change, means that women are less likely to adapt easily to climate change and more likely to suffer greater damages, economic and otherwise, than men.

While considering the applicability of Amartya Sen's concept of 'missing women,' in our analysis of the differential impact of climate change, it becomes increasingly clear that both phenomena reflect highly entrenched structural inequalities and systemic failures within societies worldwide, and specifically within Asian countries. Sen's argument, which highlights the disproportionate number of women who are essentially 'missing' due to gender biases and discriminatory practices, can be crucial in studying the intersection of the impacts of climate change with women's rights. Both issues, while distinct in their manifestations, stem from similar root causes—namely, entrenched gender biases, social neglect, and the unequal distribution of resources and opportunities. To address these intertwined issues, it is imperative that both legal frameworks and social mindsets evolve to reflect a heightened climate consciousness. The current legal and policy scenario often fail to adequately account for the differential impacts of climate change on various demographic groups, particularly women. Legal reforms should integrate gender-sensitive approaches into climate policies, ensuring that women's needs and contributions are recognized and supported. This includes creating legal mechanisms that protect women from climate-induced displacement, providing them with access to resources and technologies to adapt to environmental changes, and incorporating their voices in decision-making processes. All of this, we believe, will be useful in climate action.

Socially, there needs to be a shift in mindset towards recognizing the link between climate justice and gender equality. Education and advocacy efforts should focus on raising awareness about how climate change disproportionately affects women and the necessity of inclusive solutions. Societal attitudes must evolve to challenge traditional gender norms that exacerbate women's vulnerabilities and to support policies that promote gender equity in climate resilience and adaptation strategies. In conclusion, addressing the issue of "missing women" through the lens of climate change requires a comprehensive approach that aligns legal reforms with social transformation. By updating laws to be more inclusive and reflective of gender-specific impacts of climate change, and by fostering a societal mindset that values and supports women in the context of environmental challenges, we can work towards a more equitable and resilient future. The interconnected nature of these issues calls for a holistic response that not only addresses the immediate needs of those most affected but also transforms the underlying structures that perpetuate inequality. In doing so, we can create a world where both gender equity and climate resilience are prioritized, ensuring a more just and sustainable future for all.

REFERENCES

Afridi, F., Mahajan, K., & Sangwan, N. (2021). *The Gendered Effects of Climate Change: Production Shocks and Labor Response in Agriculture*. Indian Statistical Institute.

Desai, B., & Mandal, M. (2021). Role of Climate Change in Exacerbating Sexual and Gender Based Violence against Women: A New Challenge for International Law. *Environmental Policy and Law*, 51(3), 137–157. DOI: 10.3233/EPL-210055

International Labor Organization. (2017). Green Initiative. *Policy Brief.*

International Labor Organization. (2018). *The employment impact of climate change adaptation.*

International Labor Organization. (2018). *Greening with Jobs.*

Klasen, S., & Pieters, J. (2013). What Explains the Stagnation of Female Labor Force Participation in Urban India? *The World Bank Economic Review*, 29(3), 2015.

Ministry of Finance, Government of India. (2018). *Economic Survey 2017-18.* https://www.indiabudget.gov.in/budget2018-2019/economicsurvey2017-2018/index.html

Mitra, E., & Berlinger, J. (2024, May 29*). Indian capital of Delhi breaks all-time heat record, as authorities impose water rationing. CNN.* https://edition.cnn.com/2024/05/29/weather/india-heat-record-water-rationing-intl/index.html

Nikore, M.. (2021). India's missing working women: Tracing the journey of women's economic contribution over the last seven decades, and during. *Journal of International Women's Studies*. 19.

OWSA. (2024). *The Climate Change Crisis is Unjust for Rural Women: FAO Gender Expert.* One World South Asia.

Qian, N. (2008). Missing women and the price of tea in China: the effect of sex-specific earnings on sex imbalance. *The Quarterly Journal of Economics*, 123(3), 1251–1285. DOI: 10.1162/qjec.2008.123.3.1251

Ravi, A. (2021, Nov 11). *Abandonment, poverty, sexual exploitation: Why Indian women pay a higher price for climate crisis.* Down to Earth. https://www.downtoearth.org.in/interviews/climate-change/abandonment-poverty-sexual-exploitation-why-indian-women-pay-a-higher-price-for-climate-crisis-80160

Roy, M., & Venema, H. (2002). *Reducing risk and vulnerability to climate change in India: The capabilities approach.* Taylor & Francis.

Sanzgiri, V. (2024, June 27) *Amazon India Workers Say Unlawful Operations Persist Despite Alleged Acknowledgement by Company.* Media Nama. https://www.medianama.com/2024/06/223-amazon-workers-allege-unlawful-operations-persist/

Sarwar, M. G.. (2007). *Women's Rights to Land in Bangladesh: Roles, Limitations and Transformation*. Unnayan Oneshan.

. Sen, A. (1990, Dec. 20) *More Than 100 Million Women Are Missing.* The New York Review.

Shah, A., ... (2024). *Caste inequality in occupational exposure to heat waves in India.* IIM Bangalore Working Paper No. 709/2024.

Shayegh, S., & Dasgupta, S. (2022). *Climate change, labor availability and the future of gender inequality in South Africa*. Taylor & Francis.

Somanathan, E., Somanathan, R., Sudarshan, A., & Tewari, M. (2021). The Impact of Temperature on Productivity and Labor Supply: Evidence from Indian Manufacturing. *Journal of Political Economy*, 129(6), 1797–1827. DOI: 10.1086/713733

United Nations Development Program. (2016). Gender, climate change and food insecurity.

United Nations Development Program. (2021). Green Jobs and Eco-Entrepreneurship opportunities for women in India.

United Nations Women. (2022). *State of Gender Equality and Climate Change in Banglades*h.

Vasavada, S. (2024, Mar. 26) *Women Farmers' Land Ownership and Climate Resilience*. India Development Review.

Venugopal, V.. (2016). Occupational Heat Stress Profiles in Selected Workplaces in India. *International Journal of Environmental Research and Public Health*. PMID: 26729144

World Economic Forum. (2023). The future of jobs is green: How climate change is changing labor markets.

Chapter 20
Climate Change and Feminization of Agriculture in India:
Mapping the Opportunities and Challenges

Subrata S. Satapathy
https://orcid.org/0000-0002-0086-3506
Christ Academy Institute of Law, India

ABSTRACT

The feminization of agriculture in India, driven by male migration to urban areas, intersects with the growing challenges of climate change, creating a complex landscape of both opportunities and challenges. Women, increasingly responsible for agricultural work, face heightened vulnerabilities due to their reliance on natural resources and limited access to adaptive technologies, credit, and decision-making power. Climate change exacerbates these challenges, intensifying workloads and threatening food security as crop yields decline and traditional farming practices become less viable. However, this convergence also presents opportunities for empowering women through targeted interventions such as land ownership, access to climate-resilient technologies, and involvement in decision-making processes. By recognizing and supporting women's critical roles in agriculture India can enhance agricultural productivity, promote gender equality, and build climate resilience in rural communities.

DOI: 10.4018/979-8-3693-6069-9.ch020

INTRODUCTION AND CONCEPTUAL FRAMEWORK

Climate change represents one of the most profound and pressing challenges of the 21st century. Characterized by rising global temperatures, changing precipitation patterns, and increasing frequency of extreme weather events, climate change poses significant threats to the natural environment and human societies. The causes of climate change are multifaceted, involving the accumulation of greenhouse gases (GHGs) in the atmosphere due to human activities such as fossil fuel combustion, deforestation, and industrial processes. The Intergovernmental Panel on Climate Change (IPCC) warns that without substantial reductions in GHG emissions, global temperatures could rise by more than 1.5°C above pre-industrial levels, leading to severe and irreversible impacts on ecosystems and human livelihoods. In recent decades, the phenomenon of the feminization of agriculture has gained increasing attention. This term refers to the growing participation of women in agricultural activities, a trend that has been driven by multiple factors including male outmigration, changes in rural labor markets, and shifts in household structures. In many developing countries, men often migrate to urban areas or abroad in search of better employment opportunities, leaving women to take on greater responsibilities in farming and rural livelihoods. Consequently, women have become the primary agricultural producers, making critical contributions to food production and household nutrition.

A GENDERED ANALYSIS: LINKING CLIMATE CHANGE, FEMINIZATION OF AGRICULTURE, AND FOOD SECURITY

Women have a critical role in agricultural output, particularly in small-scale and subsistence farming, where they often perform the majority of the farm labour and greatly increase the food security of households. The term "feminisation" of agriculture refers to the trend of an increasing number of women engaging in farming operations in India. In rural India, where agriculture is the primary source of employment, women have traditionally played a significant role in agricultural duties like planting, weeding, and harvesting crops as well as taking care of livestock. However, due to a variety of socioeconomic circumstances, including as migration, urbanisation, and changing demography, males are leaving rural regions in greater numbers in pursuit of work in cities, leaving women to assume farming tasks. Understanding the interconnectedness of climate change, feminization of agriculture, and food security requires a gendered analysis. Women play a crucial role in agriculture and food systems, yet they often face gender-specific challenges that can exacerbate their vulnerability to climate change and food insecurity. This aims to explore the synergies and trade-offs between these three critical issues through a gendered lens, highlighting the importance of empowering women and promoting gender equality in the pursuit of sustainable and resilient agricultural systems. A gendered analysis of climate change, the feminization of agriculture, and food security reveals how these interconnected issues disproportionately affect women, particularly in rural areas, and how they contribute to broader socio-economic inequalities. This analysis helps to understand the unique challenges women face and the strategies that can be employed to enhance food security and resilience in the face of climate change.

FEMINIZATION OF AGRICULTURE

The term "feminization of agriculture" refers to the increasing involvement of women in agricultural activities, often due to the migration of men to urban areas in search of work or as a result of changing socio-economic conditions. This trend has significant implications for food security, rural livelihoods, and gender equality. In small-scale and subsistence farming, where they often perform the majority of the farm labour and greatly increase household food security, women are extremely important to agricultural productivity. The increasing trend of women engaging in agricultural pursuits is referred to as the "feminisation" of agriculture in India. In the past, women have played a significant role in agricultural chores including planting, weeding, and harvesting crops in addition to taking care of livestock in India's rural areas where agriculture is the primary source of employment. However, males are leaving rural areas in greater numbers in search of work in cities due to a variety of socioeconomic issues, such as urbanisation, migration, and changing demography, leaving women to assume farming tasks.

1. Increased Responsibilities: As men migrate or seek employment in other sectors, women are left to manage farms and households. This shift often leads to women taking on a greater share of agricultural work, including planting, harvesting, and managing resources like water and soil. However, despite their increased responsibilities, women often lack access to the same resources, technologies, and decision-making power as men.
2. Limited Access to Resources: Women farmers frequently have limited access to essential resources such as land, credit, agricultural inputs (seeds, fertilizers), and extension services. Cultural norms and legal barriers often restrict women's land ownership rights, making it difficult for them to invest in or benefit from their agricultural activities. This lack of access undermines their productivity and limits their ability to adapt to climate change.

3. Impact on Food Security: As primary caregivers and food providers, women play a crucial role in ensuring household food security. However, climate change exacerbates the challenges they face in this role. Changes in temperature, precipitation patterns, and the frequency of extreme weather events can reduce crop yields and increase food prices, making it harder for women to provide adequate nutrition for their families.

The feminization of agriculture also means that women are increasingly responsible for the cultivation of subsistence crops, which are vital for local food security. However, their limited access to resources and decision-making power hampers their ability to sustain and improve agricultural productivity.

According to Food and Agriculture Organization (FAO), women make up 43% of the global agricultural labour force and this number is increasing to 50% in some areas. In context of India, women contribute up to 30% of the agricultural labour force. The National Sample Survey Office (NSSO)'s recent statistics show that women's involvement in agricultural activities in India grew from 12.8% in 2005-06 to 13.9% in 2012-13. According to the 10th Agriculture Census (2015–16), there were 14% more female operational holdings in the nation in 2015–16 than there were in 2010–11. As they do not take into consideration the unpaid work that women perform on family farms, these statistics might understate the extent to which women participate in agriculture. According to a survey by the National Bank for Agriculture and Rural Development (NABARD), women are essential in the care of livestock, with 64% of them working in animal husbandry in rural areas. Due to several factors, including the decline in non-farm employment opportunities in rural areas, greater access to education and training, and the acceptance of women's rights to land ownership, women's participation in agriculture has greatly increased in recent years. Additionally, the COVID-19 pandemic has made these issues even worse for women farmers, who now have more work to do as a result of school closures, childcare responsibilities, and difficulty accessing markets and inputs. However, there are still many obstacles that women must overcome in order to increase their productivity and income, such as their restricted access to resources, technology, and credit.

CLIMATE CHANGE AND GENDER: DIFFERENTIAL IMPACTS AND ADAPTATION

The impacts of climate change are not gender neutral. Women, particularly those in rural and marginalized communities, are disproportionately affected by climate-related hazards due to existing gender inequalities in access to resources, information, and decision-making power. For instance, women often have less access to land, credit, and agricultural inputs, limiting their ability to adapt to changing climatic conditions. Additionally, women's responsibilities for household food security and water collection can increase their workload and stress during periods of climatic stress, such as droughts or floods. Despite these challenges, women also exhibit remarkable resilience and adaptive capacities. They often employ diverse strategies to cope with climate variability, such as diversifying crops, conserving seeds, and engaging in collective action. Recognizing and supporting women's adaptive strategies is essential for building climate-resilient agricultural systems. Climate change and gender intersect in complex ways, leading to differential impacts on men, women, and gender-diverse people. These differences arise from varying social roles, responsibilities, and levels of vulnerability, which can influence how individuals experience and adapt to climate change.

FEMINIZATION OF AGRICULTURE IN INDIA: OPPORTUNITIES

While the feminization of agriculture in India presents several challenges, it also offers significant opportunities for the empowerment of women, enhancement of agricultural productivity, and overall rural development. By recognizing and supporting women's roles in agriculture, India can unlock substantial economic and social benefits. Here are some key opportunities.

1) Empowerment through Ownership and Decision-Making: Encouraging and facilitating women's ownership of land can empower them to make decisions regarding crop choices, resource management, and agricultural practices. Legal reforms and policies that ensure equal inheritance rights and access to land can be pivotal in this regard. Promoting women's participation in agricultural cooperatives, self-help groups (SHGs), and decision-making bodies can strengthen their influence and leadership in rural communities. This can lead to more inclusive and equitable agricultural development.
2) Increased Agricultural Productivity: Providing women with better access to agricultural inputs such as seeds, fertilizers, irrigation, and credit can significantly boost agricultural productivity. When women farmers have the same access to resources as men, they can produce higher yields and contribute more effectively to food security. Offering targeted agricultural training programs for women can enhance their skills in modern farming techniques, sustainable agriculture, and climate-resilient practices. This can lead to improved farm management and higher productivity.

3) Economic Independence and Livelihood Diversification: Women in agriculture can be encouraged to take up agribusiness activities such as food processing, dairy farming, and agro tourism. Support through microfinance, training, and market access can help women develop these enterprises, leading to economic independence and job creation in rural areas. By engaging in value-added activities such as organic farming, packaging, and marketing, women can increase their income and create new market opportunities. This not only boosts their economic status but also enhances the overall value chain in agriculture.
4) Improved Food Security and Nutrition: Women's involvement in all stages of food production—from planting to harvesting to processing—positions them as key players in ensuring household food security. By empowering women in agriculture, food availability and diversity can be improved, leading to better nutrition outcomes for families. Women can be encouraged to grow nutrient-rich crops, such as vegetables, pulses, and millets, which can improve household nutrition and contribute to healthier diets in rural areas.
5) Advancing Gender Equality: As women take on more prominent roles in agriculture, traditional gender norms can be challenged, leading to greater gender equality. This shift can promote more equitable sharing of household responsibilities and decision-making power. Promoting gender equality through education and awareness campaigns can help change societal perceptions of women's roles in agriculture. This can lead to greater respect for women's contributions and more support from male counterparts.
6) Adoption of Sustainable Agricultural Practices: Women often have a deep understanding of local ecosystems and sustainable practices. By promoting their knowledge and leadership in sustainable agriculture, practices such as organic farming, crop diversification, and conservation agriculture can be more widely adopted. Women's involvement in agriculture offers an opportunity to develop and implement climate adaptation strategies that are sensitive to local contexts. Supporting women in adopting climate-resilient practices can enhance the resilience of entire communities to climate change.
7) Strengthening Rural Communities: Women's participation in community-based organizations, such as farmer groups and SHGs, can strengthen social cohesion and collective action in rural areas. These groups can become platforms for knowledge exchange, mutual support, and advocacy for better services and infrastructure. By ensuring that women's voices are included in rural development plans, the needs of all community members can be better addressed. This leads to more inclusive and balanced development outcomes.
8) Innovation and Technology Adoption: Providing women with access to digital technologies, such as mobile apps for agriculture, online marketplaces, and e-learning platforms, can empower them to innovate and adopt new practices.

Digital tools can also help women access market information, weather forecasts, and agricultural advice, enhancing their productivity and resilience. Developing and promoting agricultural technologies that are specifically designed for women's needs—such as lightweight tools or labor-saving devices—can improve efficiency and reduce the physical burden of farming.

9) Enhancing Social Status and Well-being: Recognizing women as farmers in official records and government schemes can elevate their social status and ensure they receive the benefits and support they deserve. This recognition can also lead to greater respect and autonomy within households and communities. Empowering women in agriculture can lead to better overall well-being, including improved health, education, and economic security for themselves and their families. This contributes to the broader development goals of reducing poverty and improving quality of life in rural areas.
10) Policy and Institutional Support: Developing and implementing gender-responsive agricultural policies that address the specific needs and challenges faced by women can create a more enabling environment for their participation and success in agriculture. Strengthening institutions that support women in agriculture, such as women's cooperatives, rural development agencies, and agricultural extension services, can provide the necessary resources, training, and advocacy to empower women and improve agricultural outcomes.

The feminization of agriculture in India is not just a challenge but also a significant opportunity to transform rural economies, enhance food security, and advance gender equality. By investing in women farmers and addressing the barriers they face, India can harness the potential of women to drive sustainable agricultural development and contribute to broader social and economic progress.

FEMINIZATION OF AGRICULTURE IN INDIA: CHALLENGES AND ISSUES

The intersection of the feminization of agriculture and climate change in India presents a unique set of challenges that disproportionately affect women in rural areas. These challenges can exacerbate existing inequalities and create new barriers to sustainable development. Here are the key challenges:

1) Increased Vulnerability to Climate Change: Women farmers, who are primarily responsible for subsistence farming and household food security, are highly dependent on natural resources. Climate change, through erratic rainfall, extreme weather events, and changing seasons, directly threatens their livelihoods.

Women often lack access to climate-resilient seeds, irrigation technologies, and other adaptive practices, making it harder for them to cope with the impacts of climate change.

2) Intensified Workload: Climate change can lead to the degradation of land, water scarcity, and reduced agricultural productivity. This forces women to work longer hours to secure water, firewood, and food for their families, further intensifying their already significant workload. As climate impacts reduce agricultural viability, men are more likely to migrate to cities for work, leaving women to manage not only the household but also the agricultural tasks. This increases their physical and mental stress.
3) Limited Access to Information and Resources: Women often have less access to climate-related information, extension services, and agricultural training due to social norms, mobility restrictions, and lower literacy levels. This limits their ability to implement adaptive strategies and technologies. Women typically have less access to financial resources, credit, and insurance, which are critical for adopting climate-resilient practices or recovering from climate-induced losses.
4) Health and Nutritional Risks: Climate change can reduce crop yields and affect food availability, leading to malnutrition and food insecurity. Women and children are often the first to suffer from food shortages. The increased workload due to climate impacts can lead to physical exhaustion and increased exposure to health hazards. For instance, women may have to walk longer distances to fetch water, which exposes them to risks like dehydration and injuries.
5) Lack of Recognition and Support: Despite their crucial role in agriculture, women are often not recognized as key stakeholders in climate adaptation policies and programs. This lack of recognition means that their specific needs and challenges are not adequately addressed in national and local climate action plan.
6) Exclusion from Decision-Making: Women are underrepresented in decision-making bodies at all levels, from household to community to national, which limits their influence on policies related to agriculture and climate change.
7) Social and Cultural Barriers: Traditional gender norms and roles often confine women to certain types of agricultural work, limiting their ability to diversify their livelihoods in response to climate change. These norms also restrict their mobility and access to markets, further exacerbating their vulnerability. Lower levels of education among rural women hinder their ability to understand and respond to climate-related risks effectively. This also limits their participation in capacity-building programs aimed at climate adaptation.
8) Psychological Stress and Mental Health Issues: The compounded stress of managing agricultural work, household responsibilities, and dealing with the impacts of climate change can lead to significant psychological stress. Women may experience anxiety, depression, and other mental health issues, which are

often overlooked and untreated in rural areas. As men migrate for work, women are often left isolated, which can exacerbate feelings of loneliness and increase the mental health burden.

9) Impact on Childcare and Education: The increased time spent on agricultural and household tasks due to climate change can reduce the time women have for childcare and education. This can have long-term impacts on the health and development of children. In some cases, girls may be taken out of school to help with household or farm work, particularly during times of climate stress, perpetuating the cycle of poverty and limiting future opportunities for empowerment.
10) Increased Gender-Based Violence: Scarcity of resources such as water and land can lead to conflicts, increasing the risk of gender-based violence. Women fetching water or firewood may face harassment or violence, especially when they have to travel further due to resource scarcity. As men migrate and women are left behind, they may face increased vulnerability to exploitation and violence, with fewer social protections available in rural areas.
11) Lack of Gender-Responsive Climate Policies: Climate policies and adaptation programs often lack a gender perspective, failing to address the specific needs of women in agriculture. There is a need for gender-responsive policies that consider women's roles, vulnerabilities, and potential contributions to climate resilience.

The challenges at the intersection of the feminization of agriculture and climate change in India are complex and multifaceted. Addressing these challenges requires a comprehensive approach that includes empowering women with knowledge, resources, and decision-making power; recognizing their contributions to agriculture; and ensuring that climate policies are gender-responsive and inclusive. Only by addressing these challenges can sustainable and equitable development be achieved in the face of climate change.

REFERENCES

Ali, D. A., Deininger, K., & Goldstein, M. (2014). Environmental and gender impacts of land tenure regularization in Africa: Pilot evidence from Rwanda. *Journal of Development Economics*, 110, 262–275. DOI: 10.1016/j.jdeveco.2013.12.009

Asfaw, S., & Maggio, G. (2018). Gender, weather shocks and welfare: Evidence from Malawi. *The Journal of Development Studies*, 54(2), 271–291.

Behrman, J. A., Bryan, E., & Goh, A. (2014). *Gender, climate change, and group-based approaches to adaptation. International Food Policy Research Institute Policy Note. Bishop-Sambrook, C. (2003). Labour saving technologies and practices for farming and household activities in Eastern and Southern Africa: labour constraints and the impact of HIV/AIDS on rural livelihoods in Bondo and Busia Districts, Western Kenya*. IFAD and FAO.

Castañeda Camey, I., Sabater, L., Owren, C., & Boyer, A. E. (2020). Gender-based violence and environment linkages: The violence of inequality. Wen, J. (ed.). Gland, Switzerland: IUCN. 272pp Chanamuto, Nicola J.C. and Stephen J.G. Hall (2015) Gender equality, resilience to climate change, and the design of livestock projects for rural livelihoods. Gender & Development 23(3): 515-530, DOI: DOI: 10.1080/13552074.2015.1096041

Chanana-Nag, N., & Aggarwal, P. K. (2020). Woman in agriculture, and climate risks: Hotspots for development. *Climatic Change*, 158(1), 13–27.

Rao, N., Mishra, A., Prakash, A., Singh, C., Qaisrani, A., Poonacha, P., Vincent, K., & Bedelian, C. (2019). A qualitative comparative analysis of women's agency and adaptive capacity in climate change hotspots in Asia and Africa. *Nature Climate Change*, 9(12), 964–971. DOI: 10.1038/s41558-019-0638-y

Scheelbeek, P. F. D., Bird, F. A., Tuomisto, H. L., Green, R., Harris, F. B., Joy, E. J. M., Chalabi, Z., Allen, E., Haines, A., & Dangour, A. D. (2018). Effect of environmental changes on vegetable and legume yields and nutritional quality. *Proceedings of the National Academy of Sciences of the United States of America*, 115(26), 6804–6809. DOI: 10.1073/pnas.1800442115 PMID: 29891659

Slavchevska, V., Kaaria, S., & Taivalmaa, S.-L. (2019). The Feminization of Agriculture: Evidence and Implications for Food and Water Security. The Oxford Handbook of Food, Water and Society. Sriram, Vidhya (2018). Achieving Gender Equality and Women's Empowerment in Smallholder Adaptation: Lessons from IFAD's Adaptation in Smallholder Agriculture Programme. CCAFS Info Note. Wageningen, the Netherlands: CGIAR Research Program on Climate Change, Agriculture and Food Security (CCAFS). https://ccafs.cgiar.org/publications/achieving-gender-equality-andwomen%E2%80%99s-empowerment-smallholder-adaptation-lessons-ifad%E2%80%99s#.XkxqhiOIY2w

Chapter 21
The Consideration of Children's Rights in Corporate Action:
Current Developments in Germany

Christopher Bohlens
https://orcid.org/0009-0001-9653-2743
Leuphana University Lüneburg, Germany

ABSTRACT

This chapter provides a concise overview of the integration of children's rights into corporate operations, highlighting the growing importance of this issue in the globalized business environment. As companies become more interconnected with global supply chains, they must address the direct and indirect impacts of their activities on children, particularly in areas like child labor prevention, fair wages, education, and healthcare. By embedding children's rights into their strategies, businesses not only fulfill legal and ethical obligations but also enhance their reputations and contribute positively to society, facing both challenges and opportunities in this endeavor.

INTRODUCTION

The consideration of children's rights in business activities is becoming increasingly important in today's globalized world. Companies no longer operate in isolation, but are part of a complex network of supply chains and business partners that transcend national borders. In this context, it is crucial to recognize and address the

DOI: 10.4018/979-8-3693-6069-9.ch021

impact of business decisions on the youngest and often most vulnerable members of society - children.

Children's rights and corporate action are closely linked, as companies play a key role in influencing the living conditions of children worldwide. Children's rights include the right to education, protection from exploitation, health and a safe living environment. Companies have a responsibility to ensure that their business practices respect and promote these rights. This includes the prevention of child labor in supply chains, fair wages for parents to make child labor unnecessary, and investment in education and health programs. By acting responsibly, companies can not only comply with legal regulations, but also contribute to the long-term well-being of society and strengthen their own reputation.

Children's rights, as enshrined in the UN Convention on the Rights of the Child, include the right to protection from economic exploitation, the right to education, the right to health and the right to an adequate standard of living. These rights must not only be seen as moral obligations, but must also be anchored in corporate practice. Companies have a responsibility to ensure that their business activities do not violate these rights, but actively contribute to their promotion.

The pressure on companies to face up to their social responsibility comes not only from non-governmental organizations and governments, but increasingly also from consumers and investors who demand ethical behavior and sustainable business practices. It is therefore essential that companies develop strategies and measures to respect and promote children's rights. This includes reviewing and adapting supply chains, implementing child protection policies and creating education and health initiatives.

This chapter examines how companies can integrate children's rights into their business activities, what challenges they face and what best practices exist. The aim is to contribute to raising awareness and improving the consideration of children's rights in corporate activities. The research question is therefore: How can companies effectively take children's rights into account in their business activities?

BACKGROUND

The rights of children, which are meticulously articulated and enshrined in the comprehensive framework of the United Nations Convention on the Rights of the Child, encompass a wide array of fundamental entitlements that include, but are not limited to, the crucial right to protection and safeguarding against various forms of economic exploitation that could jeopardize their well-being and future prospects, the essential right to access a quality education that empowers them with knowledge and skills necessary for personal development and societal participation, the vital right to attain and maintain good health, which is indispensable for their overall growth and development, and the fundamental right to enjoy a satisfactory standard of living that ensures their basic needs are met, thereby promoting their dignity and quality of life (Convention on the Rights of the Child, 1990; Übereinkommen über die Rechte des Kindes - VN-Kinderrechtskonvention im Wortlaut mit Materialien, 2022).

The rights that have been previously mentioned should not be understood solely as moral obligations that exist in a vacuum; instead, it is of paramount importance that these rights are integrally woven into the very fabric of the operational structures and strategic frameworks that govern corporate entities. It falls upon these corporations to ensure unequivocally that their business activities do not encroach upon or violate these fundamental rights; furthermore, they should actively engage in efforts to support, promote, and elevate their advocacy in a manner that not only respects but amplifies the significance of these rights within society (Collins & Guevara, 2014; Schilling-Vacaflor & Gustafsson, 2024; Villiers, 2023).

The driving force that encourages corporations to recognize and embrace their social responsibilities originates not only from the initiatives and pressures exerted by non-governmental organizations and governmental bodies, but is increasingly influenced by a growing cohort of consumers and investors who fervently advocate for ethical practices and sustainable methodologies in business operations. Therefore, it becomes critically essential for corporations to devise comprehensive strategies and enact effective measures that not only protect but also promote the fundamental rights of children within their operational frameworks. This obligation entails a thorough examination and subsequent modification of supply chains, the establishment of robust child protection policies, as well as the creation and implementation of educational and health initiatives aimed at fostering the well-being of young individuals (Cao et al., 2023; National Coalition Deutschland – Netzwerk zur Umsetzung der UN-Kinderrechtskonvention e.V., 2020; Petter, 2023; Winston, 2002).

THEORETICAL AND LEGAL BACKGROUND

Children's rights have historically evolved from an almost exclusive focus on protection needs to comprehensive rights to survival, development, protection and participation. The movement to recognize children's rights began in earnest in the early 20th century, leading to the adoption of the Geneva Declaration of the Rights of the Child by the League of Nations in 1924. This eventually led to the UN Convention on the Rights of the Child (UNCRC) in 1989, which recognized children as having their own rights for the first time (Fass, 2011).

The UN Convention on the Rights of the Child (1989) is the most important international agreement that comprehensively defines children's rights. It comprises 54 articles that guarantee basic rights such as the right to life, survival and development, protection from abuse and exploitation, as well as the right to education and participation. Almost all countries worldwide have ratified the convention, making it one of the most widely accepted human rights documents (Levy et al., 2022).

International laws and regulations play a crucial role in protecting children's rights in business activities. In addition, the impact of business activities on children's rights is increasingly recognized and companies are encouraged to acknowledge negative impacts and commit to upholding and promoting children's rights (Kolieb, 2020). Efforts will be made to eradicate harmful child labor practices worldwide, with a focus on aligning international legal instruments to protect children's labor rights and freedoms, emphasizing the importance of ILO conventions and global cooperation in the elimination of child labor (Erdem Türkelli, 2020; Hallsworth, 2022; ILO, 2021; United Nations Global Compact, 2024a, 2024b).

ILO Convention No. 138, of the International Labor Organization (ILO), also known as the Minimum Age Convention of 1973, sets the minimum age for admission to employment or work. The main objective of this convention is to eliminate child labor by setting the minimum age for legal work at 15 years, with an exception for developing countries, which can set the minimum age at 14 years. For hazardous work, the minimum age is set at 18. This convention aims to protect the development of children and ensure they have access to education (Hesketh et al., 2006; ILO, 2024; Übereinkommen 138, n.d.).

ILO Convention No. 182 was adopted in 1999. It obliges the signatory states to take immediate measures to eliminate the worst forms of child labor. These include slavery, forced labor, trafficking in children, prostitution, pornographic performances, illegal activities such as drug trafficking and work that endangers the health, safety or morals of children. The aim is to protect the rights and well-being of children worldwide (ILO, 2024; Klinger et al., 2016; Übereinkommen 182, n.d.).

The UN Guiding Principles on Business and Human Rights (UNGP) (UN & UNHR, 2011; UNHR & United Nations Global Compact, 2014) were adopted by the UN Human Rights Council in 2011. They are based on three pillars. The state duty to protect - states must protect against human rights violations by third parties, including companies. The corporate responsibility - companies should respect human rights and avoid negative impacts of their activities. Access to remedy - victims of human rights violations must have access to effective remedies, both at state and corporate level (Brune, 2020; Business & Human Rights Resource Centre, 2024b, 2024a; Wapler, 2019; Westra, 2014).

The Principles on Children's Rights and Business (2012) were developed by UNICEF, the UN Global Compact and Save the Children to help companies assess and manage the impact of their business activities on children. These ten principles include a commitment to combating child labor, ensuring safe and healthy working conditions for young workers and promoting the well-being of children in the communities in which companies operate (Fottrell, 2021; Geschäftsstelle Deutsches Global Compact Netzwerk (DGCN), 2012; UN Global Compact Netzwerk Deutschland e. V., 2022).

UNICEF's “Children's Rights in Sustainability Reporting” (2014) is designed to help companies respect and promote children's rights through their reporting (GRI - Global Reporting Initiative) and communications. Companies can support children's rights through their policies, processes and operations in the workplace, marketplace and community (UNICEF, 2014a).

Further guidance with workbooks, policies and codes of conduct and tools is available (UNICEF, 2014b; UNICEF & Danish Institute for Human Rights, 2013; UNICEF & Save the Children Deutschland e.V., 2013).

The OECD Guidelines for Multinational Enterprises contain recommendations for responsible business conduct, including respect for human rights, which include children's rights. These guidelines call on companies to carry out due diligence to identify, prevent and mitigate adverse impacts on human rights. This includes respecting children's rights along the entire supply chain. Implementing these guidelines helps companies not only to meet their legal obligations, but also to fulfill their social responsibility (BMWK-Bundesministerium für Wirtschaft und Klimaschutz, 2024; OECD, 2019, 2023, 2024).

Efforts are being made to eliminate harmful child labor practices worldwide, with a focus on aligning international legal instruments to protect children's labor rights and freedoms, emphasizing the importance of ILO conventions and global cooperation in the elimination of child labor (Hallsworth, 2022).

In 2020, UNICEF announced a child protection by design standard for the use of data by technology companies (Hartung, 2020). A joint position paper by several NGOs in 2020 makes recommendations on how the German government and Bundestag can protect children's rights (Appel et al., 2020).

The new EU recommendation, adopted in April 2024, aims to strengthen child protection systems in the member states and thus fulfill the Children's Rights Strategy 2021. It promotes the development of integrated systems using EU instruments. Protection from child labor is also manifested (European Commission, 2024a, 2024b).

DISCUSSION

In order to effectively guarantee that the exploitative practice of child labor is not utilized within their operations, businesses can adopt and enforce comprehensive ethical policies that delineate the minimum age requirements for employment in emerging markets, which should be meticulously designed to consider and integrate both universalist ethical principles that emphasize the inherent rights of all individuals and relativist ethical perspectives that recognize and respect the diverse cultural contexts in which these businesses operate (Gupta et al., 2010).

Furthermore, organizations operating within the corporate landscape have the capacity to formulate comprehensive codes of conduct that incorporate rigorous and systematic monitoring mechanisms, which are specifically aimed at ensuring adherence to minimum age requirements, implementing thorough oversight of all involved parties, and establishing clear and effective sanctions for any infractions that may occur, all while being cognizant of the necessity to refrain from imposing excessively harsh regulations that could inadvertently exacerbate the precarious circumstances faced by child laborers (Kolk & van Tulder, 2002).

A study shows that companies worldwide must promote and protect children's rights by actively integrating them into their activities (Schudak & Mahlmann, 2017). An analysis showed that 48% of German companies consider children's rights to be relevant mainly in developing and emerging countries and consider them to be less important for Germany. Many companies do not understand how their activities affect children's rights. Children's rights are relevant in supply chains, corporate activities, marketing and product design. Few companies consider children's rights holistically, often focusing on combating child labor, family-friendly workplaces and social projects. Measures in product design and marketing are rare. More than half of the companies do not have adequate management systems to ensure children's rights. However, sectors with a high risk of child labor often show good commitment. A comprehensive awareness of the importance of children's rights and their positive impact on business success is largely lacking. Companies see it as the responsibility of politicians to anchor children's rights more firmly. Children's rights are not explicitly mentioned in the National Action Plan for Business and Human Rights (2016-2020) (Auswärtiges Amt & Interministeriellen Ausschusses Wirtschaft und Menschenrechte, 2017) Small and medium-sized enterprises often rate their commitment higher than large corporations and refute the assumption that they lack resources for CSR issues (Schudak & Mahlmann, 2017).

The Alliance 8.7 (named after the SDG goal) (Statistisches Bundesamt, 2024, p. 8) is a global multi-stakeholder partnership founded in 2016 with the aim of eliminating forced labor, modern slavery, human trafficking and child labor by 2030. With 217 partners, it works through four action groups at a global level.

These groups focus on tackling the root causes of these human rights violations in different contexts: conflict and humanitarian crises, supply chains, migration, and rule of law and governance (Alliance 8.7, 2024, p. 7).

On the 10th anniversary (2022) of the Child Rights Principles for Business, UNICEF emphasizes that companies play a central role in protecting and promoting children's rights. These principles offer concrete recommendations for action to integrate children's rights into business activities and along the supply chain. Despite existing challenges, many companies have shown a positive impact on children's rights through targeted measures such as living wages and family-friendly workplaces. Compliance with children's rights is considered a minimum requirement and companies should consider their impact along the entire value chain in order to minimize both direct and indirect negative impacts on children and make positive contributions (Much, 2022).

The Otto Group (a German trading and service company with 40.000 employees based in Hamburg, which operates worldwide in the retail, financing and logistics and mail order sectors) reports from its experience that it systematically identifies its greatest need for action based on potentially negative impacts as the basis of its CR strategy. An important social goal is that only employees of the legal minimum age are involved in the production of all products sold. The Otto Group recognizes that child labour-free production is only possible through the strengthening of children's rights and preventative approaches. Exchange with business partners is essential in order to pass on requirements and raise awareness of social standards, e.g. through training. Achieving the goals requires continuous exchange and improvement of practices with partners and suppliers. An important learning process is the step-by-step approach and the identification of internal multipliers. Cooperation with partners with expertise in children's rights and local structures offers additional security and progress for the Otto Group (Much, 2022).

Ten years after the launch of the Children's Rights and Business Principles (CRBPs), UNICEF, Save the Children and the UN Global Compact have published reports (2022) to review progress and outline future steps to integrate children's rights into responsible and sustainable business practices. Human rights are essential to sustainability and require a justice and equity-based approach to environmental action. Children, who are severely affected by climate change, are at the center of this vision as their development shapes the future of society. Companies must comply with human rights and environmental due diligence that focuses on children's rights and their potential negative impacts. The effective participation of children in the relevant processes is crucial. Cooperation between companies, governments and organizations is fundamental to achieving the Sustainable Development Goals. Voluntary action alone is insufficient; mandatory due diligence and legislative action are necessary to achieve a transformative impact and fully protect children's rights (Kang, 2021; Mawson, 2022; Mawson et al., 2022; UNICEF, 2011, 2011).

While the Child Rights and Business Principles (CRBP) offers a comprehensive and well-structured framework that encourages businesses to actively uphold and respect the fundamental rights of children, it is crucial to acknowledge that significant challenges continue to persist in the practical implementation of these principles, particularly in sectors characterized by intricate and multifaceted supply chains as well as in geographical regions marked by a pronounced lack of effective regulatory oversight and governance. In order to effectively tackle and mitigate these formidable challenges, it is imperative that there is a unified and collaborative effort that involves not only businesses but also governmental entities and representatives of civil society, all working in concert to ensure that the rights of children are given the utmost priority and rigorously safeguarded across all facets and activities of business operations (Bueno & Kaufmann, 2022; Collins, 2023).

A comprehensive study conducted in the year 2022 meticulously investigates the multifaceted role that the rights of children play within the operational frameworks and corporate social responsibilities of companies situated in the geographical regions of Switzerland and Liechtenstein. More specifically, this scholarly inquiry scrutinizes various aspects including the level of awareness exhibited by these corporations, the range of activities they undertake, the myriad challenges they encounter, and the potential opportunities that exist for the effective promotion of children's rights throughout their entire value chain, which encompasses all stages from production to consumption. It is noteworthy that the majority of these companies do not specifically address the unique rights and needs of children within their corporate policies and practices; however, they do express a broader commitment to the principles of human rights in general, which serves as a foundational ethical framework guiding their operations (GCBHR & CIDE, 2022).

Another Stakeholder engagement and child participation discussion from Nordic companies shows that. It is crucial for businesses to increase their awareness and provide comprehensive in-house training on child rights and safeguarding across all levels of the organization. This training should be integrated into the company's culture and policies to ensure that every employee understands the importance of protecting children's rights and the role they play in safeguarding these rights. By doing so, businesses can create a safer environment for children and ensure that their operations align with ethical standards.

A comprehensive guide that has been meticulously crafted by the Global Child Forum is specifically tailored for businesses that possess a keen interest in actively incorporating children into their decision-making frameworks, effectively elucidating the myriad benefits that accrue not only to the children involved but also to the corporations themselves. Furthermore, it underscores the critical importance of children's active participation in governance processes as a means for them to assert and exercise their inherent rights, while simultaneously offering businesses invaluable and innovative perspectives that can significantly enhance their operational strategies. This guide has been developed through a collaborative effort that includes contributions from esteemed corporate leaders, academic researchers, and the voices of young individuals, ensuring a well-rounded and inclusive approach to the subject matter. Although the primary audience for this guide is sustainability teams operating within multinational corporations, it is equally advantageous for smaller enterprises that seek to adopt similar practices. The relevance of this guide extends to various industries that have both direct and indirect interactions with children, particularly in their capacity as consumers, thereby encompassing a wide range of sectors that impact young people's lives (Collins, 2023).

Companies have the capacity to significantly enhance and fortify the rights of children through the implementation and active engagement in corporate social responsibility (CSR) initiatives, thereby leading to a considerable expansion of their social influence and impact within the communities they operate. Among the various measures that can be adopted in this context, it is crucial to include practices such as the avoidance of child labor, the provision of comprehensive support for young employees and their parents, along with the assurance of safe and conducive working conditions, as well as the delivery of secure and reliable products to consumers. A particular emphasis is placed on the promotion of education and the holistic development of children, which not only plays a vital role in fostering social sustainability but also serves to enhance and elevate the corporate image and overall attractiveness of the company in the eyes of various stakeholders (Save the Children Deutschland e.V., 2024; UNICEF, 2024).

In June 2024 an article discusses the challenges of identifying products associated with child labor and the role of the recently implemented German Supply Chain Act ("Lieferkettengesetz") and upcoming EU regulations in addressing this issue. Child labor remains widespread globally, particularly in agriculture, where about 70% of child labor occurs. Common products linked to child labor include textiles, cocoa, coffee, tea, and certain minerals used in electronics and cosmetics. Consumers can exert pressure on companies by critically evaluating their supply chains and opting for products with certifications like Fairtrade. However, these labels are not foolproof guarantees against child labor. The German Supply Chain Act, effective since 2023, mandates that large companies ensure human rights are respected throughout their supply chains. From 2024, it will apply to companies with over 1,000 employees. This law aims to hold companies accountable for human rights violations, including child labor, within their supply chains. The EU is also moving forward with similar legislation, although it has been slightly weakened during negotiations. The EU law, once passed, will extend the responsibility of large companies to ensure their products do not involve child labor or environmental harm. These laws represent steps toward more ethical supply chains, but their effectiveness will depend on rigorous enforcement and broader application across more industries (Winkler, 2024).

SOLUTIONS AND RECOMMENDATIONS

The examination of the rights of children within the realm of corporate practices represents a significant area of inquiry that is garnering a growing level of interest and scrutiny, which inherently encompasses both advantageous aspects as well as certain limitations that merit consideration.

Strengths

- Enhancing awareness: A considerable number of corporations have started to acknowledge the critical significance of safeguarding children's rights and are consequently taking steps to implement measures that are deemed appropriate and relevant; nevertheless, it is essential to note that these initiatives frequently manifest merely as broad guidelines and non-binding recommendations rather than enforceable standards.
- Collaborative partnerships: The establishment of cooperative relationships with non-governmental organizations (NGOs) and other relevant entities aimed at eradicating child labor while simultaneously promoting educational opportunities has demonstrably yielded a positive influence on the overall situation of children's rights in various contexts.
- Transparency and accountability in reporting: An increasing number of corporations are now taking the initiative to publish comprehensive reports detailing their ongoing endeavors and the progress they have made in relation to the promotion and protection of children's rights, thereby enhancing transparency and fostering accountability in their corporate social responsibility practices.

Weaknesses

- Implementation in the supply chain: The complexities inherent in the execution of policies aimed at safeguarding the rights of children within the supply chain can often pose significant challenges, as it is frequently arduous to ascertain that suppliers operating remotely adhere to these vital rights, particularly given that the legislative framework governing such matters is relatively nascent and still evolving.

- Cultural differences: The variances that exist in the understanding and practical application of children's rights across diverse geographical regions contribute to the challenges faced in achieving a uniform and consistent implementation of these rights, thereby complicating efforts to standardize practices on a global scale.
- Lack of continuity: Many initiatives that are designed to promote and protect children's rights tend to be structured as temporary projects, which often results in their lack of sustainable integration into the broader corporate strategy; consequently, these efforts frequently rely on voluntary commitments that may not be consistently upheld or enforced.

Identification of Gaps

- The absence of thorough due diligence: A significant number of corporations fail to conduct systematic and exhaustive auditing processes on a regular basis, thereby neglecting the critical assessment of their operational and financial practices which could potentially lead to unrecognized vulnerabilities.
- Inadequate training provisions: The existing training initiatives designed for both employees and suppliers frequently fall short of being sufficiently robust and comprehensive, resulting in a lack of necessary knowledge and skills that are essential for maintaining operational efficiency and compliance.
- Resource constraints: It is a well-documented fact that small and medium-sized enterprises (SMEs) frequently encounter substantial limitations in terms of the financial and human resources required to effectively implement and sustain holistic and comprehensive strategies aimed at enhancing their operational frameworks.

Strategies To Enhance the Consideration Of Children's Rights

- Systematic due diligence obligations: The establishment and execution of extensive and thorough due diligence obligations throughout the entirety of the supply chain, which must incorporate regular and systematic audits alongside comprehensive risk assessments, is essential in ensuring accountability and adherence to established standards.
- Training and awareness-raising: The implementation of robust training programs aimed at both employees and suppliers is crucial for the purpose of significantly elevating awareness levels and deepening the understanding of children's rights and the importance of safeguarding them within all organizational practices.

- Use of technology: The adoption and integration of advanced technological solutions, such as blockchain, serve to markedly enhance transparency and improve traceability within the supply chain, thereby ensuring that all actions and processes are clear and accountable.
- Long-term partnerships: The cultivation of enduring partnerships with non-governmental organizations (NGOs) and various other entities is vital for the ongoing promotion and advancement of children's rights, facilitating a collaborative approach that can yield lasting positive impacts.
- Integrated strategies: The thorough incorporation of the principles pertaining to children's rights into the overarching corporate strategy and organizational culture is imperative, as it not only reinforces commitment but also ensures that these principles are deeply embedded within the operational framework of the entity.

Recommendations for Implementation

- Management commitment: It is of paramount importance to secure the unwavering support and steadfast commitment of senior management, as their involvement is critical for the successful execution and integration of the proposed strategies aimed at enhancing the rights of children within the organization.
- Clear guidelines and processes: The formulation and establishment of comprehensive and unambiguous guidelines and processes is essential for the seamless integration of children's rights into every facet and operational area of the business, thereby ensuring adherence to these principles across the entire organizational framework.
- Measurement and reporting: The creation and implementation of robust mechanisms dedicated to the regular measurement and detailed reporting of progress concerning the implementation of child rights principles is vital, as it will facilitate transparency and accountability in the organization's commitments.
- Continuous improvement: The introduction of a systematic approach aimed at fostering continuous improvement is necessary, which should be grounded in constructive feedback and regular evaluations of the effectiveness of the measures taken, allowing for ongoing refinement and enhancement of practices in relation to children's rights.
- Stakeholder engagement: It is imperative to ensure the active and meaningful engagement of all pertinent stakeholders, including but not limited to employees, suppliers, customers, and local communities, throughout the entire

process of implementing children's rights, as their collective input and collaboration will significantly enrich the effectiveness of these initiatives.

By synthesizing a robust framework of ethical guidelines, meticulously developed comprehensive codes of conduct, and substantial support for various global initiatives aimed at social responsibility, corporations can effectively implement proactive strategies designed to prevent and systematically eradicate the insidious issue of child labor within both their operational practices and extensive supply chains. While numerous guidelines and frameworks have been established to address this critical concern, it is often the case that these measures lack the necessary enforcement mechanisms and remain largely as voluntary self-declarations rather than obligatory standards that must be adhered to by corporations.

FUTURE RESEARCH DIRECTIONS

As we cast our gaze toward the horizon of future developments, the incorporation of children's rights within the operational practices of corporations, as well as within the expansive frameworks of social responsibility that govern their conduct, emerges as a particularly fertile and promising domain ripe for rigorous academic inquiry and investigation. Notable trends that are surfacing in contemporary discourse indicate a burgeoning emphasis on the pivotal role that businesses are expected to play in the active safeguarding and promotion of children's rights; this shift is largely propelled by an array of factors, including but not limited to significant regulatory transformations, rapid technological progressions, and an escalating set of societal expectations that demand accountability and ethical behavior from corporate entities. While the present chapter has meticulously established a foundational understanding of the current landscape regarding these issues, it is crucial to recognize that several key areas emerge that merit further scholarly exploration to enhance our comprehensive understanding and the practical implementation of effective strategies aimed at protecting children's rights.

1. Regulatory Compliance and Global Standards

A primary and salient area for prospective research endeavors pertains to the influence exerted by the continuously evolving landscape of regulatory frameworks on the corporate practices that relate directly to the rights of children. In light of the introduction of new regulatory measures, such as the European Union's Corporate Sustainability Reporting Directive (CSRD), alongside the potential establishment of international standards that govern human rights due diligence, it becomes imperative to conduct a thorough investigation into the ways in which corporations are adapting to these emerging requirements. Future research initiatives could delve into an in-depth analysis of the effectiveness of these regulations in promoting authentic corporate responsibility, as opposed to mere compliance that may lack substantive commitment, while concurrently identifying exemplary best practices that could be standardized and implemented across diverse sectors and geographical regions.

2. Technological Integration and Safeguarding

The pivotal role that technology plays in either enhancing or undermining the rights of children constitutes yet another critical and urgent area warranting extensive future research. As corporate entities increasingly embrace advanced technologies, such as blockchain to enhance supply chain transparency or artificial intelligence in the development of new products, the implications for the preservation and protection of children's rights are profound and multifaceted. Research efforts could center on investigating the potential of these innovative technologies to serve as both protective instruments and potential threats to children's well-being, with a focus on examining case studies that illustrate successful implementations while also identifying inherent risks that necessitate proactive mitigation strategies. Furthermore, the ethical considerations that arise from the utilization of big data and the operation of digital platforms in relation to the privacy and safety of children present an additional rich and complex domain that requires thorough scholarly exploration.

3. Corporate Social Responsibility (CSR) and ESG Impact

As the criteria encompassing Environmental, Social, and Governance (ESG) considerations increasingly come to occupy a central position in the strategic planning and operational execution of corporations, the imperative to integrate the rights of children within these frameworks is poised to gain heightened significance and urgency. Future research could undertake a critical analysis of the extent to which children's rights are being effectively incorporated into the metrics of ESG assessment, as well as the consequential impact that this integration has on the overall performance and public reputation of corporations. Longitudinal studies could be designed to rigorously assess whether companies that prioritize the rights of children within their ESG strategies experience tangible and measurable benefits, such as enhanced investor confidence or increased consumer loyalty, and how these factors collectively contribute to the long-term success and sustainability of these corporate entities.

4. Cultural and Regional Variability in Implementation

The multifaceted and practical challenges associated with the implementation of initiatives aimed at promoting children's rights across a wide array of diverse cultural and regional contexts necessitate further comprehensive study and investigation. In particular, scholarly research could delve into the specific barriers and obstacles that various companies encounter in distinct geographic areas around the globe, such as differing legal standards that govern children's rights, cultural attitudes and perceptions towards children that vary significantly from one region to another, and the economic constraints that may hinder effective implementation. By conducting comparative studies, researchers might uncover valuable insights into how multinational corporations can strategically adjust and tailor their operational strategies to respect and align with local cultural contexts while simultaneously upholding and maintaining a steadfast commitment to the universally recognized principles of children's rights.

5. Long-Term Impact of Corporate Initiatives on Children's Welfare

There exists a pressing need for more extensive and rigorous research into the long-term impacts and consequences of corporate initiatives on the welfare of children within various communities. While it is true that many corporations engage in a plethora of short-term projects aimed at addressing children's welfare, the sustainability, efficacy, and overall effectiveness of these initiatives over an extended period of time remain significantly less understood and documented in the academic literature. Future empirical studies could concentrate on conducting longitudinal assessments of various corporate programs that are designed with the intention of improving children's education, health, and overall social conditions, thereby providing critical insights into which specific approaches yield the most enduring and beneficial outcomes for children over time.

6. New Paradigms in Corporate Responsibility

The viability and potential efficacy of emerging paradigms that place the rights and welfare of children at the very core of corporate responsibility represents yet another area that is ripe for extensive exploration and scholarly inquiry. This exploration encompasses the examination of innovative models of corporate governance that prioritize ethical considerations and social impact alongside the traditional objective of profit maximization. Research could systematically explore how these newly developed models operate in practice, meticulously identifying both the strengths and potential pitfalls or challenges inherent in such approaches. Furthermore, the development of novel metrics and analytical tools for assessing the impact of corporate decisions on children's rights could also emerge as a highly valuable and significant area of study within the broader context of corporate social responsibility.

The future landscape of research within the domain of children's rights and corporate responsibility is undeniably rich with a multitude of possibilities and opportunities for scholarly exploration. By systematically investigating these emerging trends and actively addressing the gaps that have been identified in the current body of knowledge, scholars and researchers can significantly contribute to the development of more effective, impactful, and sustainable practices within the corporate sector. The insights gleaned from this research endeavor will not only aid corporations in fulfilling their ethical obligations and responsibilities but will also ensure that their operational activities positively contribute to the overall welfare and well-being of children on a global scale.

CONCLUSION

In summary, it can be conclusively articulated that the thoughtful and deliberate consideration of children's rights within the framework of corporate activities assumes a pivotal and fundamental role in both the overarching concept of sustainable development and the ethical responsibilities that companies are obligated to uphold in today's society. The intricate and expansive global networking of supply chains, as well as the multitude of business partners that companies engage with, necessitates that these corporations commit themselves even more profoundly to the protection and proactive promotion of the rights of children. This commitment encompasses a variety of essential measures, particularly focusing on the prevention of child labor, ensuring fair and equitable remuneration for parents, and making substantial investments in the realms of education and health programs that benefit children.

Historical developments, including the adoption of the UN Convention on the Rights of the Child along with various international agreements such as the conventions established by the International Labour Organization (ILO), have collectively established a robust foundation for the comprehensive recognition and protection of children's rights in contemporary society. In this context, companies are increasingly being urged to actively integrate these fundamental rights into their business practices, thereby fulfilling not only their legal obligations but also addressing the social expectations that have been set forth by various stakeholders. The UN Guiding Principles on Business and Human Rights, alongside the Principles on Children's Rights and Business Conduct, provide invaluable guidance that can assist companies in navigating this complex landscape.

The study conducted reveals that, despite a notable increase in awareness and numerous initiatives aimed at addressing this critical issue, a significant number of companies continue to encounter formidable challenges when it comes to the practical implementation of children's rights. There exist considerable disparities in both the perception and execution of children's rights across different geographical regions as well as varying sizes of companies, with small and medium-sized enterprises facing particular difficulties in securing the necessary resources required for the implementation of comprehensive measures. Furthermore, it is important to note that many of these commitments are often voluntary in nature, which can complicate their execution.

To enhance the consideration of children's rights, it is imperative that systematic due diligence obligations are established and maintained throughout the entire supply chain, accompanied by regular audits and comprehensive risk assessments that are aimed at identifying potential areas of concern. Additionally, the establishment of training programs for both employees and suppliers, along with the incorporation of modern technologies such as blockchain, can significantly improve the levels of transparency and traceability within corporate practices. Moreover, developing long-term partnerships with non-governmental organizations (NGOs) and other relevant entities, as well as firmly embedding the principles of children's rights within the corporate strategy and organizational culture, are also of paramount importance.

The recommendations for effective implementation underscore the necessity for a strong commitment from senior management, the establishment of clear guidelines and processes, the creation of mechanisms for the regular measurement and reporting of progress, and the ongoing involvement of relevant stakeholders throughout the process. By engaging in a coordinated and sustainable commitment to these principles, companies can not only contribute meaningfully to the protection and enhancement of children's rights but also bolster their own reputation and ensure long-term success within the business landscape.

Overall, this chapter elucidates the fact that the integration of children's rights into corporate activities transcends mere moral obligation, evolving into a strategic necessity that cannot be overlooked. Consequently, companies that adopt a proactive and responsible approach to these issues can not only fulfill their legal requirements but also make a substantial and positive contribution to the overall well-being of society as a whole.

REFERENCES

Alliance 8.7. (2024). *Alliance 8.7*. Alliance 8.7. https://www.alliance87.org/

Appel, P., Erlenstedt, M., Kippenberg, J., Kulakow, L., Lüdemann-Dundua, A., Pflüger, F., & Wichmann, S. (2020, October). *Kinderrechte und unternehmerische Sorgfaltspflichten—Wie Bundesregierung und Bundestag Kinderrechte in globalen Lieferketten jetzt schützen sollten—Positionspapier* (Deutsches Komitee für UNICEF e.V., Human Rights Watch Verein zur Wahrung der Menschenrechte e.V., Kindernothilfe e.V., Plan International Deutschland e.V., Save the Children Deutschland e.V., terre des hommes Deutschland e. V., & World Vision Deutschland e.V., Eds.). https://media.business-humanrights.org/media/documents/positionspapier-kinderrechte.pdf

Auswärtiges Amt, & Interministeriellen Ausschusses Wirtschaft und Menschenrechte. (2017, September). *Nationale Aktionsplan – Umsetzung der UN-Leitprinzipien für Wirtschaft und Menschenrechte (2016-2020)*. https://www.auswaertiges-amt.de/blob/297434/8d6ab29982767d5a31d2e85464461565/nap-wirtschaft-menschenrechte-data.pdf

BMWK-Bundesministerium für Wirtschaft und Klimaschutz. (2024). *OECD-Leitsätze für multinationale Unternehmen* [Artikel]. OECD-Leitsätze für multinationale Unternehmen. https://www.bmwk.de/Redaktion/DE/Artikel/Aussenwirtschaft/oecd-leitsaetze-fuer-multinationale-unternehmen.html

Brune, J. (2020). *Menschenrechte und transnationale Unternehmen: Grenzen und Potentiale des UN Framework for Business and Human Rights* (1st ed.). Nomos. DOI: 10.5771/9783748908746

Bueno, N., & Kaufmann, C. (2022). *Guiding Principles on Business and Human Rights* (SSRN Scholarly Paper 4530382). DOI: 10.4337/9781789903621.guiding.principles.business

Business & Human Rights Resource Centre. (2024a). *UN Guiding Principles*. Business & Human Rights Resource Centre. https://www.business-humanrights.org/en/big-issues/un-guiding-principles-on-business-human-rights/

Business & Human Rights Resource Centre. (2024b). *UN-Leitprinzipien*. Business & Human Rights Resource Centre. https://www.business-humanrights.org/de/schwerpunkt-themen/un-leitprinzipien/

Cao, Y., Lawson, B., & Pil, F. K. (2023). Social Sustainability and Human Rights in Global Supply Chains. *International Journal of Operations & Production Management*, 44(1), 370–390. DOI: 10.1108/IJOPM-10-2022-0670

Collins, T. (2023). *Global Child Forum Corporate Guide: Children's Participation: How to involve children in decision-making* [Report]. Toronto Metropolitan University. DOI: 10.32920/23913060.v1

Collins, T., & Guevara, G. (2014). Some Considerations for Child Rights Impact Assessment (CRIAs) of Business. *Revue Générale de Droit*, 44(1), 153–192. DOI: 10.7202/1026401ar

Convention on the Rights of the Child. (1990). https://www.ohchr.org/en/instruments-mechanisms/instruments/convention-rights-child

Erdem Türkelli, G. (2020). *Children's Rights and Business: Governing Obligations and Responsibility*. DOI: 10.1017/9781108681841

European Commission. (2024a, April 23). *Recommendation on developing and strengthening integrated child protection systems in the best interests of the child | European Commission—C(2024) 2680 final*. https://commission.europa.eu/document/36591cfb-1b0a-4130-985e-332fd87d40c1_en

European Commission. (2024b, April 30). *Justice & Consumers—Recommendation on developing and strengthening integrated child protection systems in the best interests of the child*. https://ec.europa.eu/newsroom/just/items/828836/en

Fass, P. S. (2011). A Historical Context for the United Nations Convention on the Rights of the Child. *The Annals of the American Academy of Political and Social Science*, 633(1), 17–29. DOI: 10.1177/0002716210382388

Fottrell, D. (2021). *Revisiting Children's Rights: 10 Years of the UN Convention on the Rights of the Child*. Brill.

GCBHR, & CIDE. (2022, August 31). *Addressing Children's Rights in Business. An Assessment from Switzerland and Liechtenstein*. https://gcbhr.org/insights/2022/08/addressing-childrens-rights-in-business-an-assessment-from-switzerland-and-liechtenstein-publication

Geschäftsstelle Deutsches Global Compact Netzwerk (DGCN). (2012, August). *Kinderrechte und unternehmerisches Handeln—Grundsätze zum Schutz und zur Förderung von Kinderrechten durch Unternehmen*. https://www.unicef.de/informieren/materialien/kinderrechte-unternehmerisches-handeln/194576

Gupta, S., Pirsch, J., & Girard, T. (2010). An Empirical Examination of a Multinational Ethical Dilemma: The Issue of Child Labor. *Journal of Global Marketing*, 23(4), 288–305. DOI: 10.1080/08911762.2010.504519

Hallsworth, K. (2022). Problems of international legal regulation of protection of labor rights of children and adolescents. *Vìsnik*, 121(121), 5–5. DOI: 10.17721/1728-2195/2022/2.121-1

Hartung, P. (2020, November). *The children's rights-by-design standard for data use by tech companies*. UNICEF. https://www.unicef.org/globalinsight/media/1286/file/%20UNICEF-Global-Insight-DataGov-data-use-brief-2020.pdf

Hesketh, T., Gamlin, J., & Woodhead, M. (2006). Policy in child labour. *Archives of Disease in Childhood*, 91(9), 721–723. DOI: 10.1136/adc.2006.096263 PMID: 16923854

ILO. (2021, June 10). *Child Labour: Global estimates 2020, trends and the road forward.* https://www.ilo.org/publications/major-publications/child-labour-global-estimates-2020-trends-and-road-forward

ILO. (2024, August 1). *ILO Conventions on child labour | International Labour Organization.* ILO Conventions on Child Labour | International Labour Organization. https://www.ilo.org/international-programme-elimination-child-labour-ipec/what-child-labour/ilo-conventions-child-labour

Kang, J. (2021). *UNICEF Deutschland: Empfehlungen zur Dialogfassung der Deutschen Nachhaltigkeitsstrategie 2021*. UNICEF.

Klinger, R., Krajewski, M., Krebs, D., & Hartmann, C. (2016). *Verankerung menschenrechtlicher Sorgfaltspflichten von Unternehmen im deutschen Recht: Gutachten.* Amnesty International.

Kolieb, J. (2020). *Protecting the Most Vulnerable: Embedding Children's Rights in the Business and Human Rights Project* (SSRN Scholarly Paper 3834369). https://papers.ssrn.com/abstract=3834369 DOI: 10.4337/9781786436405.00026

Kolk, A., & van Tulder, R. (2002). Child Labor and Multinational Conduct: A Comparison of International Business and Stakeholder Codes. *Journal of Business Ethics*, 36(3), 291–301. DOI: 10.1023/A:1014009313508

Lem, S. S., & Naalsund, E. (2023, March). *Discussion brief – involving children and children's rights in business*. UNICEF Norway. https://www.unicef.no/sites/default/files/2023-03/unicef-norway-involving-children-and-childrens-rights-in-business-discussion-brief.pdf

Levy, S., Migacheva, K., Ramirez, L., Okorodudu, C., Cook, H., Araújo-Soares, V., Minescu, A., Livert, D., Ragin, D., & Walker, P. (2022). A human rights based approach to the global children's rights crisis: A call to action. *The Journal of Social Issues*, 78(4), 1085–1097. Advance online publication. DOI: 10.1111/josi.12563

Mawson, A. (2022). *Charting the Course: Embedding children's rights in responsible business conduct. A Brief.* UNICEF. https://www.unicef.org/media/121001/file/Charting%20The%20Course.pdf

Mawson, A., Cantú Rivera, H., & Jayasekaran, S. (2022). *Charting the Course: Embedding children's rights in responsible business conduct.* UNICEF. https://www.unicef.org/media/129596/file/Charting%20The%20Course.pdf

Much, L. (2022, October 10). *10 Jahre kinderrechtlicher Grundsätze für Unternehmen – Bilanz und Ausblick.* https://www.unicef.de/informieren/aktuelles/blog/-/jubilaeum-child-rights-business-principles/313368

National Coalition Deutschland – Netzwerk zur Umsetzung der UN-Kinderrechtskonvention e. V. (2020, August 19). Das Kinderrecht auf eine gesunde Umwelt verwirklichen. *Netzwerk Kinderrechte.* https://netzwerk-kinderrechte.de/2020/08/19/das-kinderrecht-auf-eine-gesunde-umwelt-verwirklichen-2/

OECD. (2019). *Child labour risks in global supply chains.* Ending Child Labour, Forced Labour and Human Trafficking in Global Supply Chains. https://mneguidelines.oecd.org/child-labour-risks-in-global-supply-chain.htm

OECD. (2023). *OECD-Leitsätze für multinationale Unternehmen zu verantwortungsvollem unternehmerischem Handeln.* Organisation for Economic Co-operation and Development. https://www.oecd-ilibrary.org/finance-and-investment/oecd-leitsatze-fur-multinationale-unternehmen-zu-verantwortungsvollem-unternehmerischem-handeln_abd4d37b-de

OECD. (2024). *MNE Guidelines—Organisation for Economic Co-operation and Development.* OECD Guidelines for Multinational Enterprises on Responsible Business Conduct. https://mneguidelines.oecd.org/mneguidelines/

Petter, J. (2023, June 12). Lieferkettengesetz: Analyse von »Save the Children zeigt in 18 von 20 untersuchten Lieferketten Hinweise auf Kinderarbeit. *Der Spiegel.* https://www.spiegel.de/ausland/lieferkettengesetz-und-kinderarbeit-analyse-von-save-the-children-zeigt-in-18-von-20-untersuchten-faellen-hinweise-a-09f2d164-fdc1-4503-be2f-13652eb09963

Save the Children Deutschland e. V. (2024, June). *CSR-Kommunikation.* CSR-Kommunikation. https://www.savethechildren.de/unterstuetzen/fuer-unternehmen/unsere-unterstuetzung/csr-kommunikation/

Schilling-Vacaflor, A., & Gustafsson, M.-T. (2024). Towards more sustainable global supply chains? Company compliance with new human rights and environmental due diligence laws. *Environmental Politics*, 33(3), 422–443. DOI: 10.1080/09644016.2023.2221983

Schudak, A., & Mahlmann, I. (2017, August). *Studie: Kinderrechte in deutschen Unternehmensaktivitäten—Status und Bewertung* (Geschäftsstelle Deutsches Global Compact Netzwerk (DGCN), Ed.). https://www.globalcompact.de/migrated_files/wAssets/docs/Menschenrechte/studie-kinderrechte-in-deutschen-unternehmensaktivitaeten.pdf

Statistisches Bundesamt. (2024, June). *SDG 8—Menschenwürdige Arbeit und Wirtschaftswachstum - Deutschlands SDG Indikatoren*. http://sdg-indikatoren.de/8/

Übereinkommen 138 Übereinkommen Über Das Mindestalter Für Die Zulassung Zur Beschäftigung, 1973. Retrieved June 17, 2024, from https://www.ilo.org/sites/default/files/wcmsp5/groups/public/@ed_norm/@normes/documents/normativeinstrument/wcms_c138_de.htm

Übereinkommen 182 Übereinkommen Über Das Verbot Und Unverzügliche Maßnahmen Zur Beseitigung Der Schlimmsten Formen Der Kinderarbeit, 1999. Retrieved June 16, 2024, from https://www.ilo.org/sites/default/files/wcmsp5/groups/public/@ed_norm/@normes/documents/normativeinstrument/wcms_c182_de.htm

Übereinkommen über die Rechte des Kindes - VN-Kinderrechtskonvention im Wortlaut mit Materialien. (2022). https://www.bmfsfj.de/resource/blob/93140/fe59de84a8fc3a6ffc61e8a5559cac9d/uebereinkommen-ueber-die-rechte-des-kindes-data.pdf

UN, & UNHR. (2011). *Guiding principles on business and human rights: Implementing the United Nations "protect, respect and remedy" framework*. United Nations. DOI: 10.18356/9ceabfd3-en

UN Global Compact Netzwerk Deutschland e. V. (2022, May 17). *Unternehmerisches Handeln muss Kinderrechte schützen – Zehn Jahre Grundsätze zum Schutz und zur Förderung von Kinderrechten durch Unternehmen*. https://www.globalcompact.de/news/meldung/unternehmerisches-handeln-muss-kinderrechte-schuetzen

UNHR, & United Nations Global Compact. (2014, June). *The UN Guiding Principles on Business and Human Rights: Relationship to UN Global Compact Commitments*. https://www.business-humanrights.org/en/latest-news/pdf-the-un-guiding-principles-on-business-and-human-rights-relationship-to-un-global-compact-commitments/

UNICEF. (2011). *Children's Rights and Business Principles | UNICEF*. https://www.unicef.org/documents/childrens-rights-and-business-principles

UNICEF. (2014a, April). *Children's Rights in Sustainability Reporting—A guide for integrating children's rights into the GRI reporting framework*. UNICEF. https://www.business-humanrights.org/en/latest-news/pdf-childrens-rights-in-sustainability-reporting-a-guide-for-integrating-childrens-rights-into-the-gri-reporting-framework/

UNICEF. (2014b, August). *Children Are Everyone's Business Workbook 2.0—A guide for integrating children's rights into policies, impact assessments and sustainability reporting*. UNICEF. https://www.business-humanrights.org/en/latest-news/pdf-children-are-everyones-business-workbook-20-a-guide-for-integrating-childrens-rights-into-policies-impact-assessments-and-sustainability-reporting/

UNICEF. (2024, June 11). *CSR - Verantwortung übernehmen*. UNICEF Österreich. https://unicef.at/mitmachen/als-unternehmen-helfen/csr-verantwortung-uebernehmen/

UNICEF, & Danish Institute for Human Rights. (2013, December). *Children's Rights in Impact Assessments: A guide for integrating children's rights into impact assessments and taking action for children*. https://resource-centre.savethechildren.net/document/childrens-rights-impact-assessments-guide-integrating-childrens-rights-impact-assessments/

UNICEF, & Save the Children Deutschland e.V. (2013, September). *Children's Rights in Policies and Codes of Conduct. A tool for companies*. UNICEF. https://resource-centre.savethechildren.net/document/childrens-rights-policies-and-codes-conduct-tool-companies/

United Nations Global Compact. (2024a). Business & Human Rights Navigator. *Business & Human Rights Navigator*. https://bhr-navigator.unglobalcompact.org/issues/kinderarbeit/weiterfuehrende-informationen/?lang=de

United Nations Global Compact. (2024b). Business & Human Rights Navigator engl. *Business & Human Rights Navigator*. https://bhr-navigator.unglobalcompact.org/issues/child-labour/#further-guidance

Villiers, C. (2023). A game of cat and mouse: Human rights protection and the problem of corporate law and power. *Leiden Journal of International Law*, 36(2), 415–438. DOI: 10.1017/S0922156522000632

Wapler, F. (2019). Umsetzung und Anwendung der Kinderrechtskonvention in Deutschland. *Recht der Jugend und des Bildungswesens*, 67(3), 252–273. DOI: 10.5771/0034-1312-2019-3-252

Westra, L. (2014). *Child Law: Children's Rights and Collective Obligations*. Springer. DOI: 10.1007/978-3-319-05071-3

Winkler, E. (2024, June 12). *Kinderarbeit: Konsumentinnen können Druck auf Unternehmen ausüben | MDR.DE* [MDR]. UNICEF zu Kinderarbeit: Konsumentinnen können Druck auf Unternehmen ausüben. https://www.mdr.de/nachrichten/deutschland/wirtschaft/kinderarbeit-produkte-erkennen-lieferkettengesetz-100.html

Winston, M. (2002). NGO Strategies for Promoting Corporate Social Responsibility. *Ethics & International Affairs*, 16(1), 71–87. DOI: 10.1111/j.1747-7093.2002.tb00376.x

Compilation of References

Aayog, N. (2022). *Decoding government support to women entrepreneurs in India*. Women Entrepreneurship Platform, Microsave.net https://www.niti.gov.in/sites/default/files/2023-03/Decoding-Government-Support-to-Women-Entrepreneurs-in-India.pdf

Abraham, R., Lahoti, R., & Swaminathan, H. (2021). *Childbirth and women's labour market transitions in India* (WIDER Working Paper No. 2021/128). UNU-WIDER.

Afridi, F., Mukhopadhyay, A., & Sahoo, S. (2012). Female labour force participation and child education in India: the effect of the national rural employment guarantee scheme.

Afridi, F., Mahajan, K., & Sangwan, N. (2021). *The Gendered Effects of Climate Change: Production Shocks and Labor Response in Agriculture*. Indian Statistical Institute.

Agarwal, B. (2010). *Gender and Green Governance: The Political Economy of Women's Presence Within and Beyond Community Forestry*. Oxford University Press. DOI: 10.1093/acprof:oso/9780199569687.001.0001

Agudo, P. A. (n.d). *Climate change and the human rights to water and sanitation*. Office of the United Nations High Commissioner for Human Rights. https://www.ohchr.org/sites/default/files/2022-01/climate-change-2-friendlyversion.pdf

Ahmad, T., Zhang, D., Huang, C., Zhang, H., Dai, N., Song, Y., & Chen, H. (2021). Artificial intelligence in sustainable energy industry: Status Quo, challenges and opportunities. *Journal of Cleaner Production*, 289, 125834. DOI: 10.1016/j.jclepro.2021.125834

Al Jazeera. (2023, May 22). *Climate change causes 2m deaths in 50 years; poor suffer most: UN*. Al Jazeera. https://www.aljazeera.com/news/2023/5/22/climate-change-causes-2m-deaths-in-50-years-poor-suffer-most-un

Alam, , MAli, , M. F. (2022). Health and wellbeing and quality of life in the changing urban environment: Challenges and government response in Fiji island. *Journal of Positive School Psychology, 6*(9), 2853-2863.

Albuquerque, C. D. (2010). Climate Change and the Human Rights to Water and Sanitation. Position Paper by the Independent expert on the human rights to safe drinking water.

Alcorn, J. B. (1989). Process as resource: The traditional agricultural ideology of Bora and Huastec resource management and its implications for research. *Adv Econ Botany*, 7, 63–77.

Ali, A., Septyanto, A. W., Chaudhary, I., Al Hamadi, H., Alzoubi, H. M., & Khan, Z. F. (2022, February). Applied Artificial Intelligence as Event Horizon Of Cyber Security. In *2022 International Conference on Business Analytics for Technology and Security (ICBATS)* (pp. 1-7). IEEE. DOI: 10.1109/ICBATS54253.2022.9759076

Ali, D. A., Deininger, K., & Goldstein, M. (2014). Environmental and gender impacts of land tenure regularization in Africa: Pilot evidence from Rwanda. *Journal of Development Economics*, 110, 262–275. DOI: 10.1016/j.jdeveco.2013.12.009

Alkire, S., Meinzen-Dick, R., Peterman, A., Quisumbing, A., Seymour, G., & Vaz, A. (2013). The women's empowerment in agriculture index. *World Development*, 52, 71–91. DOI: 10.1016/j.worlddev.2013.06.007

Alliance 8.7. (2024). *Alliance 8.7*. Alliance 8.7. https://www.alliance87.org/

Al-Matary, A., & Ali, J. (2014). Controversies and considerations regarding the termination of pregnancy for Foetal Anomalies in Islam. *BMC Medical Ethics*, 15(1), 10. DOI: 10.1186/1472-6939-15-10 PMID: 24499356

Anastasi, S., Madonna, M., & Monica, L. (2021). Implications of embedded artificial intelligence-machine learning on safety of machinery. *Procedia Computer Science*, 180, 338–343. DOI: 10.1016/j.procs.2021.01.171

Angelopoulos, A., Michailidis, E. T., Nomikos, N., Trakadas, P., Hatziefremidis, A., Voliotis, S., & Zahariadis, T. (2019). Tackling faults in the industry 4.0 era—A survey of machine-learning solutions and key aspects. *Sensors (Basel)*, 20(1), 109. DOI: 10.3390/s20010109 PMID: 31878065

Angelou, M. (1978.). Still I Rise. Poets.org. Retrieved April 1, 2023, from https://poets.org/poem/still-i-rise

Anna, G. J. (1989). The Supreme Court, Privacy and Abortion. *The New England Journal of Medicine*, 321(17), 1200–1203. DOI: 10.1056/NEJM198910263211720 PMID: 2677728

Ansari, H., & Shetty, S. (2023). *Women can accelerate climate action. idr*https:// idronline.org/article/gender/women-can-accelerate-climate-action/

Anyanwu, A., Olorunsogo, T., Abrahams, T. O., Akindote, O. J., & Reis, O. (2024). Data confidentiality and integrity: a review of accounting and cybersecurity controls in superannuation organizations. *Computer Science & IT Research Journal*, 5(1), 237–253. DOI: 10.51594/csitrj.v5i1.735

Aoun, A., Ilinca, A., Ghandour, M., & Ibrahim, H. (2021). A review of Industry 4.0 characteristics and challenges, with potential improvements using blockchain technology. *Computers & Industrial Engineering*, 162, 107746. DOI: 10.1016/j. cie.2021.107746

APA PsycNet. (n.d.). Retrieved April 16, 2023, from https://psycnet.apa.org/

Appel, P., Erlenstedt, M., Kippenberg, J., Kulakow, L., Lüdemann-Dundua, A., Pflüger, F., & Wichmann, S. (2020, October). *Kinderrechte und unternehmerische Sorgfaltspflichten—Wie Bundesregierung und Bundestag Kinderrechte in globalen Lieferketten jetzt schützen sollten—Positionspapier* (Deutsches Komitee für UNICEF e.V., Human Rights Watch Verein zur Wahrung der Menschenrechte e.V., Kindernothilfe e.V., Plan International Deutschland e.V., Save the Children Deutschland e.V., terre des hommes Deutschland e. V., & World Vision Deutschland e.V., Eds.). https:// media.business-humanrights.org/media/documents/positionspapier-kinderrechte.pdf

Arden, N. S., Fisher, A. C., Tyner, K., Lawrence, X. Y., Lee, S. L., & Kopcha, M. (2021). Industry 4.0 for pharmaceutical manufacturing: Preparing for the smart factories of the future. *International Journal of Pharmaceutics*, 602, 120554. DOI: 10.1016/j.ijpharm.2021.120554 PMID: 33794326

Aristotle. (1997). *Politics Book VII & VIII, translated by commentary by Richard Kraut*. Oxford Univ. Press.

Arunachala Gounder (Dead) by LRs v Punnusamy & Ors (2011) Civil Appeal No. 6659/2011

Asadollahi-Yazdi, E., Couzon, P., Nguyen, N. Q., Ouazene, Y., & Yalaoui, F. (2020). Industry 4.0: Revolution or Evolution? *American Journal of Operations Research*, 10(06), 241–268. DOI: 10.4236/ajor.2020.106014

Asfaw, S., & Maggio, G. (2018). Gender, weather shocks and welfare: Evidence from Malawi. *The Journal of Development Studies*, 54(2), 271–291.

Ashok, S., Ramaswamy, S., & Seshadri, S. (2022). A Transdisciplinary Perspective on the Adolescent Consent-Abuse Binary. *Journal of Indian Association for Child and Adolescent Mental Health*, 18(3), 210–213. DOI: 10.1177/09731342221143854

Asian Development Bank. (2020). *Civil Society Brief India.* https://www.adb.org/sites/default/files/publication/879896/civil-society-brief-india.pdf

Asian Development Bank. (2021). *Gender Equality Results Case Study Nepal Gender Equality and Empowerment of Women Project.* ISBN 978-92-9257-479-6 (Print), 978-92-9257-480-2 (e-ISBN) https://www.adb.org/sites/default/files/publication/185563/nepal-gender-equality-and-empowerment-women-project.pdf

Atapattu, S. (2015, October 16). *Human Rights Approaches to Climate Change.* Routledge. http://books.google.ie/books?id=A-e9CgAAQBAJ&pg=PT48&dq=https://doi.org/10.4324/9781315849683&hl=&cd=2&source=gbs_api

Atapattu, S. (2015). *Human rights approaches to climate change: challenges and opportunities*. Routledge.

Auswärtiges Amt, & Interministeriellen Ausschusses Wirtschaft und Menschenrechte. (2017, September). *Nationale Aktionsplan – Umsetzung der UN-Leitprinzipien für Wirtschaft und Menschenrechte (2016-2020).* https://www.auswaertiges-amt.de/blob/297434/8d6ab29982767d5a31d2e85464461565/nap-wirtschaft-menschenrechte-data.pdf

Awan, U., Sroufe, R., & Shahbaz, M. (2021). Industry 4.0 and the circular economy: A literature review and recommendations for future research. *Business Strategy and the Environment*, 30(4), 2038–2060. DOI: 10.1002/bse.2731

Badri, A., Boudreau-Trudel, B., & Souissi, A. S. (2018). Occupational health and safety in the industry 4.0 era: A cause for major concern? *Safety Science*, 109, 403–411. DOI: 10.1016/j.ssci.2018.06.012

Bag, S., & Pretorius, J. H. C. (2022). Relationships between industry 4.0, sustainable manufacturing and circular economy: Proposal of a research framework. *The International Journal of Organizational Analysis*, 30(4), 864–898. DOI: 10.1108/IJOA-04-2020-2120

Ballew, M., ... (2018), "Gender Differences in Public Understanding of Climate Change", https://climatecommunication.yale.edu/publications/gender-differences-in-publicunderstanding-of-climate-change/

Bâ, M. (1979). *So long a letter*. Les Nouvelles Éditions Africaines du Sénégal.

Bandopadhyay, S. (2013, August). Kabuliwalar Bangali Bau. Amar Boi. Retrieved April 1,2023,from https://www.amarboi.com/2013/08/kabuliwalar-bangali-bau-susmita-bandopadhay.html

Banerjee, M. (1999). A Report on the impact of Farakka Barrage on the Human Fabric: A study of the upstream and downstream areas of Farakka Barrage. On behalf of South Asian Network on Dams, River, and People (SANDRP), New Delhi.

Banerjee, D., & Rao, T. S. (2021). Sexual and Reproductive Rights: Time to Ring the Bell? *Journal of Psychosexual Health*, 3(4), 301–304. DOI: 10.1177/26318318211053900

Banerjee, N., & Misra, V. (2024). Role of Gender in Sustainable Mitigation for Climate Change in Agriculture. In *The Climate-Health-Sustainability Nexus: Understanding the Interconnected Impact on Populations and the Environment* (pp. 51–69). Springer Nature Switzerland. DOI: 10.1007/978-3-031-56564-9_3

Banerjee, S. N., & Chakraborty, P. (1983). Some observation on recent trends of shifting of Ganga between Rajmahal and Ahiron. *Journal of the Geographical Society of India*, 24, 318–332.

Bantjes, R., & Trussler, T. (1999). Feminism and the Grass Roots: Women and Environmentalism in Nova Scotia, 1980–1983. *Canadian Review of Sociology/Revue Canadienne de Sociologie, 36*(2), 179–198. DOI: 10.1111/j.1755-618X.1999.tb01274.x

Barge, S. (1997), Situation analysis of medical termination of pregnancy in Gujarat, Maharashtra, Tamil Nadu and Uttar Pradesh (Paper presentation), MTP workshop Ford Foundation, Mumbai.

Basulo-Ribeiro, J., & Teixeira, L. (2024). The Future of Healthcare with Industry 5.0: Preliminary Interview-Based Qualitative Analysis. *Future Internet*, 16(3), 68. DOI: 10.3390/fi16030068

Baxi, S. P., Navsharan. (2021, April 4). *A Case for the Release of All Women Undertrial Prisoners – The Leaflet*. Theleaflet.in. https://theleaflet.in/a-case-for-the-release-of-all-women-undertrial-prisoners/

Bayram, O., Talay, I., & Feridun, M. (2022). Can FinTech promote sustainable finance? Policy lessons from the case of Turkey. *Sustainability (Basel)*, 14(19), 12414. DOI: 10.3390/su141912414

Behrman, J. A., Bryan, E., & Goh, A. (2014). *Gender, climate change, and group-based approaches to adaptation. International Food Policy Research Institute Policy Note. Bishop-Sambrook, C. (2003). Labour saving technologies and practices for farming and household activities in Eastern and Southern Africa: labour constraints and the impact of HIV/AIDS on rural livelihoods in Bondo and Busia Districts, Western Kenya*. IFAD and FAO.

Belur, J., & Singh, B. (2015). Child sexual abuse and the law in India: A commentary. *Crime Science*, 4(1), 26. Advance online publication. DOI: 10.1186/s40163-015-0037-2

Berer, M. (2000). Making abortions safe: A matter of good public health policy and practice. *Bulletin of the World Health Organization*, 78(5), 580–592. PMID: 10859852

Berkes, F. 2001. Religious Traditions and Biodiversity, appeared in *Encyclopedia of Biodiversity*, Vol. 5, pp. 109-120.

Beverley, A. (2020). Uranium Mining, Interdisciplinarity, and Ecofeminism in Donna Smyth's Subversive Elements. *Canadian Literature*, 143, 40–59.

Bhat, A., & Bhatt, M. (2017). Skewed Sex Ratio and Women's Reproductive Rights in India: Issues and Challenges [IJSR]. *International Journal of Scientific Research*, 6, 860–863.

Bhat, P. I. (2017). "Women's Rights As Human Rights": A Dynamic Approach Towards Gender Justice. *CMR University Journal for Contemporary Legal Affairs*, 1(1), 67.

Bhowmik, D. (2022). An Introduction to Climate Fintech. *European Journal of Science. Innovation and Technology*, 2(4), 24–35.

Bhuiyan, A. B., Ali, M. J., Zulkifli, N., & Kumarasamy, M. M. (2020). Industry 4.0: Challenges, opportunities, and strategic solutions for Bangladesh. *International Journal of Business and Management Future*, 4(2), 41–56. DOI: 10.46281/ijbmf.v4i2.832

Bhullar, D., Mittal, P., & Singla, A. (2015). Child sexual abuse in India: A review. *International Journal of Ethics*, 1(2), 38–44. Advance online publication. DOI: 10.18099/ijetv.v1i2.6816

Bin Amin, S., Taghizadeh-Hesary, F., & Khan, F. (2022). Facilitating green digital finance in Bangladesh: Importance, prospects, and implications for meeting the SDGs. In *Green Digital Finance and Sustainable Development Goals* (pp. 143–165). Springer Nature Singapore. DOI: 10.1007/978-981-19-2662-4_7

Blobel, B. (2020, September). Application of industry 4.0 concept to health care. In *pHealth 2020:Proceedings of the 17th International Conference on Wearable Micro and Nano Technologies for Personalized Health* (Vol. 273, p. 23). IOS Press.

BMWK-Bundesministerium für Wirtschaft und Klimaschutz. (2024). *OECD-Leitsätze für multinationale Unternehmen* [Artikel]. OECD-Leitsätze für multinationale Unternehmen. https://www.bmwk.de/Redaktion/DE/Artikel/Aussenwirtschaft/oecd-leitsaetze-fuer-multinationale-unternehmen.html

Bokhari, S. A. A., & Myeong, S. (2023). The influence of artificial intelligence on e-Governance and cybersecurity in smart cities: A stakeholder's perspective. *IEEE Access : Practical Innovations, Open Solutions*, 11, 69783–69797. DOI: 10.1109/ACCESS.2023.3293480

Bongomin, O., Gilibrays Ocen, G., Oyondi Nganyi, E., Musinguzi, A., & Omara, T. (2020). Exponential disruptive technologies and the required skills of industry 4.0. *Journal of Engineering*, 2020, 1–17. DOI: 10.1155/2020/8090521

Books, A. (n.d.). There's Something In The Water. Retrieved April 17, 2023, from https://atlanticbooks.ca/theres-something-in-the-water

Borowski, P. F. (2021). Digitization, digital twins, blockchain, and industry 4.0 as elements of management process in enterprises in the energy sector. *Energies*, 14(7), 1885. DOI: 10.3390/en14071885

Borowski, P. F. (2021). Innovative processes in managing an enterprise from the energy and food sector in the era of industry 4.0. *Processes (Basel, Switzerland)*, 9(2), 381. DOI: 10.3390/pr9020381

Boserup, E. (1970). *Women's role in economic development.* St. Martin's Press.

Braided Way Magazine. (2018, November 3). *Women Are the First Environment.* Interview with Mohawk elder Katsi Cook. Retrieved April 1, 2023, from https://braidedway.org/women-are-the-first-environment-interview-with-mohawk-elder-katsi-cook/S

Brierley, C. (2022). *Extreme weather and climate events likely to drive increase in gender-based violence, UNDRR*https://www.preventionweb.net/news/extreme-weather-and-climate-events-likely-drive-increase-gender-based-violence

Brock, J. K. U., & Von Wangenheim, F. (2019). Demystifying AI: What digital transformation leaders can teach you about realistic artificial intelligence. *California Management Review*, 61(4), 110–134. DOI: 10.1177/1536504219865226

Brune, J. (2020). *Menschenrechte und transnationale Unternehmen: Grenzen und Potentiale des UN Framework for Business and Human Rights* (1st ed.). Nomos. DOI: 10.5771/9783748908746

Bueno, N., & Kaufmann, C. (2022). *Guiding Principles on Business and Human Rights* (SSRN Scholarly Paper 4530382). DOI: 10.4337/9781789903621.guiding.principles.business

Bunn, I. D. (2012). *The right to development and international economic law: legal and moral dimensions*. Bloomsbury Publishing.

Bureau of Police Research and Development. (2003a). *Model Prison Manual for the Superintendence and Management of Prisons in India* (p. Chapter 3). Ministry of Home Affairs.

Bureau of Police Research and Development. (2003b). *Model Prison Manual for the Superintendence and Management of Prisons in India* (p. Chapter 26).

Burns, B. & Lee, A. (2015). *Background Paper – Gender Mainstreaming, EGM/GR-CR/BP.1, UN-Women, UN DESA, UNFCCC Secretariat*https://wedo.org/wp-content/uploads/2016/08/Gender-mainstreaming-practices-in-the-context-of-climate-responses.pdf

Business & Human Rights Resource Centre. (2024a). *UN Guiding Principles*. Business & Human Rights Resource Centre. https://www.business-humanrights.org/en/big-issues/un-guiding-principles-on-business-human-rights/

Business & Human Rights Resource Centre. (2024b). *UN-Leitprinzipien*. Business & Human Rights Resource Centre. https://www.business-humanrights.org/de/schwerpunkt-themen/un-leitprinzipien/

Buteau, S. (2021). *Roadmap for digital technology to foster India's MSME ecosystem—opportunities and challenges*, Europe PMC https://europepmc.org/article/pmc/pmc8662980

Campbell, J. (2017). Gendered Environments in Canada: An Analysis of Women and Environments Magazine, from 1976 to 1997 (Master's thesis, University of Saskatchewan). Retrieved from https://harvest.usask.ca/handle/10388/8521

Campbell-Verduyn, M. (2023). Conjuring a cooler world? Imaginaries of improvement in blockchain climate finance experiments. *Environment and Planning C: Politics and Space*, 23996544231162858.

Cao, Y., Lawson, B., & Pil, F. K. (2023). Social Sustainability and Human Rights in Global Supply Chains. *International Journal of Operations & Production Management*, 44(1), 370–390. DOI: 10.1108/IJOPM-10-2022-0670

Carson, D. K., Foster, J. M., & Tripathi, N. (2013). Child sexual abuse in India: Current issues and research. *Psychological Studies*, 58(3), 318–325. DOI: 10.1007/s12646-013-0198-6

CasimirRaj, M., Director, D.B., & Nepolian, A. (2021). CHILD SEXUAL ABUSE IN INDIA. *International Journal of Approximate Reasoning*, 9, 353–360.

Castañeda Camey, I., Sabater, L., Owren, C., & Boyer, A. E. (2020). Gender-based violence and environment linkages: The violence of inequality. Wen, J. (ed.). Gland, Switzerland: IUCN. 272pp Chanamuto, Nicola J.C. and Stephen J.G. Hall (2015) Gender equality, resilience to climate change, and the design of livestock projects for rural livelihoods. Gender & Development 23(3): 515-530, DOI: DOI: 10.1080/13552074.2015.1096041

Census of India, 2011, A-11 Individual Scheduled Tribe Primary Census Abstract Data and its Appendix, Last modified July 7, 2016.

Center for Reproductive Rights. (2021), https://reproductiverights.org/our-issues/adolescent-sexual-and-reproductive-health-and-rights/

Center for Reproductive Rights. Reproductive Rights in Indian Courts, www.reproductiverights.org

Chakraborty, A., & Sutradhar, R. (2023). Unpaid Work by Women: A State-level Analysis of Evidence from Time-use Survey. *Indian Journal of Human Development*, 17(3), 461–476. DOI: 10.1177/09737030231218147

Chanana-Nag, N., & Aggarwal, P. K. (2020). Woman in agriculture, and climate risks: Hotspots for development. *Climatic Change*, 158(1), 13–27.

Chander, B., Pal, S., De, D., & Buyya, R. (2022). Artificial intelligence-based internet of things for industry 5.0. *Artificial intelligence-based internet of things systems*, 3-45.

Chaudhary, B., & Kr, R. (2018). Medical examination of child of sexual assault under the protection of children from sexual offences (POSCO), act 2012., 2, 50-53. https://doi.org/.DOI: 10.18231/j.ijfmts.2017.016

Chen, Y., & Volz, U. (2021). Scaling up sustainable investment through blockchain-based project bonds. *ADB-IGF Special Working Paper Series "Fintech to Enable Development, Investment, Financial Inclusion, and Sustainability.*

Chen, Y., Lu, Y., Bulysheva, L., & Kataev, M. Y. (2022). Applications of blockchain in industry 4.0: A review. *Information Systems Frontiers*, , 1–15. DOI: 10.1007/s10796-022-10248-7

Chingarande, D. et al. (2020). *Mainstreaming gender into National Adaptation Planning and implementation in SubSaharan Africa, Working Paper No323 CGIAR Research Program on Climate Change, Agriculture and Food Security (CCAFS)* https://cgspace.cgiar.org/server/api/core/bitstreams/16905e9b-6c8a-4a17-8450-fd27d1c475dd/content

Choudhary, V., Satapathy, S., & Sagar, R. (2019). Qualitative study on the impact of child sexual abuse: Perspectives of children, caregivers, and professionals in Indian context. *Journal of Child Sexual Abuse*, 28(4), 489–510. DOI: 10.1080/10538712.2018.1563262 PMID: 30681937

Choudhry, V., Dayal, R., Pillai, D., Kalokhe, A. S., Beier, K., & Patel, V. (2018). Child sexual abuse in India: A systematic review. *PLoS One*, 13(10), e0205086. DOI: 10.1371/journal.pone.0205086 PMID: 30300379

Christian Community Welfare Council of India v. Govt of Maharashtra, (Supreme Court of India 1996).

Chueca Vergara, C., & Ferruz Agudo, L. (2021). Fintech and sustainability: Do they affect each other? *Sustainability (Basel)*, 13(13), 7012. DOI: 10.3390/su13137012

Ciminelli, G., Schwellnus, C., & Stadler, B. (2021), "Sticky floors or glass ceilings? The role of human capital, working time flexibility and discrimination in the gender wage gap", *OECD Economics Department Working Papers*, No. 1668, OECD Publishing, Paris. DOI: 10.1787/18151973

Cioffi, R., Travaglioni, M., Piscitelli, G., Petrillo, A., & De Felice, F. (2020). Artificial intelligence and machine learning applications in smart production: Progress, trends, and directions. *Sustainability (Basel)*, 12(2), 492. DOI: 10.3390/su12020492

Clayton, A., Josefsson, C., & Wang, V. (2017), "Quotas and Women's Substantive Representation: Evidence from a Content Analysis of Ugandan Plenary Debates", *Politics & Gender*, Vol. 13/2, pp. 276-304, . https://dx.doi.org/. https://www.oxfamamerica.org/explore/research-publications/researchbackgrounderwash-gender-report-puerto-rico/DOI: 10.1017/S1743923X16000453

Climate change and water-related disasters. (n.d.). UNEP - UN Environment Programme. Retrieved April 2, 2024, from https://www.unep.org/topics/fresh-water/disasters-and-climate-change/climate-change-and-water-related-disasters

Climate Impacts Tracker Asia. (2023). Gender Equality at COP28: Advancing Climate Justice for Women and Girls. Retrieved from https://www.climateimpactstrackerasia.org/gender-equality-cop28

Climate Investment Funds. (n.d.). *Gender.* (Accessed on: 17-02-2024) https://www.cif.org/mainstreaming-gender

Cole, S., Tallman, P., Salmon-Mulanovich, G., & Rusyidi, B. (2024). Water insecurity is associated with gender-based violence: A mixed-methods study in Indonesia. *Social Science & Medicine*, 344, 116507. DOI: 10.1016/j.socscimed.2023.116507 PMID: 38340386

Collins, L. *The complicated life of the abortion pill,* New Yorker (Last updated on July 5, 2022), https://www.newyorker.com/science/annals-of-medicine/emile-baulieu-the-complicated-life-of-the-abortion-pill/amp

Collins, T. (2023). *Global Child Forum Corporate Guide: Children's Participation: How to involve children in decision-making* [Report]. Toronto Metropolitan University. DOI: 10.32920/23913060.v1

Collins, T., & Guevara, G. (2014). Some Considerations for Child Rights Impact Assessment (CRIAs) of Business. *Revue Générale de Droit*, 44(1), 153–192. DOI: 10.7202/1026401ar

Commer, C. O. (2024). When Conscious Consumerism Backfires: A Critical Study on Social Exclusion, Partnerships, and White Saviorism. *Sociology*, 14(1), 9–26.

Contagion of Covid 19 Virus in Prisons, Suo Motu Writ Petition(C) No.1/2020, (Supreme Court of India, May 2021).

Convention on the Rights of the Child. (1990). https://www.ohchr.org/en/instruments-mechanisms/instruments/convention-rights-child

Cooper, B. (2021, October 18). *"the lower the income, the higher the prevalence for abuse": Domestic violence linked to poverty*. Indianapolis Recorder. https://indianapolisrecorder.com/the-lower-the-income-the-higher-the-prevalence-for-abuse-domestic-violence-linked-to-poverty/

COP28 Launches Partnership to Support Women's Economic Empowerment and Ensure a Gender-Responsive Just Transition at COP28 Gender Equality Day," COP28 News, December 2023

Cornwall, A. (2003). Whose voices? Whose choices? Reflections on gender and participatory development. *World Development*, 31(8), 1325–1342.

Costa, E. (2024). Industry 5.0 and SDG 9: A symbiotic dance towards sustainable transformation. *Sustainable Earth Reviews*, 7(1), 4. DOI: 10.1186/s42055-024-00073-y

Craig, E. (2020). Child's play or sexual abuse? Reviewing the efficacy of the justice framework in dealing with child on child sexual abuse in the United Kingdom. *Journal of Child Sexual Abuse*, 29(6), 734–748. DOI: 10.1080/10538712.2020.1719448 PMID: 32286188

Da Costa, K. (2014). Can the observance of human rights of individuals enhance their resilience to cope with natural disasters? *Procedia Economics and Finance*, 18, 62–70. DOI: 10.1016/S2212-5671(14)00914-9

Dalyan, M., Syarifuddin, , Yulandari, , Mastang, , Suma, M., Sosrohadi, S., & Andini, C. (2024). Harmony and Sustainability: Traditional Ecological Knowledge Systems of the Kaluppini Indigenous People. *International Journal of Religion*, 5(6), 82–92. DOI: 10.61707/tdyqck03

Das, J. D., Dutta, T., & Saraf, A. K. (2007). Remote sensing and GIS application in change detection of the Barak River channel, N.E. India. *Photonirvachak (Dehra Dun)*, 35(4), 301–312. DOI: 10.1007/BF02990786

Das, R., & Samanta, G. (2023). Impact of floods and river-bank erosion on the riverine people in Manikchak Block of Malda District, West Bengal. *Environment, Development and Sustainability*, 25(11), 13595–13617. DOI: 10.1007/s10668-022-02648-1 PMID: 36118734

Das, T., Haldar, K., Sarkar, S., Borderon, K., Kienberger, D., Das Gupta, M., Kundu, S., & Guha-Sapir, D. (2017). Impact of riverbank erosion: A case study. *Australasian Journal of Disaster and Trauma Studies*, 21(2), 73–81. https://trauma.massey.ac.nz/issues/2017-2/AJDTS_21_2_Das.pdf

Datar v. Union of India: Why This case is Important, (2020) HRLN, www.reproductiverights.org/sites/crr.civicactions.net/files/documents/Datar_v_India.pdf

Datar, N. (2020), India Moves to Liberalization of Abortion Rules, *Businessline*, https://www.thehindubusinessline.com/specials/pulse/india-moves-to-liberalisation-of-abortion-rules/article31189432.ece

Datta, B., & Misra, G. (2000). Advocacy for Sexual and Reproductive Health: The Challenge in India. *Reproductive Health Matters*, 8(16), 24–34. DOI: 10.1016/S0968-8080(00)90184-8 PMID: 11424247 de la Peña Zarzuelo, I., Soeane, M. J. F., & Bermúdez, B. L. (2020). Industry 4.0 in the port and maritime industry: A literature review. *Journal of Industrial Information Integration*, 20, 100173. DOI: 10.1016/j.jii.2020.100173

Decision 1/CP.16, Cancun Agreements: Outcome on the implementation of agreed outcomes related to climate change (FCCC/CP/2010/9 & Add.1)

Decision 1/CP.18, Doha Climate Gateway (FCCC/CP/2012/8 & Add1)

Decision 1/CP.20, Lima Call for Climate Action (FCCC/CP/2014/10 & Add1)

Decision 1/CP.7, The Marrakesh Ministerial Declaration, para3 (FCC/CP/2001/13 & Add.1)

Dell'Erba, M. (2021). Sustainable digital finance and the pursuit of environmental sustainability. *Sustainable Finance in Europe: Corporate Governance, Financial Stability and Financial Markets*, 61-81.

Desai, B., & Mandal, M. (2021). Role of Climate Change in Exacerbating Sexual and Gender Based Violence against Women: A New Challenge for International Law. *Environmental Policy and Law*, 51(3), 137–157. DOI: 10.3233/EPL-210055

Devi, A. (2013.). Satyabati trilogy.

Devi, A. (n.d.). Sociology Group. Retrieved April 3, 2023, from https://www.sociologygroup.com/ashapurna-devi/

Devi, A. Feminist Writer. (2019, March 11). Feminism in India. Retrieved April 1,2023,from https://feminisminindia.com/2019/03/11/ashapurna-devi-feminist-writer-bengali/

Devi, A. The Voice of Women. Get Bengal. Retrieved April 1, 2023, from https://www.getbengal.com/details/ashapurna%20devi%20the%20voice%20of%2women

Devi, A. (1964). *First promise*. Mitra & Ghosh Publishers.

Devika Biswas v. Union of India, (2016) 10 SCC 726.

Dewan, A., Corner, R., Saleem, A., Rahman, M. M., Haider, M. R., Rahman, M. M., & Sarker, M. H. (2017). Assessing channel changes of the Ganges-Padma River system in Bangladesh using Landsat and hydrological data. *Geomorphology*, 276, 257–279. DOI: 10.1016/j.geomorph.2016.10.017

Dinesh Arokia Raj, A., Jha, R. R., Yadav, M., Sam, D., & Jayanthi, K. (2024). Role of Blockchain and Watermarking Toward Cybersecurity. In *Multimedia Watermarking: Latest Developments and Trends* (pp. 103–123). Springer Nature Singapore. DOI: 10.1007/978-981-99-9803-6_6

Diro, A., Zhou, L., Saini, A., Kaisar, S., & Hiep, P. C. (2024). Leveraging zero knowledge proofs for blockchain-based identity sharing: A survey of advancements, challenges and opportunities. *Journal of Information Security and Applications*, 80, 103678. DOI: 10.1016/j.jisa.2023.103678

Dobrich, E. O. R. (2024). Perspectives on Advancing Gender and Environmental Justice: Implications and Applications of a Conceptual Model for Supporting Women's Environmental Leadership. *Journal of Academic Perspectives Volume*, 3, 130.

Dorfleitner, G., & Braun, D. (2019). Fintech, digitalization and blockchain: possible applications for green finance. *The rise of green finance in Europe: opportunities and challenges for issuers, investors and marketplaces*, 207-237.

Dube, M. W. (2024). African Eco-feminisms: African Women Writing Earth, Gender and the Sacred. In *Ecofeminist Perspectives from African Women Creative Writers: Earth, Gender, and the Sacred* (pp. 3–33). Springer International Publishing. DOI: 10.1007/978-3-031-48509-1_1

Dutta, S. (2019). *Supporting last-mile women energy entrepreneurs: What works and what does not, Energia*https://www.energia.org/assets/2019/01/Supporting-Last-Mile-Women-Entrepreneurs.pdf

Duyck, S. et al. (n.d.). *Delivering on the Paris Promises: Combating Climate Change while Protecting Rights, Recommendations for the Negotiations of the Paris Rule Book, UNFCCC*https://unfccc.int/sites/default/files/903.pdf

Dwivedi, Y. K., Hughes, L., Ismagilova, E., Aarts, G., Coombs, C., Crick, T., Duan, Y., Dwivedi, R., Edwards, J., Eirug, A., Galanos, V., Ilavarasan, P. V., Janssen, M., Jones, P., Kar, A. K., Kizgin, H., Kronemann, B., Lal, B., Lucini, B., & Williams, M. D. (2021). Artificial Intelligence (AI): Multidisciplinary perspectives on emerging challenges, opportunities, and agenda for research, practice and policy. *International Journal of Information Management*, 57, 101994. DOI: 10.1016/j.ijinfomgt.2019.08.002

Eastin, J. (2018). Climate change and gender equality in developing states. *World Development*, 107, 289–305. DOI: 10.1016/j.worlddev.2018.02.021

Eco-Feminist Appropriations of Indigenous Feminisms and Environmental Violence. (n.d.). The Feminist Wire. Retrieved April 16, 2023, from https://thefeministwire.com/2018/10/eco-feminist-appropriations-of-indigenous-feminisms-and-environmental-violence/

El Zein, B., Elrashidi, A., Dahlan, M., Al Jarwan, A., & Jabbour, G. (2024). Nano and Society 5.0: Advancing the Human-Centric Revolution.

Elahi, K. M., John, R. R., & Rogge, J. R. (1990). *Riverbank erosion, flood and population displacement in Bangladesh: a report on the riverbank erosion impact study. Riverbank Erosion Impact Study (REIS)*. Jahangirnagar University.

Environmental racism and Black environmentalism in Canada. (n.d.). Smart Prosperity Institute. Retrieved April 1, 2023, from https://institute.smartprosperity.ca/library/research/environmental-racism-and-black-environmentalism-canada

Erdem Türkelli, G. (2020). *Children's Rights and Business: Governing Obligations and Responsibility*. DOI: 10.1017/9781108681841

Escape from Taliban. (2003). Wikipedia. Retrieved April 4, 2023, from https://en.wikipedia.org/wiki/Escape_from_Taliban

European Commission. (2024a, April 23). *Recommendation on developing and strengthening integrated child protection systems in the best interests of the child | European Commission—C(2024) 2680 final*. https://commission.europa.eu/document/36591cfb-1b0a-4130-985e-332fd87d40c1_en

European Commission. (2024b, April 30). *Justice & Consumers—Recommendation on developing and strengthening integrated child protection systems in the best interests of the child*. https://ec.europa.eu/newsroom/just/items/828836/en

Fass, P. S. (2011). A Historical Context for the United Nations Convention on the Rights of the Child. *The Annals of the American Academy of Political and Social Science*, 633(1), 17–29. DOI: 10.1177/0002716210382388

Federation of Obstetrics and Gynecological Societies of India (FOGSI). (2002), Safe abortions save lives: understanding the MTP Act (Paper Presentation), MTP Committee, Medico-Legal Committee FOGSI, New Delhi.

Felsberger, A., Qaiser, F. H., Choudhary, A., & Reiner, G. (2022). The impact of Industry 4.0 on the reconciliation of dynamic capabilities: Evidence from the European manufacturing industries. *Production Planning and Control*, 33(2-3), 277–300. DOI: 10.1080/09537287.2020.1810765

Fenwick, M., & Vermeulen, E. P. (2020). Banking and regulatory responses to FinTech revisited-building the sustainable financial service'ecosystems' of tomorrow. *Singapore Journal of Legal Studies*, (Mar 2020), 165-189.

Finance, A. T. C. C. (2015). *Industry 4.0 Challenges and solutions for the digital transformation and use of exponential technologies. Finance, audit tax consulting corporate*. Swiss.

Finn, S., Herne, M., & Castille, D. (2017, August 29). The Value of Traditional Ecological Knowledge for the Environmental Health Sciences and Biomedical Research. *Environmental Health Perspectives*, 125(8), 085006. DOI: 10.1289/EHP858 PMID: 28858824

Flint, C. G., Robinson, E. S., Kellogg, J., Ferguson, G., BouFajreldin, L., Dolan, M., Raskin, I., & Lila, M. A. (2011). Promoting wellness in Alaskan villages: Integrating traditional knowledge and science of wild berries. *EcoHealth*, 8(2), 199–209. DOI: 10.1007/s10393-011-0707-9 PMID: 21915737

Fottrell, D. (2021). *Revisiting Children's Rights: 10 Years of the UN Convention on the Rights of the Child*. Brill.

Fraga-Lamas, P., Lopes, S. I., & Fernández-Caramés, T. M. (2021). Green IoT and edge AI as key technological enablers for a sustainable digital transition towards a smart circular economy: An industry 5.0 use case. *Sensors (Basel)*, 21(17), 5745. DOI: 10.3390/s21175745 PMID: 34502637

Fury for the Sound: The Women at Clayoquot. (n.d.). Moving Images Distribution. Retrieved April 1, 2023, from https://www.movingimages.ca/store/products.php?pid=1312

Gadekar, R., Sarkar, B., & Gadekar, A. (2022). Key performance indicator based dynamic decision-making framework for sustainable Industry 4.0 implementation risks evaluation: Reference to the Indian manufacturing industries. *Annals of Operations Research*, 318(1), 189–249. DOI: 10.1007/s10479-022-04828-8 PMID: 35910040

Ganatra, B., & Hirve, S. (2002). Induced abortions in a rural community in Western Maharashtra: Prevalence and patterns. *Reproductive Health Matters*, 10(19), 76–85. DOI: 10.1016/S0968-8080(02)00016-2 PMID: 12369334

Gaur, M. K. 2006. Traditional Environmental Knowledge: Barriers in Documenting Conservation Practices in Thar Desert. *Nat. Conf.: Environment Planning & Conservation by Traditional Methods*. Dec. 22-23, 2006. Govt. College, Ajmer. pp. 125-134.

Gaur, M. K., & Gaur, H. (2004). *Combating Desertification: Building on Traditional Knowledge System of the Thar Desert Communities. Internat. J Environmental Monitoring and Assessment*. Kluwer Academic Publishers.

GCBHR, & CIDE. (2022, August 31). *Addressing Children's Rights in Business. An Assessment from Switzerland and Liechtenstein*. https://gcbhr.org/insights/2022/08/addressing-childrens-rights-in-business-an-assessment-from-switzerland-and-liechtenstein-publication

Geschäftsstelle Deutsches Global Compact Netzwerk (DGCN). (2012, August). *Kinderrechte und unternehmerisches Handeln—Grundsätze zum Schutz und zur Förderung von Kinderrechten durch Unternehmen.* https://www.unicef.de/informieren/materialien/kinderrechte-unternehmerisches-handeln/194576

ghdx.healthdata.org. https://ghdx.healthdata.org/record/india-hmis-standard-report-data-itemwise-monthly-2016-2017 (Last updated on Feb 28, 2022).

Ghosh, A., & Khaitan, N. (2017). A Womb of One's Own: Privacy and Reproductive Rights. *Engage!*

Glemarec, Y. et al. (2016). *Leveraging Co-Benefits Between Gender Equality And Climate Action For Sustainable Development*, UN Women https://unfccc.int/files/gender_and_climate_change/application/pdf/leveraging_cobenefits.pdf

Global Affairs Canada. (2017). Canada's Feminist International Assistance Policy. Retrieved April 1, 2023, from https://www.international.gc.ca/world-monde/issues_development-enjeux_developpement/priorities-priorites/policy-politique.aspx?lang=eng

Goldin, C. (1994). The U-shaped female labour force functions in economic development and economic history.

Goldin, C. (1986). The economic status of women in the early republic: Quantitative evidence. *The Journal of Interdisciplinary History*, 16(3), 375–404. DOI: 10.2307/204496

Gone, J. P. (2012). Indigenous traditional knowledge and substance abuse treatment outcomes: The problem of efficacy evaluation. *The American Journal of Drug and Alcohol Abuse*, 38(5), 493–497. DOI: 10.3109/00952990.2012.694528 PMID: 22931084

Google Trends. https://trends.google.com/trends/

Government of India (1996), The Prenatal Diagnostic Techniques (Regulation and Prevention of Misuse) Act, 1994, Gazette of India, 1996.

Government of India (2003), Medical Termination of Pregnancy Rules and Regulations, Vide GSR 485(E) and 486(E), Gazette of India, 2003.

Government of India. (1966), Report of the Shah Committee to study the question of legalization of abortion, Ministry of Health and Family Planning, 1966.

Government of India. (1996), Family Welfare Programme in India: Yearbook 1994–95, Ministry of Health and Family Welfare, 1996.

Government of India. (2000). *National Population Policy, 2000.* Department of Family Welfare, Ministry of Health and Family Welfare.

Government of India. (2009). *Annexure II, Ministry of Home Affairs.*

Government of India Annual Report 2020-21, Ministry of Micro, Small, and Medium Enterprises, https://msme.gov.in/sites/default/files/MSME-ANNUAL-REPORT-ENGLISH%202020-21.pdf

Gremaud, A., & Kochtcheeva, L. (2024). The role of non-state actors in promoting environmental justice: A comparative study of Kenya and South Africa. *GeoJournal*, 89(3), 110.

Gupta, D. N. S. (1951). *An outline of the river problems of West Bengal.* General Printers & Publishers.

Gupta, R. (2023). Industry 4.0 adaption in indian banking Sector—A review and agenda for future research. *Vision (Basel)*, 27(1), 24–32. DOI: 10.1177/0972262921996829 PMID: 36977304

Gupta, S., Pirsch, J., & Girard, T. (2010). An Empirical Examination of a Multinational Ethical Dilemma: The Issue of Child Labor. *Journal of Global Marketing*, 23(4), 288–305. DOI: 10.1080/08911762.2010.504519

guttmacher.org. https://www.guttmacher.org/news-release/2017/national-estimate-abortion-india-released (Last updated on Dec.11, 2017).

Hallsworth, K. (2022). Problems of international legal regulation of protection of labor rights of children and adolescents. *Vìsnik*, 121(121), 5–5. DOI: 10.17721/1728-2195/2022/2.121-1

Harris, A. (2018). A conversation with masterminds in blockchain and climate change. In *Transforming climate finance and green investment with blockchains* (pp. 15–22). Academic Press. DOI: 10.1016/B978-0-12-814447-3.00002-1

Hartung, P. (2020, November). *The children's rights-by-design standard for data use by tech companies.* UNICEF. https://www.unicef.org/globalinsight/media/1286/file/%20UNICEF-Global-Insight-DataGov-data-use-brief-2020.pdf

Hassan, M. K., Rabbani, M. R., & Ali, M. A. M. (2020). Challenges for the Islamic Finance and banking in post COVID era and the role of Fintech. *Journal of Economic Cooperation & Development*, 41(3), 93–116.

Hassan, N. A., Ratna, S. J., Hassan, M., & Tamanna, S. (2017). Remote Sensing and GIS for the Spatio-temporal Change Analysis of the East and the West River Bank Erosion and Accretion of Jamuna River (1995-2015), Bangladesh. *Journal of Geoscience and Environment Protection*, 5(9), 79–92. DOI: 10.4236/gep.2017.59006

Hassoun, A., Aït-Kaddour, A., Abu-Mahfouz, A. M., Rathod, N. B., Bader, F., Barba, F. J., Biancolillo, A., Cropotova, J., Galanakis, C. M., Jambrak, A. R., Lorenzo, J. M., Måge, I., Ozogul, F., & Regenstein, J. (2023). The fourth industrial revolution in the food industry—Part I: Industry 4.0 technologies. *Critical Reviews in Food Science and Nutrition*, 63(23), 6547–6563. DOI: 10.1080/10408398.2022.2034735 PMID: 35114860

Hassoun, A., Prieto, M. A., Carpena, M., Bouzembrak, Y., Marvin, H. J., Pallares, N., Barba, F. J., Punia Bangar, S., Chaudhary, V., Ibrahim, S., & Bono, G. (2022). Exploring the role of green and Industry 4.0 technologies in achieving sustainable development goals in food sectors. *Food Research International*, 162, 112068. DOI: 10.1016/j.foodres.2022.112068 PMID: 36461323

Hesketh, T., Gamlin, J., & Woodhead, M. (2006). Policy in child labour. *Archives of Disease in Childhood*, 91(9), 721–723. DOI: 10.1136/adc.2006.096263 PMID: 16923854

High Court on its Own Motion v. State of Maharashtra, Suo Moto P.I.L. No. 1 of 2016.

Hirve, S. (2004). Abortion Policy in India: Lacunae and Future Challenges. *Reproductive Health Matters*, 12(24, Supplement), 114–121. DOI: 10.1016/S0968-8080(04)24017-4 PMID: 15938164

Hoang, T. G., Nguyen, G. N. T., & Le, D. A. (2022). Developments in financial technologies for achieving the Sustainable Development Goals (SDGs): FinTech and SDGs. In *Disruptive technologies and eco-innovation for sustainable development* (pp. 1–19). IGI Global. DOI: 10.4018/978-1-7998-8900-7.ch001

Hofman, K. (2021). A defense of ecofeminism: Re-examining the Clayoquot Sound peace camp. Archived from the original on April 1, 2023. Retrieved April 1, 2023, from https://kaylahofman.com/defense-of-ecofeminism-clayoquot-sound

Hoover, E., Cook, K., Plain, R., Sanchez, K., Waghiyi, V., Miller, P., Dufault, R., Sislin, C., & Carpenter, D. O. (2012). Indigenous Peoples of North America: Environmental Exposures and Reproductive Justice. *Environmental Health Perspectives*, 120(12), 1645–1649. DOI: 10.1289/ehp.1205422 PMID: 22899635

Houde, N. (2007). The six faces of traditional ecological knowledge: Challenges and opportunities for Canadian co-management arrangements. *Ecology and Society*, 12(2), 34. DOI: 10.5751/ES-02270-120234

Howlader, N., & Rahman, M. A. (2016). Riverbank Erosion and Its Impact: A Study on Ganges River Affected Area. *Jagannath University Journal of Social Science*, 4(1-2), 73–85.

https://sustainabledevelopment.un.org/content/documents/2322UN%20Women%20Analysis%20on%20Women%20and%20SDGs.pdf

https://timesofindia.indiatimes.com/blogs/developing-contemporary-india/womens-land-rights-in-india-whats-missing-from-our-land-laws/

Human Rights Committee's View on Teitiota v. New Zealand, CCPR/C/127/D/2728/2016.

Human Rights Council. (2011) Resolution 18/22 Human rights and climate change, A/HRC/ RES/18/22, (2011). In www.ohchr.org. UN Human Rights Council. Retrieved April 2, 2024, from https://www.ohchr.org/sites/default/files/Documents/Issues/ClimateChange/A.HRC.RES.18.22.pdf

Human Rights Council. Res. 7/23, U.N. Doc. A/HRC/RES/7/23 (Mar. 28, 2008).

Human Rights Watch. World Report 2022, https://www.hrw.org/world-report/2022 (last visited July 8, 2024).

Hussainara Khatoon (iv) & Ors. v. Home Secretary. Bihar, (Supreme Court of India, 1979).

IDS Bulletin. (2020). *Gender and Energy: Opportunities for All,* Volume 51, Number 1, https://www.icrw.org/wp-content/uploads/2020/02/IDS-Bulletin-Gender-and-Energy-Opportunities-for-All.pdf

IEA. (n.d.). *The path to limiting global warming to 1.5 °C has narrowed, but clean energy growth is keeping it open.* https://www.iea.org/news/the-path-to-limiting-global-warming-to-1-5-c-has-narrowed-but-clean-energy-growth-is-keeping-it-open

IFC. (2020), Report: *Financial Inclusion for Women-owned MSMEs in India, International Finance Corporation World Bank Group*, https://www.ifc.org/en/insights-reports/2010/financial-inclusion-for-women-owned-msmes-in-india

ILO. (2014). *The gender dimension: Integrating the gender perspective in OSH policies* (ilo.org)

ILO. (2021, June 10). *Child Labour: Global estimates 2020, trends and the road forward.* https://www.ilo.org/publications/major-publications/child-labour-global-estimates-2020-trends-and-road-forward

ILO. (2021a). *Violence and harassment in the world of work: A guide on Convention.* No190 and Recommendation No206

ILO. (2022b). *Care at work: Investing in care leave and services for a more gender equal world of work.*

ILO. (2022h). *Preparing Future Generations of Women for New Job Demands: Skilling, Re-Skilling, Digitalization and Automation,* OP.3_ILO.pdf (unwomen.org)

ILO. (2023). *Achieving a just transition towards environmentally sustainable economies and societies for all.* International Labour Office.

ILO. (2024). *Gender Equality and Inclusion for a Just Transition in Climate Action a Practical Guide.* International Labour Organization. https://www.ilo.org/media/482386/download

ILO. (2024, August 1). *ILO Conventions on child labour | International Labour Organization.* ILO Conventions on Child Labour | International Labour Organization. https://www.ilo.org/international-programme-elimination-child-labour-ipec/what-child-labour/ilo-conventions-child-labour

Intergovernmental Panel on Climate Change. Climate Change 2022: Impacts, Adaptation and Vulnerability, Summary for Policymakers (2022), https://www.ipcc.ch/report/ar6/wg2/ (last visited July 8, 2024).

International Council on Human Rights Policy. Climate Change and Human Rights: A Rough Guide (2008), https://www.ichrp.org/files/reports/45/136_report.pdf (last visited July 8, 2024).

International Covenant on Civil and Political Rights (ICCPR), Article 3 (1966).

International Labor Organization. (2017). Green Initiative. *Policy Brief.*

International Labor Organization. (2018). *Greening with Jobs.*

International Labor Organization. (2018). *The employment impact of climate change adaptation.*

IPCC. 2022: Summary for Policymakers [H.-O. Pörtner, D.C. Roberts, E.S. Poloczanska, K. Mintenbeck, M. Tignor, A. Alegría, M. Craig, S. Langsdorf, S. Löschke, V. Möller, A. Okem (eds.)]. In: *Climate Change 2022: Impacts, Adaptation, and Vulnerability.* Contribution of Working Group II to the Sixth Assessment Report of the Intergovernmental Panel on Climate Change [H.-O. Pörtner, D.C. Roberts, M. Tignor, E.S. Poloczanska, K. Mintenbeck, A. Alegría, M. Craig, S. Langsdorf, S. Löschke, V. Möller, A. Okem, B. Rama (eds.)]. Cambridge University Press, Cambridge, UK and New York, NY, USA, pp. 3-33, DOI: 10.1017/9781009325844.001

IRENA. (2019). *Renewable Energy: A Gender Perspective.*

IRENA. (2022). *World Energy Transitions Outlook 2022: 1.5°C Pathway.*

Islam, M. F., & Rashid, A. N. M. B. (2011). Riverbank Erosion Displaces in Bangladesh: Need for Institutional Responses and Policy Intervention. *Bangladesh Journal of Bioethics*, 2(2), 4–19. DOI: 10.3329/bioethics.v2i2.9540

Ivanov, D., Dolgui, A., & Sokolov, B. (2019). The impact of digital technology and Industry 4.0 on the ripple effect and supply chain risk analytics. *International Journal of Production Research*, 57(3), 829–846. DOI: 10.1080/00207543.2018.1488086

Jaiwant, S. V., & Kureethara, J. V. (2023). Green Finance and Fintech: Toward a More Sustainable Financial System. In *Green Finance Instruments, FinTech, and Investment Strategies: Sustainable Portfolio Management in the Post-COVID Era* (pp. 283–300). Springer International Publishing. DOI: 10.1007/978-3-031-29031-2_12

Jakka, G., Yathiraju, N., & Ansari, M. F. (2022). Artificial Intelligence in Terms of Spotting Malware and Delivering Cyber Risk Management. *Journal of Positive School Psychology*, 6(3), 6156–6165.

Javaid, M., Haleem, A., Singh, R. P., Khan, S., & Suman, R. (2021). Blockchain technology applications for Industry 4.0: A literature-based review. *Blockchain: Research and Applications*, 2(4), 100027.

Javaid, M., Haleem, A., Singh, R. P., Suman, R., & Gonzalez, E. S. (2022). Understanding the adoption of Industry 4.0 technologies in improving environmental sustainability. *Sustainable Operations and Computers*, 3, 203–217. DOI: 10.1016/j.susoc.2022.01.008

Jayaram, V. *Hinduism and Abortions*, Hindu Website (Last updated on Sept 01, 2018, 11:35 AM), https://www.hinduwebsite.com/hinduism/h_abortions.asp

Jhunjhunwala, P(2023), Bridging the Gaps to Accelerate Women's Entrepreneurship, UNDP https://www.undp.org/india/blog/bridging-gaps-accelerate-womens-entrepreneurship

Jodoin, S., Savaresi, A., & Wewerinke-Singh, M. (2021). Rights-based approaches to climate decision-making. *Current Opinion in Environmental Sustainability*, 52, 45–53. DOI: 10.1016/j.cosust.2021.06.004

Juliana v. United States, 217 F. Supp. 3d 1224 (D. Or. 2016).

K.S. Puttaswamy v. Union of India, (2017) 10 SCC 1.

Kabeer, N. 2010. Can the MDGs provide a pathway to social justice? The challenge of intersecting inequalities. Child poverty and inequality new perspectives, 57. https://www.undp.org/content/dam/turkey/docs/Publications/mdgs/MDGs%20and%20Inequalities%20Final%20Report.pdf

Kakade, K. S., Kulkarni, A. M., & Naik, K. R. (2024). *The Role of Women Leadership in Environmental Management*. Opportunities and Challenges for Women Leaders in Environmental Management.

Kalaiarasi, H., & Kirubahari, S. (2023). Green finance for sustainable development using blockchain technology. In *Green Blockchain Technology for Sustainable Smart Cities* (pp. 167–185). Elsevier. DOI: 10.1016/B978-0-323-95407-5.00003-7

Kamla Neti (Dead) Through Legal Representatives(Appellant) vs. Special Land Acquisition Officer and Ors. (Respondents), (2023) 3 SCC 528

Kampa, D. (2020, June 27). So long a letter: So long a history—Mariama Ba, Muslimness, and women's rights. Reading Under the Olive Tree. Retrieved April 4, 2023, from https://readingundertheolivetree.com/2020/06/27/so-long-a-letter-so-long-a-history-mariama-ba-muslimness-and-womens-rights/

Kang, J. (2021). *UNICEF Deutschland: Empfehlungen zur Dialogfassung der Deutschen Nachhaltigkeitsstrategie 2021*. UNICEF.

Kapoor, A., Alcayna, T., Boer, T. d., Gleason, K., Bhandari, B., & Heinrich, D. (2021). *Climate Change Impacts on Health and Livelihoods: Fiji Assessment.*https://www.climatecentre.org/wp-content/uploads/RCRC_IFRC-Country-assessments-FIJI.pdf

Karim, A. H. M. Z. (n.d.). (2914). Flood and Riverbank Erosion Displacees: Their Indigenous Survival Strategies in Two Coastal Villages in Bangladesh. *Asian Social Science*, 10(4), 16–26.

Karkal, M. (1991). Abortion Laws and the Abortion Situation in India. *Issues in Reproductive and Genetic Engineering: Journal of International Feminist*, 4(3), 223–230. PMID: 11651217

Kasowaki, L., & Ahmet, S. (2024). *Shielding the Virtual Ramparts: Understanding Cybersecurity Essentials* (No. 11700). EasyChair.

Kaur, J. (2012). The role of litigation in ensuring women's reproductive rights: An analysis of the Shanti Devi judgement in India. *Reproductive Health Matters*, 20(39), 21–30. DOI: 10.1016/S0968-8080(12)39604-3 PMID: 22789079

Kennedy, E. H., & Kmec, J. (2018). Reinterpreting the gender gap in household pro-environmental behaviour. *Environmental Sociology*, 4(3), 299–310.

Kharb, A., Bhandari, S., Moitinho de Almeida, M., Castro Delgado, R., Arcos González, P., & Tubeuf, S. (2022, September 13). Valuing Human Impact of Natural Disasters: A Review of Methods. *International Journal of Environmental Research and Public Health*, 19(18), 11486. DOI: 10.3390/ijerph191811486 PMID: 36141766

Klasen, S., & Pieters, J. (2013). What Explains the Stagnation of Female Labor Force Participation in Urban India? *The World Bank Economic Review*, 29(3), 2015.

Klinger, R., Krajewski, M., Krebs, D., & Hartmann, C. (2016). *Verankerung menschenrechtlicher Sorgfaltspflichten von Unternehmen im deutschen Recht: Gutachten*. Amnesty International.

Knox, J. H. (2020). Constructing the human right to a healthy environment. *Annual Review of Law and Social Science*, 16(1), 79–95.

Koehler, G. (2016). Tapping the sustainable development goals for progressive gender equity and equality policy? *Gender and Development*, 24(1), 53–68. DOI: 10.1080/13552074.2016.1142217

Koengkan, M., Fuinhas, J. A., Auza, A., Castilho, D., & Kaymaz, V. (2024). Environmental Governance and Gender Inclusivity: Analyzing the Interplay of PM2. 5 and Women's Representation in Political Leadership in the European Union. *Sustainability*, 16(6), 2492.

Kolieb, J. (2020). *Protecting the Most Vulnerable: Embedding Children's Rights in the Business and Human Rights Project* (SSRN Scholarly Paper 3834369). https://papers.ssrn.com/abstract=3834369 DOI: 10.4337/9781786436405.00026

Kolk, A., & van Tulder, R. (2002). Child Labor and Multinational Conduct: A Comparison of International Business and Stakeholder Codes. *Journal of Business Ethics*, 36(3), 291–301. DOI: 10.1023/A:1014009313508

Kottis, A. P. (1990). Shifts over time and regional variation in women's labour force participation rates in a developing economy: The case of Greece. *Journal of Development Economics*, 33(1), 117–132. DOI: 10.1016/0304-3878(90)90009-Z

KS Puttaswamy v. Union of India AIR 2017 SC 4161.

Kulkarni, U. K., Kulkarni, K. V., Kokre, R. N., & Daundkar, B. B. (2016). Forensic study on child sexual abuse under POCSO act. *European Journal of Biomedical and Pharmaceutical Sciences*, 3(7), XX–XX.

Kumar, D.S., & Kumar Paswan, P. (2024). Role and importance of judiciary in protecting children from sexual offences. *Prayagraj Law Review*.

Kumar, A., Pathak, A., Kumar, S., Rastogi, P., & Rastogi, P. (2012). The problem of child sexual abuse in India laws, legal lacuna and the bill–PCSOB-2011. *Journal of the Indian Academy of Forensic Medicine*, 34(2), 169–174.

Kumar, S. H., Talasila, D., Gowrav, M. P., & Gangadharappa, H. V. (2020). Adaptations of Pharma 4.0 from Industry 4.0. *Drug Invention Today*, 14(3).

Kumar, S., & Mallipeddi, R. R. (2022). Impact of cybersecurity on operations and supply chain management: Emerging trends and future research directions. *Production and Operations Management*, 31(12), 4488–4500. DOI: 10.1111/poms.13859

Kumar, V. (2014). Role of Indigenous Knowledge in Climate Change Adaptation Strategies: A Study with Special Reference to North- Western India. *J Geogr Nat Disast*, 4, 131.

Kurniawan, T. A., Maiurova, A., Kustikova, M., Bykovskaia, E., Othman, M. H. D., & Goh, H. H. (2022). Accelerating sustainability transition in St. Petersburg (Russia) through digitalization-based circular economy in waste recycling industry: A strategy to promote carbon neutrality in era of Industry 4.0. *Journal of Cleaner Production*, 363, 132452. DOI: 10.1016/j.jclepro.2022.132452

Kusz, D. (2014). Modernisation of agriculture vs sustainable agriculture.

Kuzior, A. (2022). Technological unemployment in the perspective of Industry 4.0. *Virtual Economics*, 5(1), 7–23. DOI: 10.34021/ve.2022.05.01(1)

Kwauk, C. et al. (2019). *Girls' education in climate strategies Opportunities for improved policy and enhanced action in Nationally Determined Contribution, Global Economy & Development Working Paper 133*https://www.brookings.edu/wp-content/uploads/2019/12/Girls-ed-in-climate-strategies-working-paper-FINAL.pdf

Kwaymullina, A. (2018). In Stevens, L., Tait, P., & Varney, D. (Eds.), *You Are on Indigenous Land: Ecofeminism, Indigenous Peoples and Land Justice. In Feminist Ecologies: Changing Environments in the Anthropocene* (pp. 193–208). Springer International Publishing., DOI: 10.1007/978-3-319-64385-4_11

Larrier, R. (2000). *Francophone women writers of Africa and the Caribbean*. University Press of Florida.

Laskurain-Iturbe, I., Arana-Landín, G., Landeta-Manzano, B., & Uriarte-Gallastegi, N. (2021). Exploring the influence of industry 4.0 technologies on the circular economy. *Journal of Cleaner Production*, 321, 128944. DOI: 10.1016/j.jclepro.2021.128944

Lavanya, R. (2010). The Role of Human Rights in Climate Change Adaptation, 22 J. *Envtl.*, L, 229.

Law Commission of India. (1979). *79th Report on Congestion of Under-Trial Prisoners in India.*

Le Quere et al. (2012). *Global Carbon Budget 2012, Global Carbon Project.* https://www.globalcarbonproject.org/carbonbudget/archive/2012/CarbonBudget_2012.pdf

Leal Filho, W., Krishnapillai, M., Minhas, A., Ali, S., Nagle Alverio, G., Hendy Ahmed, M. S., & Sharifi, A. (2023). Climate change, extreme events and mental health in the Pacific region. *International Journal of Climate Change Strategies and Management*, 15(1), 20–40. DOI: 10.1108/IJCCSM-03-2022-0032

Lem, S. S., & Naalsund, E. (2023, March). *Discussion brief – involving children and children's rights in business.* UNICEF Norway. https://www.unicef.no/sites/default/files/2023-03/unicef-norway-involving-children-and-childrens-rights-in-business-discussion-brief.pdf

Leng, J., Ye, S., Zhou, M., Zhao, J. L., Liu, Q., Guo, W., Cao, W., & Fu, L. (2020). Blockchain-secured smart manufacturing in industry 4.0: A survey. *IEEE Transactions on Systems, Man, and Cybernetics. Systems*, 51(1), 237–252. DOI: 10.1109/TSMC.2020.3040789

Letzing, J., & Sung, M. (2024, April). *Is climate inaction a human rights violation?* World Economic Forum. https://www.weforum.org/agenda/2024/04/climate-progress-human-rights-violation

Levy, S., Migacheva, K., Ramirez, L., Okorodudu, C., Cook, H., Araújo-Soares, V., Minescu, A., Livert, D., Ragin, D., & Walker, P. (2022). A human rights based approach to the global children's rights crisis: A call to action. *The Journal of Social Issues*, 78(4), 1085–1097. Advance online publication. DOI: 10.1111/josi.12563

Li, B. H., Hou, B. C., Yu, W. T., Lu, X. B., & Yang, C. W. (2017). Applications of artificial intelligence in intelligent manufacturing: A review. *Frontiers of Information Technology & Electronic Engineering*, 18(1), 86–96. DOI: 10.1631/FITEE.1601885

Lim, C. H., Lim, S., How, B. S., Ng, W. P. Q., Ngan, S. L., Leong, W. D., & Lam, H. L. (2021). A review of industry 4.0 revolution potential in a sustainable and renewable palm oil industry: HAZOP approach. *Renewable & Sustainable Energy Reviews*, 135, 110223. DOI: 10.1016/j.rser.2020.110223

Limon, M. (2009). Climate Change, Human Rights and the UN: A New Role for the Office of the High Commissioner for Human Rights? 38 Brighton J. *Int'l*, L, 37.

Limon, M. (2009). Human rights and climate change: Constructing a case for political action. *Harv. Envtl. L. Rev.*, 33, 439.

Lioness News Desk. (2016). *5 Top Risks Women Entrepreneurs Face, Lioness*https://lionessmagazine.com/5-top-risks-women-entrepreneurs-face/

Liu, H., Yao, P., Latif, S., Aslam, S., & Iqbal, N. (2022). Impact of Green financing, FinTech, and financial inclusion on energy efficiency. *Environmental Science and Pollution Research International*, 29(13), 1–12. DOI: 10.1007/s11356-021-16949-x PMID: 34705207

Li, Z., Liang, X., Wen, Q., & Wan, E. (2024). The Analysis of Financial Network Transaction Risk Control based on Blockchain and Edge Computing Technology. *IEEE Transactions on Engineering Management*, 71, 5669–5690. DOI: 10.1109/TEM.2024.3364832

Lu, C., Lyu, J., Zhang, L., Gong, A., Fan, Y., Yan, J., & Li, X. (2020). Nuclear power plants with artificial intelligence in industry 4.0 era: Top-level design and current applications—A systemic review. *IEEE Access : Practical Innovations, Open Solutions*, 8, 194315–194332. DOI: 10.1109/ACCESS.2020.3032529

Lu, Y. (2019). Artificial intelligence: A survey on evolution, models, applications and future trends. *Journal of Management Analytics*, 6(1), 1–29. DOI: 10.1080/23270012.2019.1570365

Maathai, W. (2004). *The Green Belt Movement: Sharing the Approach and the Experience*. Lantern Books.

Macchiavello, E. (2023). Sustainable Finance and Fintech. A Focus on Capital Raising. A Focus on Capital Raising (May 31, 2023). Forthcoming in K. Alexander, M. Gargantini and M. Siri, *The Cambridge Handbook of EU Sustainable Finance-Regulation, Supervision and Governance*. CUP.

Madhok, S., Unnithan, M., & Heitmeyer, C. (2014). On reproductive justice: 'Domestic violence', rights and the law in India. *Culture, Health & Sexuality*, 16(8), 1025–1039. DOI: 10.1080/13691058.2014.918281 PMID: 24927257

Mahadew, R. (2021). A Human-Rights-Based Approach to Climate Change. Revue juridique de l'Océan Indien, (31), 155-168.

Majstorovic, V. D., & Mitrovic, R. (2019). Industry 4.0 programs worldwide. In *Proceedings of the 4th International Conference on the Industry 4.0 Model for Advanced Manufacturing: AMP 2019 4* (pp. 78-99). Springer International Publishing. DOI: 10.1007/978-3-030-18180-2_7

Maldonado, J., Bull Bennett, T. M., Chief, K., Cochran, P., Cozzetto, K., & Gough, B.. (2015). Engagement with indigenous peoples and honoring traditional knowledge systems. *Climatic Change*, 135(1), 111–126. DOI: 10.1007/s10584-015-1535-7 PMID: 26346998

Mamta Verma v. Union of India, W.P.(C) 627/2017

Mandal, S. (2017). Assessing the Instability and Shifting Character of the River Bank Ganga in Manikchak Diara of Malda District, West Bengal using Bank Erosion Hazard Index (BEHI), RS & GS. *European Journal of Geography*, 8(4), 6.

Mandela Rules, Rule. 11, 28, 45, 48, 58 & 81. (2016).

Manral, I., Panjrattan, C., Saini, R., Roy, S., & Singh, A. K. (2022). A multi-speciality approach to the protection of children from sexual offences act: A review. *Industrial Psychiatry Journal*, 32(1), 4–8. DOI: 10.4103/ipj.ipj_169_21 PMID: 37274576

Manurung, W., & Dame Panjaitan, J. (2023). Implementation of Legal Protection of Children and Women as Victims of Sexual Crimes. *International Journal of Social Research.*

Mary Roy, V. (1986). The State Of Kerala 1986 AIR 1011. *SCR*, (1), 371.

Mascarenhas, O. A., Thakur, M., & Kumar, P. (2024). Critical Thinking Applied to Ecofeminism. In A Primer on Critical Thinking and Business Ethics (pp. 89-118). Emerald Publishing Limited.

Mathew, L. A. (2016). A Multidisciplinary Approach to Child Protection for Sexual Abuse in India: The Law.

Mathew, S. (2021). History vs. Reality: The State of SRHR in India. *The YP Foundation*. Retrieved from https://theypfoundation.org/resources/history-vs-reality-of-srhr/

Mathews, B., Lee, X., & Norman, R. (2016). Impact of a New Mandatory Reporting Law on Reporting and Identification of Child Sexual Abuse: A Seven Year Time Trend Analysis. Queensland University of Technology (QUT) - Law & Justice Legal Studies Research Paper Series. https://doi.org/DOI: 10.1016/j.chiabu.2016.04.009

Mawson, A. (2022). *Charting the Course: Embedding children's rights in responsible business conduct. A Brief.* UNICEF. https://www.unicef.org/media/121001/file/Charting%20The%20Course.pdf

Mawson, A., Cantú Rivera, H., & Jayasekaran, S. (2022). *Charting the Course: Embedding children's rights in responsible business conduct*. UNICEF. https://www.unicef.org/media/129596/file/Charting%20The%20Course.pdf

Medical Termination of Pregnancy (Amendment) Act. 2002, No.64, Acts of the Parliament, 2002 (India).

Medical Termination of Pregnancy (Amendment) Act. 2021, No. 8, Acts of the Parliament, 2021(India).

Medical Termination of Pregnancy (MTP) Act. 1971 No. 34, Acts of the Parliament, 1971(India).

Medical Termination of Pregnancy (MTP) Act. 1971, No. 34, Acts of the Parliament, 1971(India).

Meera Santosh Pal v. Union of India, (2017) 3 SCC 462.

Meera Santosh Pal v. Union of India, 2017 AIR SC 787.

Mehrotra, S., & Sinha, S. (2017). Explaining falling female employment during a high growth period. *Economic and Political Weekly*, 52(39), 54–62.

Mercy Corps Ventures. (2022). *Understanding MSME Resilience to Climate Change*, Medium, https://medium.com/mercy-corps-social-venture-fund/understanding-msme-resilience-to-climate-change-cd50355bd31b

Meyendorf, N., Ida, N., Singh, R., & Vrana, J. (2023). NDE 4.0: Progress, promise, and its role to industry 4.0. *NDT & E International*, 140, 102957. DOI: 10.1016/j.ndteint.2023.102957

Micaroni, M. (2020). *Sustainable Finance: Addressing the SDGs through Fintech and Digital Finance solutions in EU* (Doctoral dissertation, Politecnico di Torino).

Migliorelli, M., & Dessertine, P. (2019). The rise of green finance in Europe. *Opportunities and challenges for issuers, investors and marketplaces. Cham. Palgrave Macmillan*, 2, 2019.

Ministry of Finance, Government of India. (2018). *Economic Survey 2017-18*.https://www.indiabudget.gov.in/budget2018-2019/economicsurvey2017-2018/index.html

Ministry of Finance. (2022). *Union Budget 2022-23*. Government of India.

Ministry of Home Affairs, Government of India. (2016). *Model Prison Manual, 2016, table 26.157.*

Ministry of Home Affairs. (1983). *Report of the All-India Committee on Jail Reforms 1980-83*. Government of India.

Ministry of Women & Child Development. (2018). *Report on Women in Prisons*. Government of India.

Mirza, A. T. M., Rahman, T., Islam, S., & Rahman, S. H. (2014). Coping with flood and riverbank erosion caused by climate change using livelihood resources: A case study of Bangladesh. *Climate and Development*, 7(2), 185–191. DOI: 10.1080/17565529.2014.910163

Mirza, M. M. Q. (2004). The Ganges Water Diversion: Environmental Effects and Implications — An Introduction. In Mirza, M. M. Q. (Ed.), *The Ganges Water Diversion: Environmental Effects and Implications. Water Science and Technology Library* (Vol. 49). Springer., DOI: 10.1007/978-1-4020-2792-5_1

Mithas, S., Chen, Z. L., Saldanha, T. J., & De Oliveira Silveira, A. (2022). How will artificial intelligence and Industry 4.0 emerging technologies transform operations management? *Production and Operations Management*, 31(12), 4475–4487. DOI: 10.1111/poms.13864

Mitra, A. (2019a). Women's work in response to urbanisation: Evidence from Odisha. *ANTYAJAA: Indian Journal of Women and Social Change, 4*(1), 92–106. https://journals.sagepub.com/doi/full/10.1177/2455632719836804

Mitra, E., & Berlinger, J. (2024, May 29). *Indian capital of Delhi breaks all-time heat record, as authorities impose water rationing. CNN*. https://edition.cnn.com/2024/05/29/weather/india-heat-record-water-rationing-intl/index.html

Mitra, S. (2019b). Women and unpaid work in India: A macroeconomic overview. https://slidetodoc.com/ women-and-unpaid-work-in-India-a-macroeconomic/

Mitra, S., & Mehta, S. K. (2021). It is time to compensate women's unpaid labour. *India Development Review*.https://idronline.org/its-time-to-compensate-womens-unpaid-labour/

Mittal, R. (2023), India's progressive approach towards abortion laws, *Legalbites,*https://www.legalbites.in/topics/articles/indias-progressive-approach-towards-abortion-laws-345469?infinitescroll=1

Mohanty, H., & Banerjee, D. (2021). An Analysis of Protection of Children from Sexual Offences Act, 2012 (POCSO ACT). SSRN Electronic Journal. https://doi.org/DOI: 10.2139/ssrn.3768096

Mollah, T. H., & Ferdaush, J. (2015). Riverbank Erosion, Population Migration and Rural Vulnerability in Bangladesh (A case Study on Kazipur Upazila at Sirajgonj District). *Environment and Ecology Research*, 3(5), 125–131. DOI: 10.13189/eer.2015.030502

Mondal, M., & Satpati, L. N. (2012). Morphodynamic setting and nature of bank erosion of the ichamati river in swarupnagar and baduria blocks, 24 Parganasz (N), West Bengal. *Indian Journal of Spatial Science*, 3(2), 35–43.

Montt, G. (2018). Too polluted to work? The gendered correlates of air pollution on hours worked. *IZA Journal of Labor Economics*, 7(1), 7. DOI: 10.1186/s40172-018-0067-6

Moorehead, V. D.Jr, Gone, J. P., & Damia, D. (2015). A gathering of native american healers: Exploring the interface of indigenous tradition and professional practice. *American Journal of Community Psychology*, 56(3-4), 383–394. DOI: 10.1007/s10464-015-9747-6 PMID: 26351006

Moore, N. (2011). Eco/feminism and rewriting the ending of feminism: From the Chipko movement to Clayoquot Sound. *Feminist Theory*, 12(1), 3–21. DOI: 10.1177/1464700110390592

Morgan, J., Halton, M., Qiao, Y., & Breslin, J. G. (2021). Industry 4.0 smart reconfigurable manufacturing machines. *Journal of Manufacturing Systems*, 59, 481–506. DOI: 10.1016/j.jmsy.2021.03.001

Moro-Visconti, R., Cruz Rambaud, S., & López Pascual, J. (2020). Sustainability in FinTechs: An explanation through business model scalability and market valuation. *Sustainability (Basel)*, 12(24), 10316. DOI: 10.3390/su122410316

Mota, S. A. (2002). The U.S. Supreme Court Addresses the Child Pornography Prevention Act and Child Online Protection Act in Ashcroft v. Free Speech Coalition and Ashcroft v. American Civil Liberties Union. *Federal Communications Law Journal*, 55, 4.

Mourtzis, D., Angelopoulos, J., & Panopoulos, N. (2022). A Literature Review of the Challenges and Opportunities of the Transition from Industry 4.0 to Society 5.0. *Energies*, 15(17), 6276. DOI: 10.3390/en15176276

Mrs X and Mrs Y v. Union of India, W.P. 81 of 2017.

MSME Finance Gap. (n.d.). *SME Finance Forum*.https://www.smefinanceforum.org/data-sites/msme-finance-gap

Much, L. (2022, October 10). *10 Jahre kinderrechtlicher Grundsätze für Unternehmen – Bilanz und Ausblick.* https://www.unicef.de/informieren/aktuelles/blog/-/jubilaeum-child-rights-business-principles/313368

Mukherjee, P. (2013, February 05). Cry of Help for Erosion Victims. Business Standard. Retrieved from https://www.business-standard.com/article/economy-policy/cry-of-help-for-erosion-victims-108021901020_1.html

Mutunda, S. Women subjugating women: Re-reading Mariama Ba's So long a letter and Scarlet song. Retrieved April 5, 2023, from https://escholarship.org/content/qt4j47346d/qt4j47346d_noSplash_f8ea2e0a8f21692d94777ca2fe506bd4.pdf

Naderi, N., & Tian, Y. (2022). Leveraging Blockchain Technology and Tokenizing Green Assets to Fill the Green Finance Gap. *Energy Research Letters, 3*(3).

Nahavandi, S. (2019). Industry 5.0—A human-centric solution. *Sustainability (Basel)*, 11(16), 4371. DOI: 10.3390/su11164371

Nandini Tushar Rawool v. State – W.P No.8313 of 2018.

Narayanan, S. *What Hinduism says about abortion Article*, speakingtree.in (Last updated on Oct 24, 2016, 1:15 PM), https://www.speadkingtree.in/allslides/what-hinduism-says-about-abortion-article

Nassiry, D. (2018). *The role of fintech in unlocking green finance: Policy insights for developing countries* (No. 883). ADBI working paper.

National Coalition Deutschland – Netzwerk zur Umsetzung der UN-Kinderrechtskonvention e. V. (2020, August 19). Das Kinderrecht auf eine gesunde Umwelt verwirklichen. *Netzwerk Kinderrechte.* https://netzwerk-kinderrechte.de/2020/08/19/das-kinderrecht-auf-eine-gesunde-umwelt-verwirklichen-2/

National Commission for Women. (2020, April 22). *Press Note* .

National Crime Records Bureau. (2021). *Prison Statistics India 2020.* Ministry of Home Affairs.

Nazem, N. I., & Elahi, K. M. (1990). Impact of Riverbank Erosion on Land Settlements in Bangladesh. In A.Q. M. Mahtub (Ed.), *Proceedings of the Seminar on People and Environment in Bangladesh*, Dhaka, Bangladesh: UNDP and UNFPA

Nehra, S. S., & Rajput, A. S. (2019). The Legal Personality of an Unborn Child: A Comparative Analysis of USA & India. *Amity International Journal of Juridical Sciences*, 5, 104.

Nenavath, S., & Mishra, S. (2023). Impact of green finance and fintech on sustainable economic growth: Empirical evidence from India. *Heliyon*, 9(5), e16301. DOI: 10.1016/j.heliyon.2023.e16301 PMID: 37234625

New Climate Institute. (2015). *Climate Action Tracker: INDCs lower projected warming to 2.7°C: significant progress but still above 2°C.* https://newclimate.org/news/climate-action-tracker-indcs-lower-projected-warming-to-27degc-significant-progress-but-still

Ng, W., & Acker, A. (2018), "Understanding Urban Travel Behaviour by Gender for Efficient and Equitable Transport Policies", http://www.itf-oecd.org

Nguyen, T., Gosine, R. G., & Warrian, P. (2020). A systematic review of big data analytics for oil and gas industry 4.0. *IEEE Access : Practical Innovations, Open Solutions*, 8, 61183–61201. DOI: 10.1109/ACCESS.2020.2979678

Nica, E., & Stehel, V. (2021). Internet of things sensing networks, artificial intelligence-based decision-making algorithms, and real-time process monitoring in sustainable industry 4.0. *Journal of Self-Governance and Management Economics*, 9(3), 35–47.

Nikolaou, N. (2018). *The Transition to a Green Global Economy: The Case of Green Finance and Green Bonds.* (Doctoral dissertation), University of Piraeus.

Nikore, M.. (2021). India's missing working women: Tracing the journey of women's economic contribution over the last seven decades, and during. *Journal of International Women's Studies*. 19.

Norgaard, K., & York, R. (2016). Gender Equality and State Environmentalism. *Gender & Society*, 19(4), 506–522. DOI: 10.1177/0891243204273612

Novianti, K., Warsilah, H., & Wahyono, A. (2016). Climate Change and Food Security on Coastal Community. *Jurnal Penelitian Kesejahteraan Sosial*, 15(3), 203–218.

Nowicka, W. (2011). Sexual and reproductive rights and the human rights agenda: Controversial and contested. *Reproductive Health Matters*, 19(38), 119–128. DOI: 10.1016/S0968-8080(11)38574-6 PMID: 22118146

Nurse, L. A., McLean, R. F., Agard, J., Briguglio, L. P., Duvat-Magnan, V., Pelesikoti, N., & Webb, A. (2014). *Small islands.* Cambridge University Press.

OCHCR. (2022). *Climate change exacerbates violence against women and girls*, United Nations Human Rights Office of the High Commissioner https://www.ohchr.org/en/stories/2022/07/climate-change-exacerbates-violence-against-women-and-girls

OECD (2019), "Gender Equality and Sustainable Infrastructure"

OECD (2020), *Policies and Practices to Promote Women in Leadership Roles in the Private*

OECD DAC Network on Gender Equality (GENDERNET). (2015). *Making climate finance work for women: Overview of the integration of gender equality in aid to climate change,* https://www.oecd.org/dac/gender-development/Making-Climate-Finance-Work-for-Women.pdf

OECD. (2008), *Household Behaviour and the Environment: Reviewing the Evidence,* https://www.oecd.org/environment/consumption-innovation/42183878.pdf

OECD. (2015). *The Economic Consequences of Climate Change.* OECD Publishing.

OECD. (2017), *Policy Brief on Women's Entrepreneurship,* https://www.oecd.org/cfe/smes/Policy-Brief-on-Women-s-Entrepreneurship.pdf

OECD. (2019). *Child labour risks in global supply chains.* Ending Child Labour, Forced Labour and Human Trafficking in Global Supply Chains. https://mneguidelines.oecd.org/child-labour-risks-in-global-supply-chain.htm

OECD. (2021), Entrepreneurship Policies through a Gender Lens, OECD Studies on SMEs and Entrepreneurship, OECD Publishing, Paris, DOI: 10.1787/71c8f9c9-en

OECD. (2021), *Gender and the Environment: Building Evidence and Policies to Achieve the* OECD, Gender and Sustainable Development: Maximising the Economic, Social and Environmental Role of Women. Organization for Economic Co- operation and Development (OECD) to the United Nations Commission on Sustainable Development (UNCSD) (2008). https://www.oecd.org/social/40881538.pdf

OECD. (2023). *OECD-Leitsätze für multinationale Unternehmen zu verantwortungsvollem unternehmerischem Handeln.* Organisation for Economic Co-operation and Development. https://www.oecd-ilibrary.org/finance-and-investment/oecd-leitsatze-fur-multinationale-unternehmen-zu-verantwortungsvollem-unternehmerischem-handeln_abd4d37b-de

OECD. (2024). *MNE Guidelines—Organisation for Economic Co-operation and Development.* OECD Guidelines for Multinational Enterprises on Responsible Business Conduct. https://mneguidelines.oecd.org/mneguidelines/

OECD. 2021. Gender and the environment: Building evidence and policies to achieve the SDGs.https://www.oecd-ilibrary.org/environment/gender-and-the-environment_3d32ca39-en

Offermann, L., & Foley, K. (2020). *Is There a Female Leadership Advantage?* Oxford Research Encyclopedia of Business and Management., DOI: 10.1093/acrefore/9780190224851.013.61

Office of the United Nations High Commissioner for Human Rights. (n.d.). *The impacts of climate change on the effective enjoyment of human rights*. Office of the United Nations High Commissioner for Human Rights. https://www.ohchr.org/en/climate-change/impacts-climate-change-effective-enjoyment-human-rights

OHCHR. (n.d.). *Special Rapporteur on toxics and human rights*. Retrieved April 1, 2023, from https://www.ohchr.org/EN/Issues/ToxicsAndHumanRights/Pages/SRToxicsAndHumanRightsIndex.aspx

Olawuyi, D. S., & Justice, C.Right to Development. (2015). A Perspective from Sub-Saharan Africa, 42 J. *Energy & Nat. Res.*, L, 16.

One, U. (2023). *The Role of Renewable Energy in Achieving Sustainable Development Goals*, https://utilitiesone.com/the-role-of-renewable-energy-in-achieving-sustainable-development-goals

Osai, A. (2022). Women are Hit Hardest in Disasters, So Why are Responses Too Often Gender-Blind? United Nations Development Programme. Retrieved from https://www.undp.org/documents/women-hit-hardest-disasters

Otero, M. (2023, May 17). *Six kilometers a day: Water and sanitation*. U.S. Agency for International Development. https://www.usaid.gov/six-kilometers-day

OWSA. (2024). *The Climate Change Crisis is Unjust for Rural Women: FAO Gender Expert*. One World South Asia.

OXFAM (2018), *The Weight of Water on Women: The Long Wake of Hurricane María in Puerto Rico*.

Ozili, P. K. (2023). Assessing global interest in decentralized finance, embedded finance, open finance, ocean finance and sustainable finance. *Asian Journal of Economics and Banking*, 7(2), 197–216. DOI: 10.1108/AJEB-03-2022-0029

Pandey, S., & Reddy, S. (2019). Article: Risk Factors, Outcomes, and Protective Factors of Child Sexual Abuse in India: A Conceptual Framework.

Parvin Shaikh, N. (2024). Feminism in Practice: Learning from the Barefoot "Solar Mamas". *Journal of International Women's Studies*, 26(3), 14.

Patel, S. et al. (2021). *Tackling gender inequality and climate change through the budget.* International Budget.org https://internationalbudget.org/wp-content/uploads/gender-climate-budgeting-synthesis-april-2021.pdf

Pedersen, C. S. (2018). The UN Sustainable Development Goals (SDGs) are a gift to business. *25th CIRP Life Cycle Engineering (LCE) Conference* (pp. 21-24). Copenhagen: Elsevier. doi:DOI: 10.1016/j.procir.2018.01.003

Pedroza-Gutiérrez, C., Solano, N., Fernández-Rivera-Melo, F., & Hernández-Herrera, I. (2024). What are the factors that enable or hinder gender equality? Lessons and experiences on gender dynamics in Mexican small-scale fisheries. *Marine Policy*, 159, 105944. DOI: 10.1016/j.marpol.2023.105944

Perelli, C., Cacchiarelli, L., Peveri, V., & Branca, G. (2024). Gender equality and sustainable development: A cross-country study on women's contribution to adopting climate-smart agriculture in Sub-Saharan Africa. *Ecological Economics*, 219, 108145. DOI: 10.1016/j.ecolecon.2024.108145

Petchesky, R. P. (1991). *Abortion and Women's Choice: The State, Sexuality and Reproductive Freedom, The Northeastern Series in Feminist Theory*. Northeastern University Press.

Petter, J. (2023, June 12). Lieferkettengesetz: Analyse von »Save the Children zeigt in 18 von 20 untersuchten Lieferketten Hinweise auf Kinderarbeit. *Der Spiegel*. https://www.spiegel.de/ausland/lieferkettengesetz-und-kinderarbeit-analyse-von-save-the-children-zeigt-in-18-von-20-untersuchten-faellen-hinweise-a-09f2d164-fdc1-4503-be2f-13652eb09963

Pew Research Center. (2019), *How people worldwide view climate change*, https://www.pewresearch.org/fact-tank/2019/04/18/a-look-at-how-people-around-the-worldview-climate-change/

Phadke, S. (1998), Pro-Choice or Population Control: A Study of the Medical Termination of Pregnancy Act, Government of India, 1971, http://doccenter.org.in/docsweb/writers/shilpa-phadke1.htm

Picess, A. R. (2018). Child sexual abuse in India: Law with lacunae and loopholes. *Supremo Amicus*, 5, 76.

Pillai, A. V., & Kostruba, A. (2021), Women's Reproductive Rights And Their Scope Under International Legal Frameworks, *Entrepreneurship, Economy and Law*, 8, 18-28, https://hal.science/hal-03397078.

Pittman, A. (2010). Capturing change in women's realities a critical overview of current monitoring & evaluation frameworks and approaches.

Pivoto, D. G., de Almeida, L. F., da Rosa Righi, R., Rodrigues, J. J., Lugli, A. B., & Alberti, A. M. (2021). Cyber-physical systems architectures for industrial internet of things applications in Industry 4.0: A literature review. *Journal of Manufacturing Systems*, 58, 176–192. DOI: 10.1016/j.jmsy.2020.11.017

Popov, V. V., Kudryavtseva, E. V., Kumar Katiyar, N., Shishkin, A., Stepanov, S. I., & Goel, S. (2022). Industry 4.0 and digitalisation in healthcare. *Materials (Basel)*, 15(6), 2140. DOI: 10.3390/ma15062140 PMID: 35329592

Prabha Tyagi v. Kamlesh Tyagi (2022) 2022 LiveLaw (SC) 474 | CrA 511 OF 2022

Pratiwi, H., Hutabarat, Z. S., Nabila, R., Hevinosa, A. D., Mauli, C., Syaputri, A. H., Nisa, A., Hidayatullah, M. S., Tyaningsih, D. R., Putri, D. I., Fakhira, A. L., Aprilia, M. D., & Al Zahra, F. (2024). Blue Economic Beacon: Red And White Leads The Way Through Niche Diplomacy. *International Journal of Multidisciplinary Sciences and Arts*, 3(1), 244–251. DOI: 10.47709/ijmdsa.v3i1.3947

Priyanka Shukla v. UOI & Ors., (W.P.(ST) No. 36727/2017).

Priyanka Yadav v. State W.P. No. 23307 of 2017.

Protishruti, P. (n.d.). Wikipedia. Retrieved April 2, 2023, from https://en.wikipedia.org/wiki/Prothom_Protishruti

Puaschunder, J. M. (2023). The Future of Resilient Green Finance. In *The Future of Resilient Finance: Finance Politics in the Age of Sustainable Development* (pp. 185–210). Springer International Publishing. DOI: 10.1007/978-3-031-30138-4_6

PWC. (2022). *Gender equality in climate action: Women at the core*. https://www.pwc.in/assets/pdfs/consulting/esg/gender-equality-in-climate-action.pdf

Qian, N. (2008). Missing women and the price of tea in China: the effect of sex-specific earnings on sex imbalance. *The Quarterly Journal of Economics*, 123(3), 1251–1285. DOI: 10.1162/qjec.2008.123.3.1251

Quaranta, M., & Dotti Sani, G. M. (2018). Left behind? Gender gaps in political engagement over the life course in twenty-seven European countries. *Social Politics*, 25(2), 254–286. DOI: 10.1093/sp/jxy005

Quirico, O., & Boumghar, M. (Eds.). (2015). *Climate change and human rights: an international and comparative law perspective*. Routledge.

Rabbi, H., Saifullah, A. S. M., Sheikh, M. S., Sarker, M. M., & Bhowmick, A. C. (2013). Recent study on river bank erosion and its impacts on land displaced people in Sirajgonj Riverine area of Bangladesh. *World journal of Applied Environmental chemistry, 2*(2), 36-43.

Radanliev, P. (2024). Cyber diplomacy: defining the opportunities for cybersecurity and risks from Artificial Intelligence, IoT, Blockchains, and Quantum Computing. *Journal of Cyber Security Technology*, 1-51.

Rahman, M. (2013). Impact of Riverbank Erosion Hazard in the Jamuna Floodplain Areas in Bangladesh. *Journal of Science Foundation*, 8(1-2), 55–65. DOI: 10.3329/jsf.v8i1-2.14627

Rajasthan Government. (n.d.). *Rajasthan Gramin Aajeevika Vikas Parishad (Aajeevika)* (Accessed on: 17-02-2024) https://rgavp.rajasthan.gov.in/

Rane, N. (2023). Transformers in Industry 4.0, Industry 5.0, and Society 5.0: Roles and Challenges.

Rani, S. (2022), Reproductive Rights of Women in India: An Overview, *IJCRT*, 10.

Ranjan, P. (2020). The Right to Development in a Climate-Constrained World: The Green Climate Fund as an Opportunity for Transformative Development, 22 Indian J. *Int'l*, L, 183.

Rao, N., Mishra, A., Prakash, A., Singh, C., Qaisrani, A., Poonacha, P., Vincent, K., & Bedelian, C. (2019). A qualitative comparative analysis of women's agency and adaptive capacity in climate change hotspots in Asia and Africa. *Nature Climate Change*, 9(12), 964–971. DOI: 10.1038/s41558-019-0638-y

Rath, K. C., Khang, A., & Roy, D. (2024). The Role of Internet of Things (IoT) Technology in Industry 4.0 Economy. In *Advanced IoT Technologies and Applications in the Industry 4.0 Digital Economy* (pp. 1-28). CRC Press.

Ravi, A. (2021, Nov 11). *Abandonment, poverty, sexual exploitation: Why Indian women pay a higher price for climate crisis.* Down to Earth. https://www.downtoearth.org.in/interviews/climate-change/abandonment-poverty-sexual-exploitation-why-indian-women-pay-a-higher-price-for-climate-crisis-80160

Raworth, K. (2012). A safe and just space for humanity: Can we live within the doughnut. *Oxfam Policy and Practice: Climate Change and Resilience*, 8(1), 1–26.

Ray, R. K., Chowdhury, F. R., & Hasan, M. R. (2024). Blockchain Applications in Retail Cybersecurity: Enhancing Supply Chain Integrity, Secure Transactions, and Data Protection. *Journal of Business and Management Studies*, 6(1), 206–214. DOI: 10.32996/jbms.2024.6.1.13

Reier Forradellas, R. F., & Garay Gallastegui, L. M. (2021). Digital transformation and artificial intelligence applied to business: Legal regulations, economic impact and perspective. *Laws*, 10(3), 70. DOI: 10.3390/laws10030070

Renu, R., & Chopra, G. (2019). Child sexual abuse in India and the Protection of Children from Sexual Offences (POCSO) Act 2012: A research review. *Integrated Journal of Social Sciences*, 6(2), 49–56.

Report of World Bank. (2023). *Exploring Opportunities for Women Entrepreneurs Driving Climate Solutions A Discussion Note, International Finance Corporation.* World Bank Group https://www.ifc.org/content/dam/ifc/doc/2023/exploring-opportunities-for-women-entrepreneurs-driving-climate-solutions.pdf

Res, G. A. 70/1, U.N. Doc. A/RES/70/1 (Sept. 25, 2015).

Rex v. Bourne, (1938)3 All E.R. 615.

Rheeney, A. (2023, March 28). *U.S.P. Student initiative to go before I.C.J.* Samoa Observer. https://www.samoaobserver.ws/category/samoa/102759

Rizzello, A., & Kabli, A. (2020). Social finance and sustainable development goals: A literature synthesis, current approaches and research agenda. *ACRN Journal of Finance and Risk Perspectives, 9.*

Roe v. Wade, (1973) 410 U.S. 113.

Rope, O. (Ed.). (2013, August 19). *Protecting the rights of women offenders: A job for the CEDAW Committee? (Olivia Rope – Penal Reform International).* Optional Protocol to CEDAW. https://opcedaw.wordpress.com/2013/08/19/protecting-the-rights-of-women-offenders/

Rowland, L. (2015). Speculative Solutions: The Development of Environmental and Ecofeminist Discourse in Margaret Atwood's MaddAddam. Studies in Canadian Literature. ISSN 1718-7850.

Roy, M., & Venema, H. (2002). *Reducing risk and vulnerability to climate change in India: The capabilities approach.* Taylor & Francis.

Sachs, J. D., Woo, W. T., Yoshino, N., & Taghizadeh-Hesary, F. (2019). Importance of green finance for achieving sustainable development goals and energy security. *Handbook of green finance: Energy security and sustainable development, 10,* 1-10.

Saeed, S., Altamimi, S. A., Alkayyal, N. A., Alshehri, E., & Alabbad, D. A. (2023). Digital transformation and cybersecurity challenges for businesses resilience: Issues and recommendations. *Sensors (Basel)*, 23(15), 6666. DOI: 10.3390/s23156666 PMID: 37571451

Saheli Women's Resource Centre. Reproductive Rights In The Indian Context, https://www.saheliwomen.org/health/reproductive-rights-in-the-indian-context

Sánchez-Sotano, A., Cerezo-Narváez, A., Abad-Fraga, F., Pastor-Fernández, A., & Salguero-Gómez, J. (2020). Trends of digital transformation in the shipbuilding sector. In *New Trends in the Use of Artificial Intelligence for the Industry 4.0*. IntechOpen. DOI: 10.5772/intechopen.91164

Sangeeta Thapa v. Government of NCT of Delhi & Ors. W.P.(C) 15241/2021.

Sanjaya, K., & Praveen, D. (2020). Ethics in the Criminal Justice System. In Deb, S., & Subhalakshmi, G. (Eds.), *Delivering Justice* (p. 223). Routledge., DOI: 10.4324/9780429324062-20

Santhya, K. G., & Jejeebhoy, J. S. (2007). *India Young people's sexual and reproductive health in India: Policies, programmes and realities*. Population Council.

Sanzgiri, V. (2024, June 27) *Amazon India Workers Say Unlawful Operations Persist Despite Alleged Acknowledgement by Company*. Media Nama. https://www.medianama.com/2024/06/223-amazon-workers-allege-unlawful-operations-persist/

Sarif, M. N., Siddiqui, L., Islam, M. S., Parveen, N., & Saha, M. (2021). Evolution of river course and morphometric features of the River Ganga: A case study of up and downstream of Farakka Barrage. *International Soil and Water Conservation Research*, 9(4), 578–590. Advance online publication. DOI: 10.1016/j.iswcr.2021.01.006

Sarkar, S. D. (2022, May 18). Octogenarian MLA's riverbed vigil in Malda: Samar Mukherjee's relentless effort to prevent illegal mining of soil and sand. The Telegraph online. Retrieved from https://www.telegraphindia.com/west-bengal/octogenarian-mlas-riverbed-vigil-in-malda/cid/1865585

Sarwar, M. G.. (2007). *Women's Rights to Land in Bangladesh: Roles, Limitations and Transformation*. Unnayan Oneshan.

Save the Children Deutschland e. V. (2024, June). *CSR-Kommunikation*. CSR-Kommunikation. https://www.savethechildren.de/unterstuetzen/fuer-unternehmen/unsere-unterstuetzung/csr-kommunikation/

Scheelbeek, P. F. D., Bird, F. A., Tuomisto, H. L., Green, R., Harris, F. B., Joy, E. J. M., Chalabi, Z., Allen, E., Haines, A., & Dangour, A. D. (2018). Effect of environmental changes on vegetable and legume yields and nutritional quality. *Proceedings of the National Academy of Sciences of the United States of America*, 115(26), 6804–6809. DOI: 10.1073/pnas.1800442115 PMID: 29891659

Schilling-Vacaflor, A., & Gustafsson, M.-T. (2024). Towards more sustainable global supply chains? Company compliance with new human rights and environmental due diligence laws. *Environmental Politics*, 33(3), 422–443. DOI: 10.1080/09644016.2023.2221983

Schoenmaker, D., & Volz, U. (2022). *Scaling up sustainable finance and investment in the Global South.* CEPR Press.

Schudak, A., & Mahlmann, I. (2017, August). *Studie: Kinderrechte in deutschen Unternehmensaktivitäten—Status und Bewertung* (Geschäftsstelle Deutsches Global Compact Netzwerk (DGCN), Ed.). https://www.globalcompact.de/migrated_files/wAssets/docs/Menschenrechte/studie-kinderrechte-in-deutschen-unternehmensaktivitaeten.pdf

Schulz, K. A., Gstrein, O. J., & Zwitter, A. J. (2020). Exploring the governance and implementation of sustainable development initiatives through blockchain technology. *Futures*, 122, 102611. DOI: 10.1016/j.futures.2020.102611

Schulz, K., & Feist, M. (2021). Leveraging blockchain technology for innovative climate finance under the Green Climate Fund. *Earth System Governance*, 7, 100084. DOI: 10.1016/j.esg.2020.100084

SDGs, OECD Publishing, Paris, .DOI: 10.1787/3d32ca39-en

Sector, https://www.oecd.org/corporate/OECD-G20-EMPOWER-Women-Leadership.pdf

Seth, K., & Sharma, R. (2015). Overview of Laws against online child sex abuse in India, U.K and U.S. *International Journal of Research*, 2, 1169–1177.

Seth, R., & Srivastava, R. (2017). Child Sexual Abuse: Management and prevention, and protection of children from Sexual Offences (POCSO). *Indian Pediatrics*, 54(11), 949–953. DOI: 10.1007/s13312-017-1189-9 PMID: 29217802

Shagun. (2023). *Climate crisis: Inequality hotspot study shows where women in agriculture are hit hardest, Down to Earth.* https://www.downtoearth.org.in/news/agriculture/climate-crisis-inequality-hotspot-study-shows-where-women-in-agriculture-are-hit-hardest-92849

Shah, A., ... (2024). *Caste inequality in occupational exposure to heat waves in India.* IIM Bangalore Working Paper No. 709/2024.

Sharma, D.P., & Kumar, D.A. (2023). Child Sexual Abuse Risk Factors, Outcomes, and Protective Factors in India: A Conceptual and framework model. *Innovative Research Thoughts.*

Sharma, A., & Singh, B. (2022). Measuring Impact of E-commerce on Small Scale Business: A Systematic Review. *Journal of Corporate Governance and International Business Law*, 5(1).

Sharma, G., Kumar, S., & Banga, H. (2022). protection of children from sexual offences with special reference to POSCO Act, 2012. *International Journal of Health Sciences*, 6706–6713. Advance online publication. DOI: 10.53730/ijhs.v6nS4.10317

Sharma, S. (2015). Reproductive Rights and Women's Health: A New Prospective. *International Journal of Humanities and Social Science Invention*, 4(4), 51–56.

Sharma, S., & Dwivedi, R. (2024). A survey on blockchain deployment for biometric systems. *IET Blockchain*, 4(2), 124–151. DOI: 10.1049/blc2.12063

Shayegh, S., & Dasgupta, S. (2022). *Climate change, labor availability and the future of gender inequality in South Africa*. Taylor & Francis.

Shih, C., Gwizdalski, A., & Deng, X. (2023). *Building a Sustainable Future: Exploring Green Finance*. Regenerative Finance, and Green Financial Technology.

Shiva, V. (1988). *Staying Alive: Women, Ecology, and Development*. Zed Books.

Shi, Z., Xie, Y., Xue, W., Chen, Y., Fu, L., & Xu, X. (2020). Smart factory in Industry 4.0. *Systems Research and Behavioral Science*, 37(4), 607–617. DOI: 10.1002/sres.2704

Shri Shantilal, H. S. (1967). *Report of The Committee to Study the Question of Legalization of Abortion*. Ministry of Health and Family Planning.

Sima, V., Gheorghe, I. G., Subić, J., & Nancu, D. (2020). Influences of the industry 4.0 revolution on the human capital development and consumer behavior: A systematic review. *Sustainability (Basel)*, 12(10), 4035. DOI: 10.3390/su12104035

Singh Rajawat, A., Bedi, P., Goyal, S. B., Shukla, P. K., Zaguia, A., Jain, A., & Monirujjaman Khan, M. (2021). Reformist framework for improving human security for mobile robots in industry 4.0. *Mobile Information Systems*, 2021, 1–10. DOI: 10.1155/2021/4744220

Singh, B. (2023). Blockchain Technology in Renovating Healthcare: Legal and Future Perspectives. In *Revolutionizing Healthcare Through Artificial Intelligence and Internet of Things Applications* (pp. 177-186). IGI Global.

Singh, B., & Kaunert, C. (2024). Future of Digital Marketing: Hyper-Personalized Customer Dynamic Experience with AI-Based Predictive Models. In Revolutionizing the AI-Digital Landscape (pp. 189-203). Productivity Press.

Singh, S. K., Sharma, S. K., Singla, D., & Gill, S. S. (2022). Evolving requirements and application of SDN and IoT in the context of industry 4.0, blockchain and artificial intelligence. *Software Defined Networks: Architecture and Applications*, 427-496.

Singh, S., & Kumar, D.R. (2022). Sexual Offences Against Female Children in India: A Critical Study. *legal researchd development an international refereed e Journal.*

Singh, S., Hussain, R., Shekhar, C., Acharya, R., Moore, A. M., Stillman, M., ... & Ball, H. (2018). Abortion and unintended pregnancy in six Indian states: findings and implications for policies and programs.

Singh, A. (2021). Abortion Law in India in the light of Medical Termination of Pregnancy (Amendment) Act, 2021. *Nyaayshastra Law Review*, 1(3), 1–13.

Singh, B. (2019). Profiling Public Healthcare: A Comparative Analysis Based on the Multidimensional Healthcare Management and Legal Approach. *Indian Journal of Health and Medical Law*, 2(2), 1–5.

Singh, B. (2020). Global science and jurisprudential approach concerning healthcare and illness. *Indian Journal of Health and Medical Law*, 3(1), 7–13.

Singh, B. (2022). COVID-19 Pandemic and Public Healthcare: Endless Downward Spiral or Solution via Rapid Legal and Health Services Implementation with Patient Monitoring Program. *Justice and Law Bulletin*, 1(1), 1–7.

Singh, B. (2022). Relevance of Agriculture-Nutrition Linkage for Human Healthcare: A Conceptual Legal Framework of Implication and Pathways. *Justice and Law Bulletin*, 1(1), 44–49.

Singh, B. (2022). Understanding Legal Frameworks Concerning Transgender Healthcare in the Age of Dynamism. *Electronic Journal of Social and Strategic Studies*, 3(1), 56–65. DOI: 10.47362/EJSSS.2022.3104

Singh, B. (2023). Federated Learning for Envision Future Trajectory Smart Transport System for Climate Preservation and Smart Green Planet: Insights into Global Governance and SDG-9 (Industry, Innovation and Infrastructure). *National Journal of Environmental Law*, 6(2), 6–17.

Singh, B. (2024). Legal Dynamics Lensing Metaverse Crafted for Videogame Industry and E-Sports: Phenomenological Exploration Catalyst Complexity and Future. *Journal of Intellectual Property Rights Law*, 7(1), 8–14.

Singh, M. M., Parsekar, S. S., & Nair, S. N. (2014). An epidemiological overview of child sexual abuse. *Journal of Family Medicine and Primary Care*, 3(4), 430. DOI: 10.4103/2249-4863.148139 PMID: 25657958

Sinha, I., & Singh, R. (2023). Critical Analysis of POCSO: In Light of the Rights of Adolescents. International Journal For Multidisciplinary Research. https://doi.org/DOI: 10.36948/ijfmr.2023.v05i06.10289

Slavchevska, V., Kaaria, S., & Taivalmaa, S.-L. (2019). The Feminization of Agriculture: Evidence and Implications for Food and Water Security. The Oxford Handbook of Food, Water and Society. Sriram, Vidhya (2018). Achieving Gender Equality and Women's Empowerment in Smallholder Adaptation: Lessons from IFAD's Adaptation in Smallholder Agriculture Programme. CCAFS Info Note. Wageningen, the Netherlands: CGIAR Research Program on Climate Change, Agriculture and Food Security (CCAFS). https://ccafs.cgiar.org/publications/achieving-gender-equality-andwomen%E2%80%99s-empowerment-smallholder-adaptation-lessons-ifad%E2%80%99s#.XkxqhiOIY2w

Smail, E. (2023). Gender Equality at COP28: Advancing Climate Justice for Women and Girls. Climate Impacts Tracker Asia. Retrieved from https://www.climateimpactstrackerasia.org/gender-equality-cop28

Smith College Finding Aids. (n.d.). *Mother's Milk Project*. Retrieved April 1, 2023, from https://findingaids.smith.edu/repositories/4/resources/146

Smt. Nivedita Basu v. The State of West Bengal & Ors., W.P. 2513 of 2022.

Somanathan, E., Somanathan, R., Sudarshan, A., & Tewari, M. (2021). The Impact of Temperature on Productivity and Labor Supply: Evidence from Indian Manufacturing. *Journal of Political Economy*, 129(6), 1797–1827. DOI: 10.1086/713733

Sommer, M., Ferron, S., Cavill, S., & House, S. (2015). Violence, gender and WASH: Spurring action on a complex, under-documented and sensitive topic. *Environment and Urbanization*, 27(1), 105–116. DOI: 10.1177/0956247814564528

Sorabjee, S. (2018). *Abortion: Its every women's right to choose*, Hindustan Times (Last updated on Jan 8, 2018 4:39 PM).https://www.hindustantimes.com/opinion/abortion-it-s-every-woman-s-right-to-choose/story-52qv723N3yXESybsvyHR0J.html

Sorensen, C. et al. (2018). *Climate Change and Women's Health: Impacts and Opportunities in India,* PMCID: PMC7007102 PMID: 32159002, *National Library of Medicine National Center for Biotechnology Information*https://www.ncbi.nlm.nih.gov/pmc/articles/PMC7007102/

Soubeyran, E., & Choudhary, K. (2023). *Why do we need to accelerate the mainstreaming of gender into climate action?* London School of Economics and Political Science https://www.lse.ac.uk/granthaminstitute/news/why-do-we-need-to-accelerate-the-mainstreaming-of-gender-into-climate-action/

Sqalli, Z. et al. (2021). *Why Climate Action Needs a Gender Focus,* BCG https://www.bcg.com/publications/2021/climate-action-impact-on-gender-equality

Sreenivasan, A., & Suresh, M. (2024). Start-up sustainability: Does blockchain adoption drives sustainability in start-ups? A systematic literature reviews. *Management Research Review*, 47(3), 390–405. DOI: 10.1108/MRR-07-2022-0519

Sreevas, T. V. T. K., & Kulkarni, P. V. (2024). Women leaders in environmental management. In *Opportunities and Challenges for Women Leaders in Environmental Management* (pp. 173–191). IGI Global.

Statistisches Bundesamt. (2024, June). *SDG 8—Menschenwürdige Arbeit und Wirtschaftswachstum - Deutschlands SDG Indikatoren*. http://sdg-indikatoren.de/8/

Stein, P. J., Stein, M. A., Groce, N., & Kett, M. (2023). The role of the scientific community in strengthening disability-inclusive climate resilience. *Nature Climate Change*, 13(2), 108–109. DOI: 10.1038/s41558-022-01564-6

Suchita Shrivastav and Anr. v. Chandigarh Administrator, SLP (C) 17985 of 2009.

Suchita Srivastava v. Chandigarh Administration, (2009) 14 SCR 989, (2009) 9 SCC 1.

Sundari, S. (2020). Structural changes and quality of women's labour in India. *The Indian Journal of Labour Economics : the Quarterly Journal of the Indian Society of Labour Economics*, 63(3), 689–717. DOI: 10.1007/s41027-020-00245-2 PMID: 33046952

Suresh, N. (2022, February 22). *Explained: How The Union Budget Funds India's Justice System*. Www.indiaspend.com. https://www.indiaspend.com/explainers/how-the-union-budget-funds-indias-justice-system-804821#:~:text=It%20used%20the%20allocated%20Rs

Swati Agarwal & Ors. V. Union of India, W.P.(C) 825/2019

Swiss Re Institute. (2021), *The economics of climate change: no action not an option*, https://www.swissre.com/dam/jcr:e73ee7c3-7f83-4c17-a2b8-8ef23a8d3312/swiss-re-instituteexpertise-publication-economics-of-climate-change.pdf

Tao, F., Akhtar, M. S., & Jiayuan, Z. (2021). The future of artificial intelligence in cybersecurity: A comprehensive survey. *EAI Endorsed Transactions on Creative Technologies*, 8(28), e3–e3. DOI: 10.4108/eai.7-7-2021.170285

Tapasya Umesha Pisal v. Union of India, 2018 (12) SCC 57.

Technoserve. (n.d.). *5 Ways Climate Change Disproportionately Impacts Women | Expert Analysis, Techno Serve Business Solutions to Poverty*. https://www.technoserve.org/blog/5-climate-change-impacts-women/

The Bharatiya Nyaya Sanhita (BNS). 2023, No. 45 Acts of the Parliament, 2023 (India).

The Constitution of India, List II, Entry 4. (1950).

The Medical Termination of Pregnancy (Amendment) Act, 2021 (Act 8 of 2021).

The Nordic Council. (2023). *Bridging the Green Jobs Divide* (norden.org)

The Sasawaka Peace Foundation and BloombergNEF. (2020), "Gender Diversity and Climate UN Environmental Programme. Global Mutilateral Environmental Agreements. https://www.unep.org/explore-topics/oceans-seas/what-we-do/working-regional-seas/partners/global-multilateral

The Star. (2024, April 18). PS wants more involvement of women in climate change decision-making. Retrieved from https://www.the-star.co.ke/news/2024-04-18-ps-wants-more-involvement-of-women-in-climate-change-decision-making/

The Universal Declaration of Human Rights (UDHR),GA Res 217 A, GAOR, UN Doc A/Res/ 217/A. (1948). U.N. General Assembly.

The Universal Declaration of Human Rights, Article 5 (1948). International Covenant on Civil and Political Rights (ICCPR), GA Res 2200A, GAOR, UN Doc A/Res/ 2200/A, (1966).

There's Something In The Water. (n.d.). Fernwood Publishing. Retrieved April 17, 2023, from https://fernwoodpublishing.ca/book/theres-something-in-the-water

Tindall, D. B., & Robinson, J. L. (2017). Collective action to save the ancient temperate rainforest: Social networks and environmental activism in Clayoquot Sound. *Ecology and Society*, 22(1), art40. Advance online publication. DOI: 10.5751/ES-09042-220140

Topping, K., & Barron, I. (2009). School-Based Child Sexual Abuse Prevention Programs: A Review of Effectiveness. *Review of Educational Research*, 79(1), 431–463. DOI: 10.3102/0034654308325582

Torgalkar, V. (2022). *India's women farm workers suffer job losses from climate shocks*. Thomson Reuters Foundation. https://www.context.news/climate-risks/indias-women-farm-workers-suffer-job-losses-from-climate-shocks

Torres Strait Islanders v. Australia, UN Doc CCPR/C/135/D/3624/2019.

Tripathy, B., & Mondal, T. (2019). Impact of Riverbank erosion on Human life. *Thematics Journal of Geography*, 8(09), 53–56. DOI: 10.26643/tjg.v8i9.8101

Tseng, M. L., Tran, T. P. T., Ha, H. M., Bui, T. D., & Lim, M. K. (2021). Sustainable industrial and operation engineering trends and challenges Toward Industry 4.0: A data driven analysis. *Journal of Industrial and Production Engineering*, 38(8), 581–598. DOI: 10.1080/21681015.2021.1950227

Tsosie, R. (1996). Tribal environmental policy in an era of self-determination: The role of ethics, economics, and traditional ecological knowledge. *Vermont Law Review*, 21, 225.

Tyagi, S., & Karande, S. (2021). Child sexual abuse in India: A wake-up call. *Journal of Postgraduate Medicine*, 67(3), 125–129. DOI: 10.4103/jpgm.JPGM_264_21 PMID: 34380802

U.N. General Assembly. (2011, March 16). *United Nations Rules for the Treatment of Women Prisoners and Non-custodial Measures for Women Offenders (the Bangkok Rules), GA Res 65/229, UN Doc A/RES/65/229.*

U.N. General Assembly. (2016, January 8). *United Nations Standard Minimum Rules for the Treatment of Prisoners (the Nelson Mandela Rules), GA Res A/RES/70/175* .

Übereinkommen 138 Übereinkommen Über Das Mindestalter Für Die Zulassung Zur Beschäftigung, 1973. Retrieved June 17, 2024, from https://www.ilo.org/sites/default/files/wcmsp5/groups/public/@ed_norm/@normes/documents/normativeinstrument/wcms_c138_de.htm

Übereinkommen 182 Übereinkommen Über Das Verbot Und Unverzügliche Maßnahmen Zur Beseitigung Der Schlimmsten Formen Der Kinderarbeit, 1999. Retrieved June 16, 2024, from https://www.ilo.org/sites/default/files/wcmsp5/groups/public/@ed_norm/@normes/documents/normativeinstrument/wcms_c182_de.htm

Übereinkommen über die Rechte des Kindes - VN-Kinderrechtskonvention im Wortlaut mit Materialien. (2022). https://www.bmfsfj.de/resource/blob/93140/fe59de84a8fc3a6ffc61e8a5559cac9d/uebereinkommen-ueber-die-rechte-des-kindes-data.pdf

Udeagha, M. C., & Ngepah, N. (2023). The drivers of environmental sustainability in BRICS economies: Do green finance and fintech matter? *World Development Sustainability*, 3, 100096. DOI: 10.1016/j.wds.2023.100096

UN CTCN. (2019). *The Women and Gender Constituency, Bhungroo – small women farmers owning and developing an innovative rainwater saving technology, UN Climate Technology Centre and Network UNFCCC Technology Mechanism*https://www.ctc-n.org/resources/bhungroo-small-women-farmers-owning-and-developing-innovative-rainwater-saving-technology

UN General Assembly. (2015). Transforming our world: the 2030 Agenda for Sustainable Development. https://www.un.org/ga/search/view_doc.asp?symbol=A/RES/70/1&Lang=E

UN Global Compact Netzwerk Deutschland e. V. (2022, May 17). *Unternehmerisches Handeln muss Kinderrechte schützen – Zehn Jahre Grundsätze zum Schutz und zur Förderung von Kinderrechten durch Unternehmen*. https://www.globalcompact.de/news/meldung/unternehmerisches-handeln-muss-kinderrechte-schuetzen

UN Women Watch. (1997). *Coordination Segment, Chapter IV,* Excerpt from A/52/3 https://www.un.org/womenwatch/osagi/pdf/ECOSOCAC1997.2.PDF

UN Women Watch. (2024). *Women, Gender Equality and Climate Change*https://www.un.org/womenwatch/feature/climate_change/

UN Women Watch. (n.d.). *Gender Equality and Climate Change*https://www.un.org/womenwatch/feature/climate_change/downloads/Women_and_Climate_Change_Factsheet.pdf

UN Women. (2023). *As climate change pushes millions of women into poverty, UN Women calls for a new feminist climate justice approach,*https://www.unwomen.org/en/news-stories/press-release/2023/12/as-climate-changes-pushes-millions-of-women-into-poverty-un-women-calls-for-a-new-feminist-climate-justice-approach

UN Women. (2023). In focus: *Women's leadership at the COP 28 climate conference*, https://www.unwomen.org/en/news-stories/in-focus/2023/11/in-focus-womens-leadership-at-the-cop-28-climate-conference

UN Women. (2024). *Explainer: How gender inequality and climate change are interconnected.* https://www.unwomen.org/en/news-stories/explainer/2022/02/explainer-how-gender-inequality-and-climate-change-are-interconnected

UN, & UNHR. (2011). *Guiding principles on business and human rights: Implementing the United Nations "protect, respect and remedy" framework.* United Nations. DOI: 10.18356/9ceabfd3-en

UNDESA. (2015). *Report Expert Group Meeting Implementation Of Gender-Responsive Climate Action In The Context Of Sustainable Development*https://unfccc.int/files/gender_and_climate_change/application/pdf/egmreport.pdf

UNDP. (2001). *Generating Opportunities: Case Studies on Energy and Women.* https://www.undp.org/publications/generating-opportunities-case-studies-energy-and-women

UNDP. (2019). *Ghana Gender*Https://Climatepromise.Undp.Org/Sites/Default/Files/Research_Report_Document/Undp-Ndcsp-Ghana-Gender-Analysis-Final2.Pdf

UNDP. (2022). Climate Promise Progress Report, New York, NY: UNDP. https://climatepromise.undp.org/sites/default/files/research_report_document/Climate%20Promise%20Global%20Progress%20Report%202022.pdf

UNDP. (n.d.). *Accelerating Opportunities for Women Climate Entrepreneurs.* https://www.undp.org/india/accelerating-opportunities-women-climate-entrepreneurs

UNDP. 2022. Climate Promise Progress Report, New York, NY: UNDP. https://climatepromise.undp.org/sites/default/files/research_report_document/Climate%20Promise%20Global%20Progress%20Report%202022.pdf

Une si longue lettre. (1979). Wikipedia. Retrieved March 28, 2023, from https://fr.wikipedia.org/wiki/Une_si_longue_lettre

UNFCCC. (2023). *Five Reasons Why Climate Action Needs Women,* United Nation Climate Change,https://unfccc.int/news/five-reasons-why-climate-action-needs-women

UNFCCC. (2023a). *Implementation of just transition and economic diversification strategies*

unfpa.org. (2022). *State of World Population Report.* (Last updated on Apr. 1, 2022), https://india.unfpa.org/sites/default/files/pub-pdf/1_key_insights_india_final_web.pdf

UNHR, & United Nations Global Compact. (2014, June). *The UN Guiding Principles on Business and Human Rights: Relationship to UN Global Compact Commitments.* https://www.business-humanrights.org/en/latest-news/pdf-the-un-guiding-principles-on-business-and-human-rights-relationship-to-un-global-compact-commitments/

UNICEF, & Danish Institute for Human Rights. (2013, December). *Children's Rights in Impact Assessments: A guide for integrating children's rights into impact assessments and taking action for children.* https://resource-centre.savethechildren.net/document/childrens-rights-impact-assessments-guide-integrating-childrens-rights-impact-assessments/

UNICEF, & Save the Children Deutschland e.V. (2013, September). *Children's Rights in Policies and Codes of Conduct. A tool for companies.* UNICEF. https://resource-centre.savethechildren.net/document/childrens-rights-policies-and-codes-conduct-tool-companies/

UNICEF. (2011). *Children's Rights and Business Principles | UNICEF*. https://www.unicef.org/documents/childrens-rights-and-business-principles

UNICEF. (2014a, April). *Children's Rights in Sustainability Reporting—A guide for integrating children's rights into the GRI reporting framework*. UNICEF. https://www.business-humanrights.org/en/latest-news/pdf-childrens-rights-in-sustainability-reporting-a-guide-for-integrating-childrens-rights-into-the-gri-reporting-framework/

UNICEF. (2014b, August). *Children Are Everyone's Business Workbook 2.0—A guide for integrating children's rights into policies, impact assessments and sustainability reporting*. UNICEF. https://www.business-humanrights.org/en/latest-news/pdf-children-are-everyones-business-workbook-20-a-guide-for-integrating-childrens-rights-into-policies-impact-assessments-and-sustainability-reporting/

UNICEF. (2022). *Study On The Role Of Mitigation-Adaptation Co-Benefits For Creating A More Resilient Future For All, Climate Sustainability Working Group (CSWG)* G20 Indonesia 2022 https://www.unicef.org/indonesia/media/17201/file/Study%20on%20The%20Role%20of%20Mitigation-Adaptation%20Co-Benefits%20For%20Creating%20A%20More%20Resilient%20Future%20For%20All.pdf

UNICEF. (2023, November 1). The climate-changed child. United Nations International Children's Emergency Fund. https://www.unicef.org/reports/climate-changed-child#download-the-report

UNICEF. (2024, June 11). *CSR - Verantwortung übernehmen*. UNICEF Österreich. https://unicef.at/mitmachen/als-unternehmen-helfen/csr-verantwortung-uebernehmen/

United Nation Climate Change. (2023) *Bhungroo*. https://unfccc.int/climate-action/momentum-for-change/women-for-results/bhungroo

United Nation Climate Change. (n.d.). *Introduction to Gender and Climate Change-* https://unfccc.int/gender

United Nations Conference on Environment and Development (UNCED), Rio de Janeiro, Brazil, (1992, June). Conferences Environment and sustainable development. https://www.un.org/en/conferences/environment/rio1992

United Nations Department of Economic and Social Affairs. (n.d.). *Water and food security*. United Nations. https://www.un.org/waterforlifedecade/food_security.shtml

United Nations Development Program. (2016). Gender, climate change and food insecurity.

United Nations Development Program. (2021). Green Jobs and Eco-Entrepreneurship opportunities for women in India.

United Nations Development Programme. (2024). *Mexico Case Study, Gender Equality And Women's Empowerment In Public Administration*, https://www.undp.org/sites/g/files/zskgke326/files/publications/es/MexicoFinal%20-%20HiRes.pdf

United Nations Environment Programme, Climate Change and Human Rights, https://www.unep.org/resources/report/climate-change-and-human-rights (last visited July 8, 2024).

United Nations Framework Convention on Climate Change (UNFCCC). (n.d.). Gender and Climate Change. Retrieved from https://unfccc.int/topics/gender/the-big-picture/gender-and-climate-change

United Nations Framework Convention on Climate Change, Climate Change: Impacts, Vulnerabilities, and Adaptation in Developing Countries, https://unfccc.int/resource/docs/publications/impacts.pdf (last visited July 8, 2024).

United Nations Global Compact. (2024a). Business & Human Rights Navigator. *Business & Human Rights Navigator*. https://bhr-navigator.unglobalcompact.org/issues/kinderarbeit/weiterfuehrende-informationen/?lang=de

United Nations Global Compact. (2024b). Business & Human Rights Navigator engl. *Business & Human Rights Navigator*. https://bhr-navigator.unglobalcompact.org/issues/child-labour/#further-guidance

United Nations New York. (2002). *Gender Mainstreaming: An Overview, Office of the Special Adviser on Gender Issues and Advancement of Women,*https://www.un.org/womenwatch/osagi/pdf/e65237.pdf

United Nations Office of the High Commissioner for Human Rights. Mapping Human Rights Obligations Relating to the Enjoyment of a Safe, Clean, Healthy and Sustainable Environment, https://www.ohchr.org/Documents/Issues/Environment/MappingReport/MappingHRObligationsRelationClimateChange.pdf (last visited July 8, 2024).

United Nations Special Rapporteur on Human Rights and the Environment, Human Rights Obligations Relating to the Enjoyment of a Safe, Clean, Healthy and Sustainable Environment. (2018), https://www.ohchr.org/Documents/Issues/Environment/SREnvironment/A_HRC_37_59.pdf (last visited July 8, 2024).

United Nations Women. (2022). *State of Gender Equality and Climate Change in Bangladesh*.

United Nations. (1994), Population and Development: Programme of Action adopted at the International Conference on Population and Development Cairo, 5–13 September 1994, Department of Economic and Social Affairs.

United Nations. (2009). *Women...In The Shadow of Climate Change From Nos3 & 4, VolXlvi, Special Climate Change Issue: "To Protect Succeeding Generations.* https://www.un.org/en/chronicle/article/womenin-shadow-climate-change

United Nations. (n.d.). Climate change for indigenous peoples. United Nations. https://www.un.org/development/desa/indigenouspeoples/climate-change.html

Uranium. (n.d.). The Mining Association of Nova Scotia. Retrieved April 1, 2023, from https://tmans.ca/uranium

USAID From the American People. (2022). *Gender Equality and Climate Finance.* https://www.climatelinks.org/sites/default/files/asset/document/2022-11/Gender%20Equality%20and%20Climate%20Finance_Technical%20Brief%20Nov%202022.pdf

Vahí, A. O. (2022). *COP27 and the centrality of human rights in Climate Action.* Universal Rights Group. https://www.universal-rights.org/cop27-human-rights/

Varghese, D. P., & John, S. (2021). The protection of children from sexual offences act, 2012: Strong provisions, weak implementation? *Iustitia (Bloomington, Ind.)*, 12(1), 95–123.

Vasavada, S. (2024, Mar. 26) *Women Farmers' Land Ownership and Climate Resilience.* India Development Review.

Vedam, S., Stoll, K., Tarasoff, L., Phillips-Beck, W., Lo, W., MacDonald, K., Metellus, A., Rost, M., Scott, M., Hodge, K., Korchinski, M., van der Pijl, M., Alonso, C., Clark, E., Tatum, A., Olson, R., Xie, K., Decker, M., Wenzel, K., & Hall, W. (2024). The RESPCCT Study: Community-led Development of a Person-Centered Instrument to Measure Health Equity in Perinatal Services. *Journal of Participatory Research Methods*, 5(1), 1. DOI: 10.35844/001c.94399

Venugopal, V.. (2016). Occupational Heat Stress Profiles in Selected Workplaces in India. *International Journal of Environmental Research and Public Health.* PMID: 26729144

Vijay, P., & Sharma, U. S. (2002). Conceptual Contours of Human Rights and Reproductive Freedom of Women in India. *Journal of Coastal Life Medicine*, 1(11), 594–600.

Villiers, C. (2023). A game of cat and mouse: Human rights protection and the problem of corporate law and power. *Leiden Journal of International Law*, 36(2), 415–438. DOI: 10.1017/S0922156522000632

Vineeta Sharma v. Rakesh Sharma & Ors.[(2020) 9 SCC 1]

Vogt, J. (2021). Where is the human got to go? Artificial intelligence, machine learning, big data, digitalisation, and human–robot interaction in Industry 4.0 and 5.0: Review Comment on: Bauer, M.(2020). Preise kalkulieren mit KI-gestützter Onlineplattform BAM GmbH, Weiden, Bavaria, Germany. *AI & Society*, 36(3), 1083–1087. DOI: 10.1007/s00146-020-01123-7

Wahlsten, D. (2004). Leilani Muir versus the Philosopher King: Eugenics on trial in Alberta. *Genetica*, 99(2-3), 185–198. DOI: 10.1007/BF02259522 PMID: 9463073

Wain, D. (2021). An Ecofeminist Treatment of Nourishment and Feeding in Margaret Atwood's MaddAddam Trilogy. In Dystopias and Utopias on Earth and Beyond (pp. 25–36). DOI: 10.4324/9781003152989-4

Wamba-Taguimdje, S. L., Fosso Wamba, S., Kala Kamdjoug, J. R., & Tchatchouang Wanko, C. E. (2020). Influence of artificial intelligence (AI) on firm performance: The business value of AI-based transformation projects. *Business Process Management Journal*, 26(7), 1893–1924. DOI: 10.1108/BPMJ-10-2019-0411

Wan, J., Yang, J., Wang, Z., & Hua, Q. (2018). Artificial intelligence for cloud-assisted smart factory. *IEEE Access : Practical Innovations, Open Solutions*, 6, 55419–55430. DOI: 10.1109/ACCESS.2018.2871724

Wapler, F. (2019). Umsetzung und Anwendung der Kinderrechtskonvention in Deutschland. *Recht der Jugend und des Bildungswesens*, 67(3), 252–273. DOI: 10.5771/0034-1312-2019-3-252

Westra, L. (2014). *Child Law: Children's Rights and Collective Obligations*. Springer. DOI: 10.1007/978-3-319-05071-3

WHO. (2021), India's Amended Law Makes Abortion Safer And More Accessible, https://www.who.int/india/news/detail/13-04-2021-india-s-amended-law-makes-abortion-safer-and-more-accessible

WHO. (2024), https://www.who.int/news/item/16-05-2024-sexual-and-reproductive-health-for-all-20-years-of-the-global-strategy

Wilson, K. (2006). Ecofeminism and First Nations Peoples in Canada: Linking culture, gender and nature. *Gender, Place and Culture*, 12(3), 333–355. DOI: 10.1080/09663690500202574

Wilson, K. (2018). For reproductive justice in an era of Gates and Modi: The violence of India's population policies. *Feminist Review*, 120(1), 23–39. DOI: 10.1057/fr.1982.13

Winkler, E. (2024, June 12). *Kinderarbeit: Konsumentinnen können Druck auf Unternehmen ausüben | MDR.DE* [MDR]. UNICEF zu Kinderarbeit: Konsumentinnen können Druck auf Unternehmen ausüben. https://www.mdr.de/nachrichten/deutschland/wirtschaft/kinderarbeit-produkte-erkennen-lieferkettengesetz-100.html

Winston, M. (2002). NGO Strategies for Promoting Corporate Social Responsibility. *Ethics & International Affairs*, 16(1), 71–87. DOI: 10.1111/j.1747-7093.2002.tb00376.x

Women Economic Empowerment. (n.d.). *Global Alliance For Mass Entrepreneurship* https://massentrepreneurship.org/women-economic-empowerment/

Women, U. N. (2014). Gender Equality and Sustainable Development. Retrieved from https://www.unwomen.org/en/digital-library/publications/2014/1/gender-equality-and-sustainable-development

Women, U. N. (2022). Human Rights, the Environment and Gender Equality. Retrieved from https://www.unwomen.org/en/policy-paper-human-rights-environment-gender-equality

Women, U. N. (2022, February 28). *Explainer: How gender inequality and climate change are interconnected*. UN Women. https://www.unwomen.org/en/news-stories/explainer/2022/02/explainer-how-gender-inequality-and-climate-change-are-interconnected

World Economic Forum. (2023). The future of jobs is green: How climate change is changing labor markets.

World Health Organization. (2023). *Climate change*. https://www.who.int/news-room/fact-sheets/detail/climate-change-and-health#:~:text=Research%20shows%20that%203.6%20billion,diarrhoea%20and%20heat%20stress%20alone

Writ Petition (Civil) No. 406 OF 2013, (Supreme Court of India, 2013).

Wunneburger, D.F., Olivares, M., & Maghelal, P.K. (2008). *Internal Security for Communities: A Spatial Analysis of the Effectiveness of Sex Offender Laws*.

X (minor) through her father v. State of Uttarakhand & Ors. W.P No. 201 of 2022.

X v. Principal Secretary Health and Family Welfare Department, (2022) 2022 SCC OnLine SC 1321.

Xiao, C. and A. McCright (2012), “Explaining Gender Differences in Concern about Environmental Problems in the United States”,Vol. 25/11, pp. 1067-1084, .DOI: 10.1080/08941920.2011.651191

Xiao, C. and D. Hong (2010), “Gender differences in environmental behaviors in China”.

Xiao, C., & Hong, D. (2010, September). Gender differences in environmental behaviors in China. *Population and Environment*, 32(1), 88–104. DOI: 10.1007/s11111-010-0115-z

Xing, K., Cropley, D. H., Oppert, M. L., & Singh, C. (2021). Readiness for digital innovation and industry 4.0 transformation: studies on manufacturing industries in the city of salisbury. *Business Innovation with New ICT in the Asia-Pacific: Case Studies*, 155-176.

Xu, L. D., Xu, E. L., & Li, L. (2018). Industry 4.0: State of the art and future trends. *International Journal of Production Research*, 56(8), 2941–2962. DOI: 10.1080/00207543.2018.1444806

XXXX v. State of Telangana, W.P. 24850 OF 2021.

Yan, C., Siddik, A. B., Yong, L., Dong, Q., Zheng, G. W., & Rahman, M. N. (2022). A two-staged SEM-artificial neural network approach to analyze the impact of FinTech adoption on the sustainability performance of banking firms: The mediating effect of green finance and innovation. *Systems*, 10(5), 148. DOI: 10.3390/systems10050148

Yang, Y., Su, X., & Yao, S. (2021). Nexus between green finance, fintech, and high-quality economic development: Empirical evidence from China. *Resources Policy*, 74, 102445. DOI: 10.1016/j.resourpol.2021.102445

Yonzon, K. & Bisht, P. (2023). *Unlocking Climate Financing for India's Social Enterprises: Challenges and Opportunities*, SAMBODHI https://sambodhi.co.in/unlocking-climate-financing-for-indias-social-enterprises-challenges-and-opportunities/

Yue, Y., & Shyu, J. Z. (2024). A paradigm shift in crisis management: The nexus of AGI-driven intelligence fusion networks and blockchain trustworthiness. *Journal of Contingencies and Crisis Management*, 32(1), e12541. DOI: 10.1111/1468-5973.12541

Z v. The State of Bihar and Ors. MANU/SC/1011/2017.

Zakrzewska, A. & Fitzgerald, L. (2023), *Empowering Women to Power the Green Transition, Cogito* https://oecdcogito.blog/2023/03/08/empowering-women-to-power-the-green-transition/

Zhong, R. Y., Xu, X., Klotz, E., & Newman, S. T. (2017). Intelligent manufacturing in the context of industry 4.0: A review. *Engineering (Beijing)*, 3(5), 616–630. DOI: 10.1016/J.ENG.2017.05.015

Zhou, J., Zhang, S., Lu, Q., Dai, W., Chen, M., Liu, X., . . . Herrera-Viedma, E. (2021). A survey on federated learning and its applications for accelerating industrial internet of things. *arXiv preprint arXiv:2104.10501*.

Zusman, E., Lee, S-Y., Rojas, A., & Adams, L. (2024). *Mainstreaming Gender Into Climate Mitigation Activities, Guidelines for Policy Makers and Proposal Developers*, https://www.adb.org/sites/default/files/publication/217771/gender-climate-mitigation.pdf

About the Contributors

Swati Chakraborty is a Visiting faculty (Assistant Professor), at the Institute of Legal Studies and Research, GLA University. Dr. Chakraborty is associated as an Adjunct Faculty with Concordia University, Canada. Dr. Chakraborty is a Writing Fellow, at COMPOSE, Universitas Islam International Indonesia. She is an International Fellow of KAICIID in Lisbon, Portugal. She is the Founder of "Webplatform4Dialogue"

Subrata Satapathy is an Assistant Professor in Sociology Christ Academy Institute of Law Karnataka State Law University Bangalore, Karnataka

***

Mumtaz Alam, Director for Centre of Graduate Studies and an Associate Professor in History at Fiji National University, earned his doctorate from Aligarh Muslim University. His research centres on medical history in the Pacific and Southeast Asia, particularly emphasising how colonial and postcolonial contexts have influenced these regions. Dr Alam's research ingeniously merges social and medical history by incorporating written records and oral testimonies. Health, Medicine, and Encounter of Cultures and Rethinking Nationalism: Exploring Its Roots, Varieties, and Theories were published by him in 2022 and 2023, respectively. Dr Alam has been recognised for his accomplishments as an active researcher and member of esteemed organisations such as IHC, PHA, AHA, and ANZSHM. In 2011, he was awarded the Best Paper Prize in Science and Technology by the Indian History Congress. The scope of his extensive research background includes ecotourism, smallpox vaccination in Fiji, environmental history, and climate history. As the editor-in-chief of several prestigious international journals, Dr. Alam substantially contributes to historical scholarship.

Himanshi Bhatia is an Assistant Professor at the School of Law, Graphic Era Hill University, Dehradun, India.

Anwesha Biswas is engaged as an Invited Lecturer in the Department of French at Syamaprasad College, kolkata. She holds a Master's Degree in French Language & Literature from the University of Calcutta, where she secured the First Class First position in the order of merit. She is also a member of the Indian Association of French Teachers. Her research interests include Feminism, Africain francophone literature, Black feminism, Mariama Bâ, Ashapurna Devi, Comparative literature, Gender Studies.

Christopher Bohlens is a junior researcher at the Leuphana University of Lüneburg. He is also head of the science working group at Transparency International Germany. He previously worked for several years in student self-administration, where he represented the interests of students nationwide. As an expert, he is also committed to the participation and representation of students' interests in quality assurance at universities. He also campaigns for open data and transparency laws. He studied business administration, economics, political science and law at the Hamburg University of Applied Sciences, Leuphana University and FernUni Hagen.

Saurabh Chandra is currently working as an Associate Professor with School of Law, Bennett University, India. He has completed graduation in Law (B.A.LL.B.) from Aligarh Muslim University (AMU). He has completed Post Graduation (LL.M. Business Laws) from National Law School of India University (NLSIU) Bangalore. He has completed Doctorate in Law (Ph.D.) From National Law University, Delhi. In addition to earning degrees in Law from a prestigious institution in India, he also holds a master's degree in Management and another in Journalism and Mass Communication. He is having teaching experience of 16 years. His areas of expertise include Business and Corporate Laws and he has various publications to his credit. He has rich experience of Academic Administration and has handled various posts of Academic Administration.

Sumedha Dey currently work as a Research Assistant at the premier research institute Centre for Studies in Social Sciences, Calcutta.

Barkha Dodai is an Assistant Professor of Law at Ramaiah College of Law,Bengaluru,India

Vishal Eswaran is an accomplished Senior Big Data Engineer with an impressive career spanning over 6 years. His fervor for constructing robust data pipelines, unearthing insights from intricate datasets, identifying trends, and predicting future trajectories has fueled his journey. Throughout his tenure, Vishal has lent his expertise to empower numerous prominent US healthcare clients, including CVS Health, Aetna, and Blue Cross and Blue Shield of North Carolina, with informed business decisions drawn from expansive datasets. Vishal's ability to distill intricate data into comprehensive documents and reports stands as a testament to his proficiency in managing multifaceted internal and external data analysis responsibilities. His aptitude for synthesizing complex information ensures that insights are both accessible and impactful for strategic decision-making. Moreover, Vishal's distinction extends to his role as a co-author of the book "Internet of Things - Future Connected Devices." This book not only underscores his prowess in the field but also showcases his visionary leadership in the realm of Internet of Things (IoT). His insights resonate with a forward-looking perspective, emphasizing the convergence of technology and human life. As the author of "Secure Connections: Safeguarding the Internet of Things (IoT) with Cybersecurity," Vishal Eswaran's reputation as a thought leader is further solidified. His work is a manifestation of his commitment to ensuring the security of interconnected devices within the IoT landscape, a vital consideration in our digitally driven world. Vishal's dedication to enhancing the safety and integrity of IoT ecosystems shines through in his wo

Vivek Eswaran With 8 years of experience as a Senior Software Engineer specializing in front-end development, Vivek Eswaran brings a vital perspective to securing the Internet of Things (IoT). At Medallia, Vivek played an instrumental role in crafting engaging user interfaces and optimized digital experiences. This profound expertise in front-end engineering equips them to illuminate the crucial synergy between usability and cybersecurity as IoT adoption accelerates. In the new book "Secure Connections: Safeguarding the Internet of Things with Cybersecurity," Vivek combines their real-world experiences building intuitive and secure software systems with cutting-edge insights into strengthening IoT ecosystems. Drawing parallels between front-end best practices and security imperatives, they offer readers an invaluable guide for fortifying IoT without compromising usability. As businesses and consumers continue rapidly connecting people, processes, and devices, Vivek's contribution provides timely insights. Blending user empathy with security proficiency, Vivek empowers audiences to realize the potential of IoT through resilient and human-centered systems designed for safety without friction.

Ushaa Eswaran Dr. Ushaa Eswaran is an esteemed author, distinguished researcher, and seasoned educator with a remarkable journey spanning over 34 years, dedicated to advancing academia and nurturing the potential of young minds. Currently serving as a Principal and Professor in Andhra Pradesh, India, her vision extends beyond imparting cutting-edge technical expertise to encompass the nurturing of universal human values. With a foundation in Electronics Engineering, Dr. Eswaran delved into the realm of biosensors, carving a pioneering path in nanosensor models, a remarkable achievement that earned her a well-deserved Doctorate. Her insights have been encapsulated in her acclaimed book, "Internet of Things: Future Connected Devices," offering profound insights into the evolving IoT landscape. Her expertise also finds its place in upcoming publications centered around computer vision and IoT technologies. Dr. Eswaran's commitment to literature is rooted in her unwavering passion to equip the younger generation with the latest knowledge fortified by ethical principles. Her book stands as a beacon of practical wisdom, providing a roadmap through the intricate IoT terrain while shedding light on its future societal impacts. Her forthcoming contributions unveil her interdisciplinary perspective, seamlessly integrating electronics, nanotechnology, and computing. Bolstering her scholarly contributions is her ORCID identifier, 0000-0002-5116-3403, a testament to her prolific research journey that encompasses over a hundred published papers. Dr. Eswaran thrives in merging her profound academic insights with her dedication to nurturing holistic student growth. Her tireless exploration of the dynamic interface between technology and human values continues to shape her works. As the author of "Secure Connections: Safeguarding the Internet of Things (IoT) with Cybersecurity," Dr. Ushaa Eswaran's voice emerges as a beacon of wisdom in the realm of IoT. Her work encapsulates her dedication to enhancing the interconnected world while ensuring its resilience against cyber threats.

Mahesh K. Gaur is an Indian, who is working at the ICAR-Central Arid Zone Research Institute, Jodhpur, India as a Principal Scientist, since September 2020. He specializes in aridlands geography and the application of satellite remote sensing, GIS and digital image processing for natural resources mapping, management and assessment. He also researches drought, desertification, land degradation, indigenous knowledge systems, and the socio-economic milieu of the Thar Desert as well as Cold Desert of Ladakh, India. Prior to this, he has worked as an Associate Professor for more than 13 years with the Department of Higher Education, Government of Rajasthan State, India. He has published 70 research papers in national and international research journals, 50 conference papers, 45 book chapters, five popular articles, and ten books

Kunal Avishek Gounder is a master's research student working on climate change.

Jay Gupta is a fourth year BBA LL.B.(Hons.) student from SVKM Narsee Monjee Institute of Management Studies, Bengaluru. From the starting of my academic career itself, I have developed deep passion for research paper publications. I have published my previous research papers in various esteemed journals and books. I am also a avid reader. I have huge interest in Climate Change and Environmental aspects.

Megha Gupta is an Independent Legal Practitioner. She completed her Bachelor's Degree in Law. Building upon her foundational knowledge, she pursued her Post Graduation, LL.M. in Criminal Law from Symbiosis Law School, Noida

Christian Kaunert is Professor of International Security at Dublin City University, Ireland. He is also Professor of Policing and Security, as well as Director of the International Centre for Policing and Security at the University of South Wales. In addition, he is Jean Monnet Chair, Director of the Jean Monnet Centre of Excellence and Director of the Jean Monnet Network on EU Counter-Terrorism (www.eucter.net).

Malay Mandal is a PhD Research scholar in University of Hyderabad working in the area of Gender, sexuality, visual culture and performing arts.

Keerthana Murali With over 5 years of experience as a Site Reliability Engineer at Dell, Keerthna Murali has honed an intricate expertise in maintaining and optimizing robust digital infrastructures. On the frontlines of ensuring seamless online experiences, Keerthna specialized in troubleshooting complex issues and proactively enhancing system performance and availability. These capabilities uniquely position them to tackle the critical challenge of security for rapidly emerging IoT ecosystems. In their new book "Secure Connections," Keerthna channels their real-world experiences maintaining enterprise-scale platforms into a compelling vision for building security into the foundation of IoT systems. Blending software engineering best practices with cybersecurity insights, they offer a prescient guide for developers, IT leaders, and security experts seeking to realize IoT's potential while mitigating its risks.

Vidya R Serving as Assistant Professor in Law at Christ Academy Institute of Law, Bengaluru. Prior to this served organizations like Kozhikode Chapter of ICAI, PALSAR, Kozhikode, Markaz Law College, Kozhikode, KMCT Law College, Kuttippuram in teaching law. Other experience as Placement Manager at ICFAI Academy, Hyderabad and Everonn Business Education Ltd, Chennai. Practiced as an Advocate for 10 years. Research Scholar at Shri Venkateshwara University, Gajraula, U.P LLM (Constitutional Law) from University of Kerala, Thiruvananthapuram. PGDIM from IGNOU

Shyamashree Roy is an Assistant Professor in Political Science, Sister Nibedita Government General Degree College for Girls, West Bengal Education Service, India. She completed her M.A. and M.Phil. in International Relations from Jadavpur University, Kolkata, India. Her areas of interest are International Relations theory, foreign policy, South Asian politics and Western Political Theory. She has presented papers at several national and international conferences and contributed many chapters and articles to edited volumes and journals.

Tanmeet Kaur Sahiwal is currently serving as the Chairperson at the School of Law, SVKM's NMIMS (Bengaluru campus). With over 6 years of experience as an Assistant Professor of Law, I am passionate about shaping the legal minds of the future. In addition to my academic roles, I proudly hold the position of COO (Legal) at the Human Rights Protection Group & MFP Federation, contributing to the intersection of law and human rights issues. As a dedicated academician, I am honored to serve as a PhD Supervisor and Head of the Research Committee, fostering a culture of inquiry and scholarly pursuit within the academic community.

Karun Sanjaya is an Assistant Professor at Symbiosis Law School, Nagpur and a Research Scholar at VIT School of Law, Vellore Institute of Technology, Chennai, India. He specialises in the area of Intellectual Property Laws. He is a recipient of many awards, including the prestigious Justice V.R. Krishna Iyer Memorial Merit Awards in the years 2019 and 2022. He has published over fifty research items and is a keen researcher in diverse areas of law, including intellectual property law.

Abhilash Arun Sapre is an Assistant Professor of Law at Gujarat National Law University,Gandhinagar

Smita Satapathy is an Assistant Professor in School of Law, Alliance University, Bengaluru. She holds a Ph.D. in Law from Siksha 'O' Anusandhan, (Deemed to be University), Bhubaneswar, Odisha. She completed her LL.M. from National Law University Odisha, Cuttack. Her areas of interest include Criminal Law, Human Rights, International Law and Environmental Law.

Bhupinder Singh working as Professor at Sharda University, India. Also, Honorary Professor in University of South Wales UK and Santo Tomas University Tunja, Colombia. His areas of publications as Smart Healthcare, Medicines, fuzzy logics, artificial intelligence, robotics, machine learning, deep learning, federated learning, IoT, PV Glasses, metaverse and many more. He has 3 books, 139 paper publications, 163 paper presentations in international/national conferences and seminars, participated in more than 40 workshops/FDP's/QIP's, 25 courses from international universities of repute, organized more than 59 events with international and national academicians and industry people's, editor-in-chief and co-editor in journals, developed new courses. He has given talks at international universities, resource person in international conferences such as in Nanyang Technological University Singapore, Tashkent State University of Law Uzbekistan; KIMEP University Kazakhstan, All'ah meh Tabatabi University Iran, the Iranian Association of International Criminal law, Iran and Hague Center for International Law and Investment, The Netherlands, Northumbria University Newcastle UK,

Surima Singh is a recent graduate of Christ Academy Institute of Law with a Bachelor of Laws degree. Her education has provided her with a strong foundation in legal principles, and she is now embarking on her legal career.

Subin Thomas has completed LL.M from Acharya Nagarjuna University, Guntur, Andhra Pradesh, specializing in Constitutional and Administrative Law. Currently, he is pursuing a PhD from the Alliance School of Law, Alliance University, Bangalore. His research interests include Constitutional Law, Human Rights, Minority Rights and Education.

Shilpy Verma is PhD in Economics from Centre for International Trade and Development, Jawaharlal Nehru University. Presently, she is working with the Amity School of Economics, Amity University, Noida as Assistant Professor. Previously she has worked in Public Debt Management Cell, Department of Economic Affairs, Ministry of Finance as a Consultant prior to this she has worked with the Ministry of Textiles, Government of India as a Young Professional, Department of Commerce, Ministry of Commerce and Industry as a Consultant, Indian Institute of Foreign Trade as Research Associate. she has a close to 2 years experience of teaching in Delhi University as lecturer, 4 Scopus indexed papers publications and 1 book chapter.

Index

A

B

C

D

M

O

P

R

S

T

U

V

W

www.ingramcontent.com/pod-product-compliance
Lightning Source LLC
LaVergne TN
LVHW080351011026
849161LV00009B/2263